FLORA OF TROPICAL EAST AFRICA

LEGUMINOSAE (Part 1)

Subfamily MIMOSOIDEAE

DC., Prodr. 2 : 424 (1825), as suborder or tribe *Mimoseae;* Benth. in Hook.,
Journ. Bot. 3 : 133 (1841)

J. P. M. BRENAN

Trees, shrubs, lianes or rarely herbs, often prickly or spiny. Leaves
bipinnate or (in exotic species only) simply pinnate or modified to phyllodes
or absent. Inflorescences usually spikes, racemes or heads of sessile or shortly
pedicellate, usually small or very small, regular, (3–)5(–6)-merous flowers.
Sepals with valvate or rarely imbricate (only in *Parkia* among our genera)
aestivation, often open from an early stage of bud, usually united to form a
toothed or lobed calyx, rarely free. Petals valvate in bud, free or more often
connate below into a tube. Stamens 4–10 (as many as or twice as many as
the petals) or numerous, free or adnate below to the corolla or the filaments
connate below into a tube, usually ± exserted. Anthers small, versatile,
sometimes with an apical gland. Pollen-grains sometimes simple, but
frequently compound or united. Pods and seeds various, the latter generally
marked with areoles.*

The generic order is in general that of Bentham's monumental " Revision of the
Suborder Mimoseae " in Trans. Linn. Soc. 30 : 336–664 (1875), although not necessarily
the true and final one. Dnyansagar (in Journ. Indian Bot. Soc. 34 : 362–374 (1955))
has suggested a new classification based on embryological and pollen characters, by
which the *Mimosoïdeae* would be divided primarily by the presence of either simple
or compound pollen-grains. He has, however, examined less than half the genera

* The seeds of *Mimosoïdeae* usually show on each face an area, generally more or less
elliptic or oblong in shape, bounded by a fine line which frequently appears as a fissure
in the testa. The size and shape of this area are often important taxonomically, and I
am employing the term " areole " to refer to it. The line (the " ligne de suture " of
Capitaine, the " linea fisural " of Boelcke and the " pleurogram " of Corner) is broken
opposite the micropyle, but occasionally, e.g. in *Acacia nubica*, almost continuous so
as to form nearly a circle.
The seeds of nearly all our *Mimosoïdeae* show these areoles. I have noted their
absence only in *Elephantorrhiza*, the winged-seeded genera *Newtonia* and *Piptadeniastrum*, and the giant *Entada* species (*E. pursaetha* and probably *E. gigas*). Areoles
occur in other species of *Entada*.
These areoles are taxonomically significant in another and more general way : while
they are of common occurrence in *Mimosoïdeae*, they do not seem to occur in the other
subfamilies of *Leguminosae*, except in a rather modified way in some species of *Cassia*.
The biological significance of the lines bounding the areoles is unknown. Capitaine
suggested that they might be lines of least resistance along which the often very hard
testa might split at germination ; while Corner suggested that the palisade-tissue of
the seed is differentiated from two sources in the seed and that the pleurogram marks
the meeting place of the two differential processes. Very little attention seems, however,
to have been paid to these lines. For further information see Capitaine, Contrib. Étude
Morphologique des Graines de Légumineuses, Thèse (Paris) (1912), Boelcke in Darwiniana 7 : 240–321 (1945), and Burkart, Las Leguminosas Argentinas, ed. 2 : 21
(1952) and, most important, Corner in Phytomorphology 1 : 117–150, especially
pp. 129–131 (1951).

occurring in our area, although others have been accounted for by Erdtman (Pollen Morphology and Plant Taxonomy, Angiosperms: 225–6 (1952)). Any classification on this basis would cut across most of the main tribes recognized by Bentham.

Existing classifications make it often difficult to identify the genera of this subfamily without complete material, including flower and fruit. As one or the other is so frequently absent, I have endeavoured to construct two alternative keys, one designed for flowering, the other for fruiting specimens, together with a conspectus of the various types of pod encountered in the subfamily. It must be clearly understood, however, that the keys and conspectus are artificial and apply to East Africa, and must not be relied upon in other areas or for identifying exotics.

KEY TO GENERA BASED ON VEGETATIVE AND FRUIT CHARACTERS

1. Plant armed with prickles, thorns or spines 2
   Plant unarmed (except for single minute very
   inconspicuous prickles below the nodes
   sometimes present in *Albizia harveyi*, and
   for some spinescent branches in *A. anthelmintica*) 9
2. Pod at maturity splitting transversely into segments each containing a seed 3
   Pod not splitting transversely 6
3. Leaflets in 1–2 pairs per pinna ; pod 2·5–3·7 cm.
   wide . . . . . . 2. **Entada** (sp. 6)
   Leaflets in 3–42 pairs per pinna ; pod 0·3–
   2·5 cm. wide 4
4. Valves falling away from the sutures ; mature
   stems and leaves always prickly ; herbs
   or shrubs to 4·5 m. high . . . 14. **Mimosa**
   Valves not falling away from the sutures ;
   leaves unarmed ; stems spinous when
   young, or armed only with stipular spines
   when mature ; usually trees 5 m. high or
   more 5
5. Lateral veins of leaflets invisible ; mature
   stems usually armed . . . . 16. **Acacia** (spp. 35, 36)
   Lateral veins of leaflets fine and clearly
   visible on both surfaces ; mature stems
   and leaves always unarmed ; juvenile and
   sucker shoots spinous . . . . 19. **Cathormion**
6. Pod straight, curved or falcate, but not spirally
   twisted or contorted . . . . 16. **Acacia**
   Pod spirally twisted or contorted 7
7. Pinnae and leaflets in one pair ; only the
   stipules spinescent . . . . 18. **Pithecellobium**
   Pinnae in more than one pair 8
8. Spines terminating short branchlets, plant
   otherwise unarmed ; pods clustered, indehiscent . . . . . . 12. **Dichrostachys**
   Spines replacing stipules and paired at nodes,
   or plant armed with prickles either scattered or 1–3 together at stem-nodes . 16. **Acacia**
9. Aquatic herb with creeping usually floating
   and swollen stems ; pod 1·3–2·7 (–3·8) cm.
   long, 1–1·2 cm. wide . . . 13. **Neptunia**
   Trees or shrubs 10
10. Leaflets alternate to subopposite 11
    Leaflets opposite 13

11. Pod dehiscing into 2 rather thin valves which
are spirally twisted after dehiscence and
satiny-yellow inside ; seeds scarlet.   .   **8. Adenanthera**
Pod indehiscent, woody ; seeds not brightly
coloured   12
12. The pod bluntly tetragonal or subcylindrical
in section ; petiolules 1·5–3 mm. long   .   **10 Amblygonocarpus**
The pod with a thick wing-like projection
running longitudinally along each of the
valves, the pod thus cruciform in section ;
petiolules 0·5–1 mm. long   .   .   .   **9. Tetrapleura**
13. Pod splitting transversely into segments each
containing a seed   14
Pod not splitting transversely   .   .   .   15
14. Width of pod 1·3–2 cm. ; forest tree with
capitate inflorescence   .   .   .   **19. Cathormion**
Width of pod 2·5–9 cm. ; inflorescence spicate
or spiciformly racemose   .   .   .   **2. Entada**
15. Pod dehiscent   16
Pod indehiscent   25
16. Valves of pod separating from each other
along one margin only   17
Valves of pods separating from each other along
both margins   19
17. Pod 5–8 mm. wide ; seeds small, black, unwinged ; inflorescence of paniculate or
racemose heads   .   .   .   .   **16. Acacia** (sp. 21)
Pod 13–32 mm. wide ; seeds large, brown,
conspicuously winged ; flowers in spikes
or spiciform racemes   18
18. Leaf-rhachis with a gland at the insertion of
each pair of pinnae ; funicle attached at
or near one end of the seed ; the cotyledons elongate   .   .   .   .   **5. Newtonia**
Leaf-rhachis eglandular ; funicle attached at
or near the middle of the seed ; the
cotyledons wider than long   .   .   **4. Piptadeniastrum**
19. Valves of pod woody, recurving   .   .   .   20
Valves of pod membranaceous to rigidly
coriaceous but not woody and recurving.   .   .   21
20. Pinnae always 1 pair per leaf ; inflorescence
capitate   .   .   .   .   .   **7. Xylia**
Pinnae 2–6 pairs per leaf ; inflorescence
racemose   .   .   .   .   .   **6. Pseudoprosopis**
21. Leaves reduced to simple entire phyllodes   .   **16. Acacia** (exotic spp.)
Leaves bipinnate   22
22. Pod 5–8 mm. wide ; inflorescences capitate   .   **16. Acacia** (sp. 21)
Pod 15–70 mm. wide   23
23. Inflorescence spicate (elongate axis visible in
fruit)   .   .   .   .   .   **16. Acacia** (sp. 10)
Inflorescence capitate   24
24. Seeds with endosperm, 4–5 mm. wide ; leaflets
acute at apex   .   .   .   .   **15. Leucaena**
Seeds without endosperm, 6·5–13 mm. wide (or
sometimes narrower and then leaflets
rounded to subacute at apex)   .   **17. Albizia**

25. Valves separating from the sutures, usually splitting into 2 layers ; branchlets glabrous ; leaflets acute . . . **3. Elephantorrhiza**
Valves not separating from the sutures and not splitting into layers 26
26. Pod 5–8 mm. wide ; flowers in small racemose or paniculate heads . . . . **16. Acacia** (sp. 21)
Pod 15–42 mm. wide or in diameter 27
27. Peduncles 9–35 cm. long, hanging ; pods 30–60 cm. long (including a 4–10 cm. long stipe) . . . . **1. Parkia**
Peduncles about 2–6 cm. long, not pendulous ; pods up to about 26 cm. long, sessile or with a very short stipe up to 5 mm. long . . . 28
28. Pod subcylindrical or shortly compressed ; inflorescence spicate . . . . **11. Prosopis**
Pod strongly flattened ; inflorescence capitate **17. Albizia** (sp. 15)

KEY TO GENERA BASED ON VEGETATIVE AND FLORAL CHARACTERS

1. Plant armed with prickles, thorns or spines 2
Plant unarmed (except for single minute very inconspicuous prickles sometimes present below the nodes in *Albizia harveyi*, and for some spinescent branches in *A. anthelmintica*) 9
2. Inflorescences two-coloured, upper part yellow, lower white to mauve ; spines terminating short branchlets, plants otherwise unarmed . . . . . . **12. Dichrostachys**
Inflorescences one-coloured ; plants with prickles or spinescent stipules 3
3. Flowers in globose or subglobose heads 4
Flowers in spikes or spiciform racemes 8
4. Pinnae one pair per leaf 5
Pinnae more than one pair per leaf 6
5. Leaflets in one pair ; only stipules spinescent ; flowers creamy . . . . . **18. Pithecellobium**
Leaflets in 10–26 pairs ; prickles scattered ; flowers mauve . . . . . **14. Mimosa** (sp. 5)
6. Stamens as many as or twice as many as the (3–)4–5(–6) corolla-lobes ; flowers mauve or pink . . . . . . **14. Mimosa**
Stamens numerous and indefinite ; flowers mostly cream, yellow or white 7
7. Mature stems and leaves normally prickly or spinous ; stamen-filaments free or nearly so . . . . **16. Acacia**
Mature stems and leaves always unarmed ; juvenile and sucker shoots spinous ; stamen-filaments united below into a tube **19. Cathormion**
8. Petals free ; stamens 10 ; anthers with a caducous apical gland . . . . **2. Entada** (sp. 6)
Petals generally connate into a tube ; stamens numerous and indefinite ; anthers glandular or eglandular . . . . **16. Acacia**

9. Inflorescence capitate, globose or subglobose 10
Inflorescence elongate, spicate or racemose 17
10. Aquatic herb with creeping usually floating
and swollen stems ; flowers of 2 sorts, ♀
in upper part of head, neuter with elongate
staminodes in lower part . . . 13. **Neptunia**
Trees or shrubs 11
11. Calyx 10–13 mm. long, with imbricate lobes ;
large claviform flower-heads on peduncles
9–35 cm. long . . . . . 1. **Parkia**
Calyx mostly up to 5, sometimes 7 mm. long,
with valvate lobes ; flower-heads smaller,
not claviform 12
12. Leaves reduced to simple entire phyllodes . 16. **Acacia** (exotic spp.)
Leaves bipinnate. 13
13. Pinnae in one pair per leaf 14
Pinnae in more than one pair per leaf 15
14. Stamens 10, free ; anthers glandular at apex . 7. **Xylia**
Stamens many (19–50), their filaments connate
below into a tube ; anthers eglandular at
apex . . . . 17. **Albizia**
15. Anthers hairy ; stamens 10, free ; petals free. 15. **Leucaena**
Anthers glabrous ; stamens numerous and
indefinite ; petals normally connate below
into a tube 16
16. Central flower or flowers of inflorescence
usually differing from and often larger
than the rest ; stamens united below into ⎰17. **Albizia***
an included or exserted tube . . ·⎱19. **Cathormion***

Central flowers of inflorescence not different
from the rest ; stamens free . . . 16. **Acacia**
17. Leaflets alternate to subopposite 18
Leaflets opposite 20
18. Anthers eglandular at apex, even in bud ;
petiolules 1·5–3 mm. long . . . 10. **Amblygonocarpus**
Anthers with a usually caducous apical gland. . . . 19
19. Leaflets 0·3–1·3 cm. wide ; calyx ± puberulous 9. **Tetrapleura**
Leaflets 1·2–2·3 cm. wide ; calyx usually
glabrous . . . . . . 8. **Adenanthera**
20. Rhachis of leaves with a gland at the insertion
of each pair or the top 1–5 pairs of pinnae . . . 21
Rhachis of leaves eglandular 23
21. Stamens numerous and indefinite ; petals
connate below . . . . . 16. **Acacia** (sp. 10)
Stamens 10 ; petals free or connate below . . . 22
22. Leaflets (5–)7–15 pairs per pinna, 4–10
(–14) mm. wide . . . . . 11. **Prosopis**
Leaflets (1–)2–3 pairs per pinna, or up to
19 pairs but then only 1–3 mm. wide . 5. **Newtonia**
23. Very tall tree of lowland rain-forest and riverine
forest, up to 50 m. high, with buttressed
base ; young branchlets ± densely pubescent . . . . 4. **Piptadeniastrum**

* These two genera are separated principally by their pods : see generic descriptions
and key to fruiting specimens. *Cathormion* is a shrub or tree of fresh-water swamp
forest, in our area recorded from Uganda.

Shrubs, lianes or small trees to 10(–15) m. high,
without buttresses ; the trees occurring
in woodland or bushland, not in forest   24
24. Petals connate below ; branchlets and racemes
glabrous ; leaflets numerous ; shrub or
small tree 1–7 m. high .   .   .   .   **3. Elephantorrhiza**
Petals free or almost so   25
25. Outside of petals glabrous .   .   .   .   **2. Entada**
Outside of petals normally ± puberulous   .   **6. Pseudoprosopis**

CONSPECTUS OF POD DIFFERENCES IN EAST AFRICAN GENERA OF
MIMOSOIDEAE

In all genera except *Acacia* the pod is usually rather constant in form.

A. Pod dehiscing into two separate valves:
   a. Valves spirally twisted :
     Adenanthera (spiral only after dehiscence)
     Pithecellobium
   b. Valves not spiral, woody :
     Acacia
     Pseudoprosopis
     Xylia
   c. Valves not spiral, papery to rigidly coriaceous :
     Acacia
     Adenanthera (spiral only after dehiscence)
     Albizia
     Leucaena
     Neptunia

B. Pod dehiscing into two valves which remain attached to one another along one margin:
     Acacia (*A. mearnsii*)
     Newtonia
     Piptadeniastrum

C. Pod splitting transversely into segments each containing a seed:
     Acacia (*A. kirkii, A. xanthophloea*)
     Cathormion
     Entada
     Mimosa

D. Pod indehiscent, not splitting transversely:
   a. Valves separating from sutures, usually splitting into two layers :
     Elephantorrhiza
   b. Valves spiral or contorted, but neither separating from sutures nor splitting :
     Acacia
     Dichrostachys
   c. Valves flattened, but neither spiral nor contorted nor separating from sutures nor splitting :
     Acacia
     Albizia
   d. Valves not flattened but convex, angled or winged, neither spiral nor contorted nor separating from sutures nor splitting :
     Acacia
     Amblygonocarpus
     Parkia
     Prosopis
     Tetrapleura

A number of exotic species of *Mimosoïdeae* are cultivated in our area ; most are mentioned under their appropriate genera, but in addition, however, there are the five following trees, all South American :—

Samanea saman (Jacq.) Merr. (*Pithecellobium saman* (Jacq.) Benth., *Enterolobium saman* (Jacq.) Prain), the Rain Tree, native of Central and South America, is planted as a shade tree in gardens and townships (T.T.C.L. : 348 ; Dale, Introd. Trees Uganda : 63 : 1953). It resembles an *Albizia*, particularly *A. versicolor* (see remarks on p. 147), but has a thick straight indehiscent pod.

Enterolobium contortisiliquum (Vell.) Morong (*E. timboüva* Mart.) is cultivated in Kenya. It resembles an *Albizia* or *Pithecellobium*, but has many very acute leaflets, and a thick indehiscent pod twisted almost into a circle.

Inga edulis Mart. (*I. vera* sensu T.T.C.L. : 345, *non* Willd.) has been tried as a shade for coffee in Tanganyika ; it has simply pinnate leaves with a winged rhachis, and pods with very thickened sulcate margins.

Stryphnodendron obovatum Benth. (*S. barbatimao* sensu T.T.C.L. : 348, *non* Mart.) is near *Amblygonocarpus* and *Tetrapleura*, with spicate flowers, glandular anthers and an indehiscent or slowly dehiscent pod which is not winged nor nearly as thickened as that of *Amblygonocarpus*. It has been grown at Amani.

Calliandra surinamensis Benth., a small tree or shrub, has leaves with one pair of pinnae, flowers in heads, and stamens red above, white below and connate in their lower part into an exserted column. It is cultivated in Kenya and Zanzibar (U.O.P.Z. : 163 (1949)). The similar *C. haematocephala* Hassk., with broader lanceolate leaflets and crimson flowers, is or may be expected in cultivation in our area.

1. PARKIA

R. Br. in Denh. & Clapp., Trav., app. : 234 (1826)

Trees, without spines or prickles. Leaves bipinnate ; leaflets ± numerous ; petiole usually glandular on its upper side. Inflorescences capitate, shortly claviform (with a globose apical part abruptly narrowed into a ± short cylindrical neck) or (but not in the African species) globose or constricted in the middle ; heads stalked, solitary or paniculate. Flowers in upper part of heads ♀, in lower part ♂ or neuter. Calyx infundibuliform or long-tubular, gamosepalous, with 4–5 imbricate segments, 2 larger and 2–3 smaller, the mouth of the calyx being thus irregular. Corolla with 5 petals, which are free, or ± united, not much exceeding the calyx. Stamens 10, all fertile, their filaments connate below into a tube, to which the petals may be also adnate ; anthers eglandular. Ovary usually stipitate. Pods oblong to linear, straight or curved, dehiscent or not, usually ± thick and often woody, or somewhat fleshy when living. Seeds ellipsoid to ellipsoid-oblong, ± compressed or flattened.

A genus of about 40 species, widely distributed through the tropics ; about 7 species in Africa and Madagascar, more in Asia and America.

P. javanica (Lam.) Merr. (*P. roxburghii* G. Don) from tropical Asia is cultivated in Tanganyika (*Greenway* 717 !) and Uganda (Dale, Introd. Trees Uganda: 53 (1953)); it has 16–30 pairs of pinnae and 30–80 pairs of leaflets per pinna, the leaflets small, 1–2 mm. wide, subfalcate and acute or subacute at apex.

Parkia filicoïdea [*Welw. ex*] *Oliv.*, F.T.A. 2 : 324 (1871) ; Crété in Trav. Lab. Médic. École Sup. Pharmacie Paris 7 (4) : 42 (1910) ; L.T.A. : 781 (1930) ; T.S.K. : 66 (1936) ; Bogdan in Nature in E. Afr., No. 4 : 11 (1947) ; T.T.C.L. : 346 (1949) ; I.T.U., ed. 2 : 227 (1952) ; Gilb. & Bout. in F.C.B. 3 : 193 (1952). Type : Angola, Cuanza Norte, Pungo Andongo, *Welwitsch* 1787 (LISU, lecto., BM, K., isolecto. !)

Tree 8–30(–35) m. high ; crown spreading, flat ; bark scaly or smooth, grey to yellow-brown. Young branchlets glabrous to puberulous. Leaves : petiole on upper side usually with 2 narrow ± collateral glands ; rhachis puberulous ; pinnae 4–11(–14, fide I.T.U. ed. 2) pairs ; leaflets 11–17 pairs (to 28 pairs in juvenile or coppice leaves), oblong, rounded at apex, asymmetrically rounded or subtruncate at base, mostly 1·2–3·2(–3·4) cm. long, 5–12(–14) mm. wide, glabrous except for puberulence on margins near base,

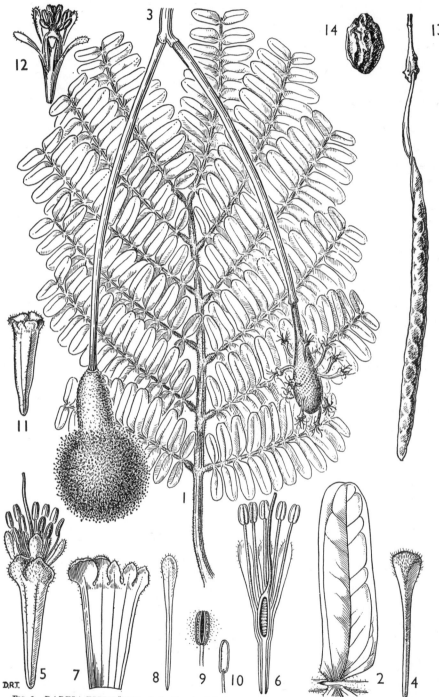

FIG. 1. *PARKIA FILICOÏDEA*—**1**, leaf, × ½; **2**, leaflet, upper side, × 2; **3**, inflorescences, × ½; **4**, bract from base of flower, × 3; **5**, ☿ flower from upper part of capitulum, × 3; **6**, longitudinal section of ☿ flower, calyx removed, × 3; **7**, calyx opened out, × 3; **8**, petal, × 3; **9**, anther, front view, filament cut off, × 4; **10**, anther, back view, filament cut off, × 4; **11**, neuter flower from lower part of capitulum, × 3; **12**, neuter flower, calyx removed, × 3; **13**, pod × ½; **14**, seed, × 1. 1–3, from *Stolz* 1676; 4–12, from *Faulkner, Pretoria No.* 11; 13, from *Semsei in F.H.* 2906; 14, from *Purves* 209.

2 longitudinal nerves more distinct than the others.   Peduncles about
9–35 cm. long.   Heads pendent, claviform, up to about 8·8 cm. long and
7·5 cm. wide, brick-red to reddish-pink, with a strong, pungent smell.
Bracteoles linear, enlarged at apex, up to 15 mm. long.   Hermaphrodite
flowers : pedicel 3–4 mm. long ; calyx 10–13 mm. long, glabrous or nearly
so except for the lobes, which are densely tomentellous outside ; the larger
lobes 1·7–2·5 mm. long, rounded ; petals adnate to staminal tube for about
2–4 mm. at base, above this free for about 10 mm. and linear-spathulate,
puberulous at apex which is about 0·6–0·75 mm. wide.   Pods 30–60 cm. long
(including 4–10 cm. long stipe), glabrous or nearly so, 1·5–2·8(–3·5) cm.
wide ; sutures straight or ± constricted between the seeds.   Fig. 1.

UGANDA.   Bunyoro District : Budongo Forest, May 1932, *Harris* H86 & B18 *in*
   F.H. 719 ! ; Mengo District : Entebbe, Nambigiruwa Swamp, Jan. 1932, *Eggeling*
   156 *in F.H.* 368 !
KENYA.   Mombasa, Mar. 1876, *Hildebrandt* 1975 ! ; " Coast forests," *Battiscombe* 1 !
TANGANYIKA.   Lushoto District : Sigi–Kwamkuyu Rivers, 15 Jan. 1931, *Greenway*
   2806 ! ; E. Mpwapwa, 9 Aug. 1933, *Hornby* 532 ! ; Morogoro District : Turiani,
   30 Oct. 1934, *E. M. Bruce* 78 ! ; Rungwe District : Bulambia. 13 Nov. 1912, *Stolz*
   1676 !
DISTR.   U2, 4 ; K7 ; T3, 5–8 ; Belgian Congo, Portuguese East Africa, Nyasaland,
   Northern Rhodesia and Angola ; recorded from W. Africa, but probably always in
   error
HAB.   Lowland rain-forest and riverine forest ; 250–1370 m.

SYN.   *P. hildebrandtii* Harms in E.J. 26 : 261 (1899) ; Crété in Trav. Lab. Mat. Médic.
   École Sup. Pharmacie Paris 7 (4) : 42, 43 (1910).   Type : Kenya, Mombasa,
   Hildebrandt 1975 (B, holo. †, BM, K, iso. !)
   P. bussei Harms in E.J. 33 : 154 (1902) ; Crété in Trav. Lab. Mat. Médic.
   École Sup. Pharmacie Paris 7 (4) : 45, 46 (1910) ; Gilb. & Bout. in F.C.B. 3 :
   144 (1952).   Types : Tanganyika, Rungwe District, Kiwira R., by the Wugu
   Hills, *Goetze* 1487 (B, syn. †) & Njombe/Songea Districts, Ruhuhu region,
   Busse 896 (B, syn. †, EA, isosyn. !)

NOTE.   I do not consider *P. bussei* and *P. filicoïdea* specifically separable, although
   they may perhaps be considered distinct varieties.   There is no evidence for the
   occurrence in our area of the plant with rather narrow linear-subcylindric pods constricted between the seeds, which is considered by Gilbert & Boutique in F.C.B. 3 :
   141–2 (1952) to correspond with true *P. filicoïdea*, unless *Brown* 193 (Uganda, Mengo
   District, Mawokota) proves to be it ; the pods of this gathering are, however, very
   old and broken, making their true dimensions difficult to ascertain.
      Pentaclethra macrophylla Benth. is recorded from E. Africa (P.O.A. B : 191, 471 &
   C : 196 (1895)) where it is said to have been collected by Hildebrandt near Mombasa.
   True *Pentaclethra* is unknown in our area, and the above mentions are almost certainly
   due to misidentification of *Parkia filicoïdea*, whose foliage has a superficial resemblance
   to that of *Pentaclethra macrophylla*.

2. ENTADA

Adans., Fam. Pl. 2·: 318 (1763)
Pusaetha [L. ex] O. Ktze., Rev. Gen. 1 : 204 (1891)
Entadopsis Britton in N. Amer. Fl. 23 : 191 (1928)

Trees, shrubs, suffrutices or lianes ; prickles absent or sometimes present.
Leaves bipinnate ; pinnae each with one to many pairs of leaflets.   Inflorescences of spiciform racemes or spikes, which are axillary or supra-
axillary, solitary or clustered and often ± aggregated.   Flowers ☿ or ♂.
Calyx gamosepalous, with 5 teeth.   Petals 5, free or nearly so (or ± connate
in species not occurring in our area), separated from ovary-base by a very
short perigynous zone composed of stamens adnate to an apparent corolla-
tube.   Stamens 10, fertile ; anthers with a usually very caducous apical
gland.   Pods straight or curved, flat or rarely spirally twisted, sometimes
very large ; at maturity the valves (but not the sutures) splitting trans-

versely into 1-seeded segments from which the outer layer (exocarp) of the pod-wall normally peels off, the inner layer (endocarp) persisting as a closed envelope round the seed ; the segments falling away from the sutures, which persist as a continuous but empty frame. Seeds (in the African species at least) ± compressed, mostly elliptic or subcircular, deep brown, smooth.

A genus of about 30 species, widespread and mainly tropical ; about 18 in Africa and Madagascar ; only about 4 in America.

Leaf-rhachis ending in a forked tendril ; pods gigantic,
    woody or very stiff, 0·4–2 m. long ; seeds 3·5–5 cm.
    wide ; large lianes with pinnae in not more than 2
    pairs per leaf and creamy to yellowish or greenish
    flowers :
    Flowers on distinct slender pedicels 1–1·5(–2) mm.
        long ; racemes supra-axillary ; pods spirally
        twisted, less woody than in sp. 2 .   .   .   1. *E. gigas*
    Flowers sessile or nearly so (pedicel to 0·5 mm.) ;
        spikes axillary ; pods straight or sometimes
        curved, but not spirally twisted, woody .   .   2. *E. pursaetha*
Leaf-rhachis not ending in a tendril ; pods at most
    0·4 m. long and usually much smaller, papery to
    subcoriaceous ; seeds 0·7–1·4 cm. wide :
    Scattered hooked prickles on stems and sometimes
        petioles and leaf-rhachides ; leaflets obovate-
        orbicular or suborbicular ; calyx pubescent ;
        shrub or tree .   .   .   .   .   6. *E. rotundifolia*
    Scattered hooked prickles absent ; plants unarmed
        except for spinescent stipules in sp. 10 ; leaflets
        ± oblong ; calyx glabrous (or slightly pubescent
        in sp. 4) :
        Flowers yellowish to whitish ; plants (except
            sp. 7) erect and without tendril-like pinnae ;
            pods straight :
            Climbing shrub or tree ; pedicels 0·3–0·75 mm.
                long ; racemes (including peduncle) 3–8 cm.
                long .   .   .   .   7. *E. leptostachya*
            Erect shrubs or trees ; pedicels 0·5–2 mm. long ;
                racemes (including peduncle) 4·5–17 cm.
                long :
                Leaflets 1–3(–3·5) mm. wide, mostly in 22–55
                    pairs per pinna, usually ± pubescent,
                    midrib excentric ; small tree.   .   .   5. *E. abyssinica**
                Leaflets 3–16 mm. wide, in 8–24 pairs per
                  pinna, midrib subcentral from shortly
                  above base :
                  Leaflets ± pubescent at least beneath ;
                    inflorescence-axes ± pubescent ; stipe
                    of pod about 1·5–3·5 cm. long ; shrub
                    1·2–1·8 m. high. .   .   .   .   4. *E. bacillaris*
                Leaflets glabrous or occasionally puberu-
                  lous ; inflorescence-axes glabrous or
                  nearly so ; stipe of pod 0·2–1·5 cm.
                  long ; shrub or small tree 1·2–10 m.
                  high .   .   .   .   .   .   3. *E. africana**

* There is evidence of hybrids occurring between *E. abyssinica* and *E. africana;* see note under the latter species (p. 13).

Flowers (or stamen-filaments) purple to red (colour
   in sp. 10 uncertain but most probably purple
   or red) ; slender lianes ; pinnae in 1–2(–3)
   pairs per leaf, one or more pinnae sometimes
   spirally twisted or tendril-like ; pods falcately
   curved :
  Young   branchlets   and   inflorescence-axes
     glabrous ; stipules minute, inconspicuous,
     not divergent nor spinescent ; pedicels
     1–1·5 mm. long ; stipe of pod 1–2·5 cm.
     long :
    Lateral nerves and veins of leaves distinctly
      raised and easily visible at least beneath ;
      leaflets 4–5(–8) pairs per pinna ; stamen-
      filaments about 3 mm. long   .   .   .   8. *E. stuhlmannii*
    Lateral nerves and veins of leaves not or
      scarcely visible on lower surface ; leaflets
      9–18 pairs per pinna ; stamen-filaments
      about 4–6·5 mm. long   .   .   .   9. *E. wahlbergii*
  Young branchlets and inflorescence-axes pubes-
    cent or puberulous ; stipules divergent,
    spinescent ; pedicels 0·25–0·5 mm. long ;
    stipe of pod 2·5–3·8 cm. long   .   .   10. *E. spinescens*

1. **E. gigas** (*L.*) *Fawc. & Rendle*, Fl. Jamaic. 4 (2) : 124 (1920) ; L.T.A. :
785 (1930), pro parte ; Brenan in K.B. 1955 : 164 (1955) ; Consp. Fl.
Angol. 2 : 257 (1956) ; F.W.T.A., ed. 2, 1 : 491 (1958). Type : Jamaica,
Patrick Browne (location unknown).

Large liane up to 25 m. high, unarmed. Young branchlets subglabrous to
puberulous or perhaps sometimes pubescent. Rhachis of leaves with (1–)2
pairs of pinnae, and ending in a forked tendril ; leaflets (3–)4(–5) pairs,
elliptic to obovate-elliptic, often asymmetric, 1·8–8 cm. long, 0·8–4 cm.
wide, emarginate at the obtuse or rounded apex, glabrous above except for
the puberulous midrib, glabrous also beneath except near base of leaflet and
(sometimes) for some pubescence along midrib. Spike-like racemes arising
from stem about 3–5 mm. above leaf-axils, solitary, 8–25 cm. long, ± pubescent, on a peduncle 1·5–6 cm. long ; pedicels 1–1·5(–2) mm. long, slender.
Flowers creamy to greenish or yellowish. Calyx somewhat puberulous or
glabrous, 1–1·25 mm. long. Petals 2·5–3 mm. long. Stamen-filaments
3·5–6 mm. long. Pods gigantic, less woody than in *E. pursaetha*, twisted
into a single or double lax spiral, with the sides also often twisted, 0·4–
1·2 m. long and 7·5–12 cm. wide ; outer layer of pod falling away to expose
the thick, chartaceous, somewhat flexible inner layer. Seeds hard, about
4–5·5 cm. in diameter.

UGANDA. Ankole District : Ruizi R., 4 Apr. 1951, *Jarrett* 400 ! ; Entebbe, Oct. 1905,
   Brown 328 !
DISTR. U2, 4 ; Central and West Africa, also in Central America, the West Indies and
   Colombia
HAB. Uncertain, but probably in riverine and lowland rain-forest ; the cited specimens
   were collected at about 1310 and 1183 m. respectively

SYN. *Mimosa gigas* L., Fl. Jamaic. : 22 (1759)
    Entada scandens (L.) Benth. subsp. *planoseminata* De Wild., Pl. Bequaert. 3 :
      85 (1925). Types : Belgian Congo, Kwilu, *Sapin* (BR, syn.) & Eala, *Goossens*
      (BR, syn.)
    E. scandens (L.) Benth. subsp. *umbonata* De Wild., Pl. Bequaert. 3 : 86 (1925).
      Type : Belgian Congo, Kitobola, *Flamigni* 469 (BR, holo., K. iso. !)
    E. planoseminata (De Wild.) Gilb. & Bout. in F.C.B. 3 : 221 (1952)
    E. umbonata (De Wild.) Gilb. & Bout. in F.C.B. 3 : 222 (1952)

NOTE. A specimen from Tanganyika (*Conrads in E.A.H.* 10904 ! from Mwanza District, Ukerewe Is. or Mwanza–Musoma area) is probably *E. gigas*, but is without flowers or fruits. If correct, this is an extension of range to T1.

2. **E. pursaetha** *DC.*, Prodr. 2 : 425 (1825) & Mém. Leg. : 421 (1826) ; Brenan in K.B. 1955 : 164 (1955) ; F.W.T.A., ed. 2, 1 : 490 (1958). Type : cultivated in Mauritius, *Delessert* (G–DC, lecto., K, photo. !)

Large liane said to reach 50 m. or more in length, unarmed. Young branchlets glabrous (but see note below). Rhachis of leaves with (1–)2 pairs of pinnae, and ending in a forked tendril ; leaflets 3–4(–5) pairs, elliptic to obovate-elliptic, 2·5–9 cm. long, 1·1–4 cm. wide, emarginate at the obtuse or rounded apex, glabrous or nearly so except for puberulence on midrib above and near base of leaflet beneath. Spikes axillary on lateral branches, which are sometimes leafless and abbreviated, the spikes thus aggregated ; spikes 7–23 cm. long, ± pubescent, on peduncles 1–8·5 cm. long ; pedicels to about 0·5 mm. long, or flowers sessile. Flowers creamy to yellow or greenish-yellow. Calyx glabrous (but see note below), about 1·25 mm. long. Petals about 2·5 mm. long. Stamen-filaments about 6 mm. long. Pods gigantic, woody, straight or sometimes curved, but not twisted, about 0·5–2 m. long and 7–15 cm. wide ; outer woody layer of pod falling away to expose the woody, rigid inner layer. Seeds hard, about 5 cm. long and 3·5–5 cm. wide.

UGANDA. Masaka District : Lake Nabugabo, Aug. 1935, *Chandler* 1388 !
KENYA. Kwale District : Kirao, Jan. 1930, *R. M. Graham* NN 757 *in F.H.* 2227 !
TANGANYIKA. Tanga District : Magunga Estate, 24 Nov. 1952, *Faulkner* 1078 ! ; Lushoto District : Pendeni–Longuza, 29 May 1917, *Peter* K338 ! ; Rungwe District : Rutengo, 26 Oct. 1910, *Stolz* 434 ! & Mabiembe, 7 Jan. 1933, *R. M. Davies* 807 !
ZANZIBAR. Pemba, Makongwe Is., 16 Dec. 1930, *Greenway* 2735 !
DISTR. **U**4 ; **K**7 ; **T**3, 7, 8 ; **P** ; widely distributed in tropical Africa and extending to South Africa ; also from India to China, the Philippines, Guam and north Australia
HAB. Lowland rain-forest & evergreen bushland in *Chlorophora-Albizia* woodland ; from near sea-level to about 600 m.

SYN. *Adenanthera gogo* Blanco, Fl. Filip. : 353 (1837). No type-specimen extant
    Entada gogo (Blanco) I.M. Johnst. in Sargentia 8 : 137 (1949)
    [*Entada gigas* sensu L.T.A. : 785 (1930), pro parte ; T.S.K. : 65 (1936) ; Bogdan in Nature in E. Afr., No. 4 : 12 (1947) ; U.O.P.Z. : 242 (1939) ; Gilb. & Bout. in F.C.B. 3 : 220 (1952) ; *non* (L.) Fawc. & Rendle]
    [*Entada phaseoloïdes* sensu T.T.C.L. : 344 (1949), *non* (L.) Merr.]

NOTE. For remarks on the identity of *E. phaseoloïdes* and *E. gigas*, names wrongly used for *E. pursaetha* in East Africa, see Johnston in Sargentia 8 : 135–138 (1949).
    E. pursaetha normally has stems and calyces glabrous ; however, a West African variant, not so far known from our area, has them pubescent.
    A specimen without flowers or fruits (Uganda, NW. Ankole District, *Dawe* 449 !) is probably *E. pursaetha*; if so, **U**2 must be added to its range.

3. **E. africana** *Guill. & Perr.* in Fl. Seneg. Tent. 233 (1832) ; L.T.A. : 786 (1930) ; Brenan in K.B. 1955 : 165 (1955) ; F.W.T.A., ed. 2, 1 : 491, fig. 156 (1958). Types : Senegal, Tiélimane, Cayor, *Leprieur* (G, syn. !, K, photo. !) ; Gambia, Albreda, *Perrottet* (G, syn. !, K, photo. !)

Shrub or small tree 1·2–10 m. high, unarmed, usually with very rough bark. Young branchlets glabrous. Leaves variable ; pinnae (1–)3–9 pairs ; tendrils absent ; leaflets (8–)10–24 pairs, oblong-elliptic, obovate-oblong or linear-oblong, 0·9–4·5 cm. long, 0·3–1·3(–1·5) cm. wide, rounded and some-times slightly mucronate at apex, glabrous or occasionally ± puberulous on both surfaces ; midrib ± central from shortly above leaflet-base. Racemes shortly supra-axillary, 1–4 together, sometimes aggregated on short lateral shoots, 6·5–15 cm. long (including the 0·6–2(–5) cm. long peduncle) ; axis glabrous or subglabrous, rarely pubescent ; pedicels 1(–1·5) mm. long. Flowers yellowish to whitish, sweetly scented. Calyx glabrous, 0·75–1·25 mm. long. Petals 1·5–3 mm. long. Stamen-filaments 4–5 mm. long.

Pods up to about 38 cm. long, 5–7·3 cm. wide, straight or nearly so, sub-coriaceous, joints conspicuously umbonate in centre. Seeds 12 mm. long, 9–10 mm. wide.

Uganda. Acholi District : Awach, Mar. 1935, *Eggeling* 1663 !
Distr. U1 ; Senegal and the Sudan to the French Cameroons, Belgian Congo and Uganda
Hab. Uncertain

Syn. *E. sudanica* Schweinf., Reliq. Kotschyanae : 8 (1868) ; L.T.A. : 789 (1930) ; I.T.U., ed. 2 : 225 (1952). Types : Sudan, Fazoghli, *Cienkowski* 252 (?LE or W, syn.) & Metemma, Gallabat, *Schweinfurth* 1891, 1935 (B, syn.†, K, isosyn. !) ; Nigeria, Nupe, *Barter* 1056 (syn. where ?, K, isosyn. !)
     Pusaetha africana (Schweinf.) O. Ktze., Rev. Gen. 1 : 204 (1891)
     Pusaetha sudanica (Schweinf.) O. Ktze., Rev. Gen. 1 : 204 (1891)
     Entadopsis sudanica (Schweinf.) Gilb. & Bout. in F.C.B. 3 : 204 (1952)

Note. *Greenway & Eggeling* 7232 (EA !, K !) from Uganda, West Nile District, Payida, has 3–9 pairs of pinnae, 14–22 pairs of leaflets 1–2·4 cm. long, 0·25–0·75 cm. wide and puberulous on both sides, the midrib in most leaflets being subcentral, but in some nearer the upper margin ; the inflorescence-axes are sparingly pubescent. It is difficult to know whether to refer this to *E. africana* or *E. abyssinica*, and I suspect it to be a hybrid between these two. Observations in the field are desired on their behaviour when growing together. Mr. F. White informs me that another gathering intermediate between these two species has been made in Uganda, Acholi District, Imatong Mts., *Eggeling* 806 (FHO). This specimen has 17 pairs of pinnae and about 30 pairs of rather large leaflets.

4. **E. bacillaris** *F. White* in Bol. Soc. Brot., sér. 2, 33 : 5 (1959). Type : Northern Rhodesia, Abercorn District, Kambole, *Richards* 9986 (K, holo. !)

Shrub 1·2–1·8 m. high. Branchlets pubescent ; young parts clothed with spreading golden indumentum. Leaves with 3–4 pairs of pinnae ; tendrils absent ; leaflets 8–13 pairs, oblong-elliptic, (1·3–)2·0–3·9(–4·6) cm. long, (0·5–)1–1·6 cm. wide, rounded or subtruncate at apex, obliquely rounded to subtruncate at base, pubescent beneath, ± pubescent to sub-glabrous above ; midrib subcentral, at least in upper part of leaflet. Racemes axillary, 1–3 together, 8–15 cm. long (including the 1·3–4 cm. long peduncle) ; axis pubescent. Pedicels 1–1·5 mm. long. Calyx glabrous or slightly pubescent at apex of lobes only, 1–2 mm. long. Corolla greenish-white to yellow, about 3·5–4 mm. long, lobes 2·5–3·5 mm. long. Stamen-filaments 5–6 mm. long. Pods 26–37 cm. long, 8–9 cm. wide, slightly falcate, sub-coriaceous ; stipe about 1·5–3·5 cm. long ; joints slightly umbonate in centre. Seeds about 12–13 mm. long and 9 mm. wide.

Tanganyika. Mpanda District : Karema, 15 Apr. 1936, *B. D. Burtt* 6056 !
Distr. T4 ; Northern Rhodesia
Hab. Rocky hills on coast-line of Lake Tanganyika clothed with *Brachystegia longifolia* and tall grass ; 840 m.

Syn. *E. nana* Harms var. *pubescens* R. E. Fr. in Schwed. Rhod.-Kongo-Exped. 1911–12, 1 : 64 (1914). Type : Northern Rhodesia, between Katwe and Abercorn, *Fries* 1215 (UPS, holo., K, photo. !)

5. **E. abyssinica** [*Steud. ex*] *A. Rich.*, Tent. Fl. Abyss. 1 : 234 (1847) ; L.T.A. : 789 (1930) ; T.S.K. : 65 (1936) ; Bogdan in Nature in E. Afr.; No. 4 : 12 (1947) ; T.T.C.L. : 344 (1949) ; I.T.U., ed. 2 : 225 (1952), Consp. Fl. Angol. 2 : 260 (1956) ; F.W.T.A., ed. 2, 1 : 491 (1958). Type : Ethiopia, Shire, " Dschogardi," *Schimper* 520 (P, syn., K, isosyn. !)

Small tree 2·7–10(–15) m. high, unarmed ; crown spreading, flat or rounded ; bark rough or smooth. Young branchlets glabrous or sometimes ± pubescent. Leaves with (1–)2–20(–22) pairs of pinnae ; tendrils absent ; leaflets (15–)22–55 pairs, mostly linear-oblong, (0·3–)0·4–1·2(–1·4) cm. long, 0·1–0·3(–0·35) cm. wide, rounded to obtuse and slightly mucronate at apex,

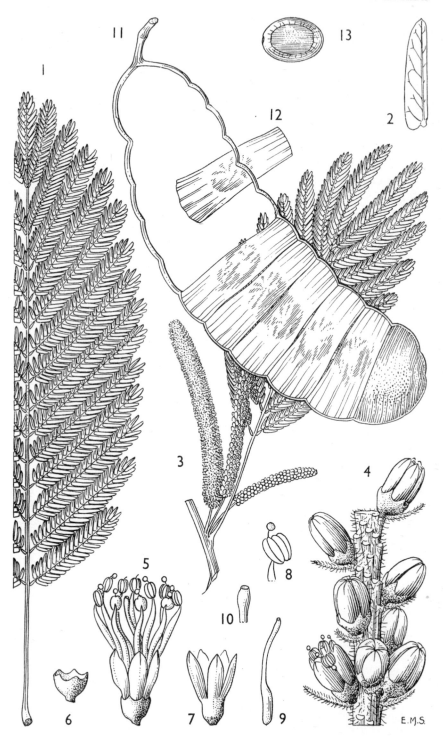

FIG. 2. *ENTADA ABYSSINICA*—**1**, leaf, × ⅔; **2**, leaflet, × 4; **3**, part of flowering branch, × ⅔; **4**, part of inflorescence, × 8; **5**, flower, ×8; **6**, calyx, × 8; **7**, petals, ×8; **8**, anther, × 16; **9**, ovary, × 8; **10**, top of style and stigma, × 10; **11**, pod, part fallen away, × ⅔; **12**, envelope of endocarp containing seed, × ⅔; **13**, seed, × 1½. 1, 2, 11–13, from *Semsei* 865; 3–10, from *Lugard* 600.

± appressed-pubescent on both surfaces, sometimes becoming glabrous above, rarely quite glabrous ; midrib starting at upper corner of the sub-truncate to rounded base, running obliquely but nearer upper margin. Racemes shortly supra-axillary, 1–4 together, 7–16 cm. long (including the 0·4–1·5 cm. peduncle) ; axis pubescent, sometimes subglabrous. Flowers creamy-white fading yellowish, sweetly scented ; pedicels 0·5–1 mm. long. Calyx glabrous, 0·75–1 mm. long. Petals 1·5–2 mm. long. Stamen-filaments 3·5–4 mm. long. Pods about 15–39 cm. long, (3·8–)5–7·5(–9) cm. wide, straight or nearly so, subcoriaceous ; joints less umbonate in centre than in *E. africana*. Seeds 10–13 mm. long, 8–10 mm. wide. Fig. 2.

UGANDA. West Nile District : Payida, Dec. 1931, *Brasnett* 324 ! ; Mbale District : Bugishu, Bugesege, 5 May 1941, *A. S. Thomas* 3852 ! ; Mengo District : 21 km. on Kampala–Bombo road, Apr. 1931, *Snowden* 2048 !
KENYA. Uasin Gishu District : Soy, 16 Apr. 1943, *Bally* 2467 ! ; Elgon, *Lugard* 600 !
TANGANYIKA. Shinyanga, *Koritschoner* 1648 ! ; Ufipa District : Sumbawanga–Pito, 25 Nov. 1949, *Bullock* 1937 ! ; Rungwe District : Tukuyu–Masoko road, 20 Feb. 1933, *R. M. Davies* 844 !
DISTR. **U**1–4 ; **K**2, 3, 5 ; **T**1–8 ; Sierra Leone and Eritrea to Angola, Southern Rhodesia and Portuguese East Africa
HAB. *Brachystegia-Julbernardia* woodland and wooded grassland ; 430–2290 m.

SYN. *Pusaetha abyssinica* ([Steud. ex] A. Rich.) O. Ktze., Rev. Gen. 1 : 204 (1891)
      Entadopsis abyssinica ([Steud. ex] A. Rich.) Gilb. & Bout. in F.C.B. 3 : 208 (1952)

VARIATION. Uganda and Kenya specimens have glabrous branchlets, but in Tanganyika, from Kondoa and Tabora Districts southwards, a variety or form with pubescent branchlets also occurs and extends into other territories south of the Flora area. Examples are : Kigoma District : Usinga, *Shabani* 3 ! ; Tabora District : without locality, *Wallace* 25 ! ; Ufipa District : Milepa, *Bullock* 3881 ! and Iringa, *Lynes* I.g.25 ! (all K, last in EA). The leaflets are normally pubescent, but may rarely be almost or quite glabrous. This is shown by *Anderson* 829 ! from Tanganyika, Lindi District, Nachingwea (EA, K).

6. **E. rotundifolia** *Harms* in E.J. 33 : 153 (1902) ; L.T.A. : 786 (1930) ; T.T.C.L. : 345 (1949). Types : Tanganyika, Lushoto District, Mazinde, by the Kisiwani road, *Busse* 362 (B, syn. †, K, isosyn. !) & Mazinde, *Holst* 3888 (B, syn. †) & without locality *Busse* 1143 (B, syn. †)

Shrub or small tree up to 9 m. high, with scattered hooked prickles on the stems and sometimes also on the petioles and leaf-rhachides. Young branchlets puberulous, quickly becoming glabrous and grey. Leaves with 1 pair of pinnae ; tendrils absent ; leaflets 1–2 pairs, obovate-orbicular or suborbicular, mostly 1·5–5 cm. in diameter, broadly rounded at apex, glabrous except for some pubescence near base beneath (sometimes above also). Spikes mostly axillary and paired, often aggregated on short lateral shoots, 3–6 cm. long (including peduncle). Flowers white, small, sessile. Calyx pubescent, 1 mm. long. Petals about 2·25 mm. long. Stamen-fila-ments about 5 mm. long. Pods about 9–18 cm. long and 2·5–3·7 cm. wide, flat, subcoriaceous. Seeds (not mature) about 9 × 7 mm.

TANGANYIKA. Pare District : Kisiwani, 1 Feb. 1936, *Greenway* 4559 ! ; Lushoto District : Mshwamba, 3 Jan. 1930, *Greenway* 2022 !
DISTR. **T**3 ; not known elsewhere
HAB. Deciduous bushland ; associated with *Acacia, Dobera, Salvadora* and *Balanites*, locally common, and a saline soil indicator ; 60–800 m.

SYN. *Entadopsis rotundifolia* (Harms) Gomes Pedro in Bol. Soc. Est. Moçamb., No. 92 : 10 (1955)

7. **E. leptostachya** *Harms* in E.J. 53 : 456 (1915) ; L.T.A. : 787 (1930) ; Bogdan in Nature in E. Afr., No. 4 : 12 (1947). Types : Kenya, Machakos District, Kibwezi, *Scheffler* 120 (B, syn. †, K, isosyn. !) & *Scheffler* 494 (B, syn. †) & Teita District, Voi, *Braun* 1540 (B, syn. †, EA, K, isosyn. !)

Climbing shrub or tree 3–6 m. high, unarmed. Young branchlets glabrous

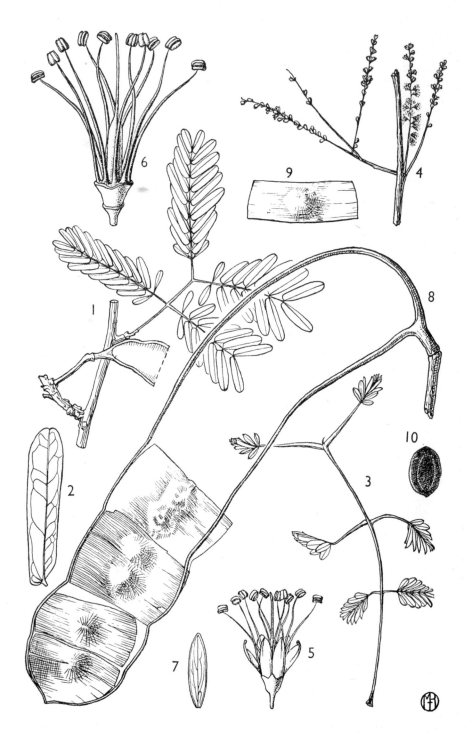

FIG. 3. *ENTADA LEPTOSTACHYA*—**1,** part of branch with leaf, × ⅔; **2,** leaflet, × 2; **3,** leaf with thickened modified pinnae for climbing, × ⅔; **4,** flowering shoot, × ⅔; **5,** flower, × 7; **6,** flower, calyx and petals removed, × 10; **7,** petal, × 7; **8,** pod, part fallen away, × ⅔; **9,** envelope of endocarp containing seed, × ⅔; **10,** seed, × 1. 1, 2, 4–9, from *Gillett* 13121; 3, from *Milne-Redhead & Taylor* 7254; 10, from *Dale* 3637.

or nearly so. Leaves with 2–4 pairs of pinnae ; tendrils absent, except for modified ± hooked pinnae on leaves on long shoots : leaflets 7–10 pairs, narrowly oblong or oblanceolate-oblong, 0·9–2·5(–3·8) cm. long, 0·3–0·9 (–1·4) cm. wide, rounded and sometimes emarginate at apex, asymmetric at base, puberulous on both surfaces, sometimes subglabrous or glabrous. Spike-like racemes axillary, solitary or 2–3 together, often aggregated on short lateral leafless shoots, 3–8 cm. long (including peduncle), glabrous or nearly so ; pedicels 0·3–0·75 mm. long. Flowers yellow (? or white), sweetly scented. Calyx glabrous, 0·5–0·75 mm. long. Petals about 2 mm. long. Stamen-filaments about 2·5–4 mm. long. Pods 17–23 cm. long and 4·3–8·4 cm. wide, flat, subcoriaceous. Seeds about 11–14 mm. long and 9 mm. wide. Fig. 3.

KENYA. Northern Frontier Province : Dandu, 9 May & 20 July 1952, *Gillett* 13121 ! ; Masai District : Garabani–Emali Hill, 6 Mar. 1930, *van Someren* 6 ! ; Teita District : between Voi and Mackinnon Road, Mar. 1937, *Dale in F.H.* 3637 & *in C.M.* 14017 ! ; Kwale District : between Samburu and Mackinnon Road, 31 Aug. 1953, *Drummond & Hemsley* 4075 !
TANGANYIKA. Pare District : Kisiwani, 8 Nov. 1955, *Milne-Redhead & Taylor* 7254 !
DISTR. **K**1, 4, 6, 7 ; **T**3 ; Somaliland Protectorate
HAB. Dry scrub with trees & deciduous bushland ; 350–1520 m.

SYN. *Entadopsis leptostachya* (Harms) Cuf., Enum. Pl. Aeth. : 210 (1955)

NOTE. According to Harms (see reference above), *E. leptostachya* is occasionally erect, and may have up to 4 pairs of pinnae, 11 pairs of leaflets, and pods up to 30 cm. long.

8. **E. stuhlmannii** (*Taub.*) *Harms* in V.E. 3 (1) : 401 (1915) ; L.T.A. : 788 (1930) ; T.T.C.L. : 345 (1949) ; Brenan in K.B. 1955 : 170 (1955). Types : Tanganyika, Uzaramo District, *Stuhlmann* 6845, 6939, 6965, 7114 & Bagamoyo District, *Stuhlmann* 7197 (all B, syn. †)

Slender woody climber, unarmed, said to have a tuberous root. Young branchlets glabrous, often flexuous. Leaves with 2(–3) pairs of pinnae ; one or more of the pinnae, usually terminal, sometimes modified to a tendril, or spirally twisted at base and bearing leaflets above ; leaves of scrambling shoots may have the terminal pair of pinnae leafless and tendril-like, and the lower pair much reduced ; leaflets 4–5(–8) pairs, obovate- or oblanceolate-oblong, or sometimes narrowly oblong, 1–2·7 cm. long, 5–14 mm. wide, rounded to subtruncate and mucronate or not at apex, asymmetric at base, glabrous, lateral nerves and venation distinctly raised and easily visible at least beneath and often on both surfaces. Stipules minute, inconspicuous, not divergent nor spinescent. Racemes axillary, solitary, often aggregated, 3·5–6 cm. long (not including the usually 1–2 cm. long peduncle), glabrous ; pedicels 1–1·5 mm. long. Flowers purple or brownish-red. Calyx glabrous, 1 mm. long. Petals 2·5 mm. long. Stamen-filaments about 3 mm. long. Pods about 12–20 cm. long (to 30 cm. long, *fide* Taubert), 2·7–4·3 cm. wide, subcoriaceous, falcately curved, with a stipe 1·5–2·5 cm. long. Mature seeds not seen, probably similar to those of *E. wahlbergii*.

TANGANYIKA. Rufiji District, without locality, 17 Jan. 1931, *Musk* 117 & Utete, 2 Dec. 1955, *Milne-Redhead & Taylor* 7540 ! ; Lindi District : Mtange, 23 Mar. 1943, *Gillman* 1383 ! & Nachingwea, 5 Mar. 1953, *Anderson* 853 !
DISTR. **T**6, 8 ; Portuguese East Africa
HAB. " Coastal scrub," deciduous bushland and scattered-tree grassland ; 15–1600 m.

SYN. *Pusaetha stuhlmannii* Taub. in P.O.A. C : 196 (1895)
      [*Entada wahlbergii* sensu L.T.A. : 788 (1930) ; F.T.A. 2 : 326 (1871), pro parte, quoad spec. Meller ; *non* Harv.]
      Entadopsis stuhlmannii (Taub.) Gomes Pedro in Bol. Soc. Est. Moçamb. No. 92 : 10 (1955)

9. **E. wahlbergii** *Harv.* in Fl. Cap. 2 : 277 (1862) ; Brenan in K.B. 1955 : 169 (1955) ; F.W.T.A., ed. 2, 1 : 492 (1958). Type : South Africa, " Cape of Good Hope," *Wahlberg* (S, holo. !)

Slender woody climber to 3 m. high or more, unarmed.   Young branchlets glabrous, flexuous.   Leaves with (1–)2(–3) pairs of pinnae ; one or more of the pinnae, usually terminal, sometimes modified to a tendril, or spirally twisted at base and bearing leaflets above ;   leaflets 9–18 pairs, narrowly oblong or oblanceolate-oblong, 0·8–1·6 cm. long, 1·5–4·25 mm. wide, rounded and usually mucronate at apex, asymmetric at base, glabrous, lateral nerves not or scarcely visible beneath.   Stipules minute, inconspicuous, to 1·5 mm. long, not divergent nor spinous.   Racemes axillary, solitary, often aggregated on short leafless shoots or occupying terminal parts of shoots, 3–6 cm. long (not including the 4–10(–35 ?) mm. long peduncle), glabrous.   Flowers dark purple or red, on pedicels 1–1·5 mm. long.   Calyx glabrous, 1–1·5 mm. long.   Petals 3–3·5 mm. long.   Stamen-filaments 4–6·5 mm. long.   Pods about 11–23 cm. long, 2·9–3·8 cm. wide, flat, subcoriaceous, falcately curved, with a stipe 1–2 cm. long.   Seeds about 10–11 mm. long and 7–8 mm. wide.

UGANDA.   West Nile District : Nebbi, Sept. 1940, *Purseglove* 1055 ! & Ajugopi [? Adzugopi], 29 Apr. 1932, *Hancock in Tothill* 1115 ! ;   Bunyoro District :   Bukumi flats, near Butiaba, 1955, *Dawkins* 888 ! & Lake Albert rift, Sept. 1938, *Hancock* 4A !
TANGANYIKA.   Ufipa District :   Kasanga, 14 June 1957, *Richards* 10096 ! & 15 June 1957, *Richards* 10113 !
DISTR.   U1, 2 ;   T4 ;   Portuguese Guinea and French Sudan to Nigeria ;   Belgian Congo, the Sudan, ? Northern Rhodesia, Natal.
HAB.   Wooded grassland ;   610–1070 m.

SYN.   *Pusaetha wahlbergii* (Harv.) O. Ktze., Rev. Gen. 1 : 204 (1891)
    E. flexuosa Hutch. & Dalz., F.W.T.A. 1 : 356 (1928) & in K.B. 1928 : 401 (1928).   Type : Nigeria, Nupe, *Barter* 991 (K, holo. !)
    Entadopsis flexuosa (Hutch. & Dalz.) Gilb. & Bout. in F.C.B. 3 : 206 (1952)
    Entadopsis wahlbergii (Harv.) Gomes Pedro in Bol. Soc. Est. Moçamb. No. 92 : 10 (1955)

NOTE.   The form of this variable and widespread species occurring in Uganda has 10–18 pairs of leaflets per pinna, the former 2–3·8 mm. wide ;   pedicels about 1 mm. long ;   petals 3·25–3·5 mm. long, without a minute appendage at apex bent inwards and downwards.   For further discussion of the variation of this species see Brenan in K.B. 1955 : 169–170 (1955).
    The Tanganyika specimens are flowering when leafless.   In the absence of foliage and ripe pods their identification is not beyond doubt.

10. **E. spinescens** *Brenan* in K.B. 1955 : 168 (1955).   Type : Tanganyika, Mpwapwa District, near Gulwe, *B. D. Burtt* 4639 (K, holo. !)

Slender woody climber up to 3·6 m. high or more, unarmed except for stipules.   Young branchlets pubescent or puberulous, often flexuous. Leaves with 1–2 pairs of pinnae ;   one or more of the pinnae sometimes modified to a tendril, or spirally twisted at base and bearing leaflets above ; leaflets 12–17 pairs, narrowly oblong, mostly 1–1·8 cm. long and 2–4 mm. wide (the lowest of each pinna often smaller), rounded and obtuse or mucronate at apex, asymmetric at base, glabrous or with some hairs on midrib and margins ;   lateral nerves not or scarcely visible beneath.   Stipules divergent, rigid, subconical, spinescent, ± pubescent, 1·5–3·5 mm. long. Racemes axillary, solitary, 3–5 cm. long (not including the 1·5–3 cm. long peduncle) ;   axis ± pubescent or puberulous ;   pedicels 0·25–0·5 mm. long. Flower-colour uncertain, probably as in *E. wahlbergii*.   Calyx glabrous, about 1 mm. long.   Petals 3 mm. long.   Stamen-filaments 3–3·5 mm. long. Pods about 13–17 cm. long, flat, subcoriaceous, falcately curved, with a stipe 2·5–3·8 cm. long.   Seeds (unripe) about 8 mm. long and 5 mm. wide.

TANGANYIKA. Dodoma District : Manyoni–Kilimatinde road, 16 Dec. 1935, *B. D. Burtt* 5389 ! ; Mpwapwa, 2 Mar. 1937, *Hornby* 398 ! ; Mpwapwa District : Gulwe, 22 Jan. 1933, *B. D. Burtt* 4639 !
DISTR.   T5 ; not known elsewhere
HAB.   Deciduous bushland and tall deciduous thickets ; 910–1220 m.

SYN.   [*E. flexuosa* sensu T.T.C.L. : 344 (1949), *non* Hutch. & Dalz.]

Imperfectly known species
11. E. sp.

Bushy shrub up to 3 m. high, with decumbent, unarmed branches. Young branchlets glabrous. Leaves with 4–5 pairs of pinnae ; no tendrils ; leaflets 11–12 pairs, oblong, 1·6–2·8 cm. long, 6–9 mm. wide, thinly appressed-pubescent especially beneath, midrib becoming subcentral ; lateral nerves easily visible. Flowers unknown. Pods about 15 cm. long and 6·5 cm. wide.

ZANZIBAR.   Pemba, Ras Domoni, 17 Dec. 1930, *Greenway* 2747 !
DISTR.   P ; see note below
HAB.   In sand near saline marsh

NOTE.   Near *E. leptostachya* but with more numerous leaflets and more pubescent rhachides to the pinnae. This may prove to be *E. kirkii* Oliv., hitherto known only from Portuguese East Africa, but the material is not adequate for certainty.
It is possible that *Schlieben* 5469 ! (Tanganyika, Lindi District, Lake Lutamba) may be the flowering condition of this. Some of the pinnae are here twisted and becoming tendril-like, and the flowers are reddish-yellow, in denre spiciform racemes whose rhachides are glabrous. More material is wanted.

3. ELEPHANTORRHIZA
Benth. in Hook., Journ. Bot. 4 : 344 (1841)

Small trees, shrubs or suffrutices, unarmed. Leaves bipinnate, pinnae mostly with many pairs of leaflets. Inflorescences of spiciform racemes which are axillary, solitary or clustered, often ± aggregated. Flowers normally ⚥. Calyx gamosepalous, with 5 teeth. Petals 5, connate below, free above. Stamens 10, fertile, free, not adnate to corolla ; anthers with a usually very caducous apical gland. Pods straight or somewhat curved, not spirally twisted ; at maturity the valves separating from the persistent sutures, but not splitting into segments ; the outer layer (exocarp) of the pod-wall often peeling off the inner layer (endocarp), the layers remaining intact or breaking irregularly. Seeds ± compressed.

A genus of about 9 species restricted to Africa south of the equator.

E. goetzei (*Harms*) *Harms* in V.E. 3 (1) : 400 (1915) ; L.T.A. : 802 (1930) ; T.T.C.L. : 344 (1949). Type : Tanganyika, Rufiji District, *Goetze* 82 (B, holo. †, K, iso. !)

Shrub or small tree 1–7 m. high, deciduous ; bark dark dull brown or red. Young branchlets glabrous, becoming blackish. Leaves up to 42 cm. long, glabrous or nearly so ; pinnae 4–31 pairs ; leaflets 12–48 pairs, linear to narrowly oblong, 3·5–12 mm. long, 0·7–3 mm. wide, glabrous ; midrib starting at distal corner of leaf-base, gradually becoming almost central in the leaf ; proximal side of base rounded and almost auriculate ; apex acute and nearly symmetrical ; lateral nerves and veins not or scarcely visible. Racemes glabrous, 5–22 cm. long ; pedicels 1–2 mm. long. Flowers variously described as yellow, or with brownish-violet petals and yellow stamens.* Calyx 1–1·5 mm. long. Petals 2·5–3 mm. long. Stamen-filaments about 5 mm. long. Pods linear-oblong, up to 44 cm. long and 2–3 cm. wide, the seeds showing as bumps at intervals. Seeds lenticular, 16–20 mm. long, 14–18 mm. wide and 10–12 mm. thick. Fig. 4, p. 20.

* Collectors' notes on flower-colour in this species leave it uncertain whether there is variation in the colour.

2—2

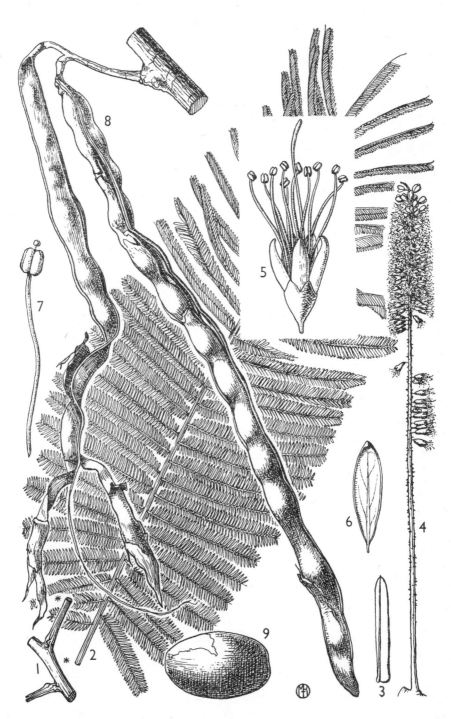

Fig. 4. *ELEPHANTORRHIZA GOETZEI*—1, part of branch with petiole bases, × ½; 2, leaf (detached from petiole base of 1), × ½; 3, leaflet, × 6; 4, flowering raceme, × 1½; 5, flower, × 9; 6, petal, × 15; 7, anther, × 15; 8, pods, × 1; 9, seed, × 3. 1–3, from *Milne-Redhead & Taylor* 9549; 4–7, from *Andrada* 1452; 8–9, from *G. Jackson* 1418.

TANGANYIKA. Rufiji District : S. of the R. Rufiji, Nov. 1898, *Goetze* 82 ! ; Songea District : Nangurukuru Hill, about 26 km. E. of Songea, 8 Apr. 1956, *Milne-Redhead & Taylor* 9549 ! ; Kilwa District : Mtumbati Valley, *Crosse-Upcott* 152 !
DISTR. T6, 8 ; Portuguese East Africa, Nyasaland, Northern (*fide* Greenway) and Southern Rhodesia ; possibly in Angola
HAB. Deciduous woodland ; at edge of rocky outcrop on hillside in Songea District ; 250–1020 m.

SYN. *Piptadenia goetzei* Harms in E.J. 28 : 397 (1900)

4. PIPTADENIASTRUM

Brenan in K.B. 1955 : 179 (1955)

Tree, tall, unarmed. Leaves bipinnate, pinnae each with many pairs of leaflets ; rhachis of leaf without glands ; pinnae often alternate. Flowers small, ☿, in often aggregated, spiciform racemes. Calyx gamosepalous, 5-toothed, glabrous outside except for base. Petals 5, free, glabrous outside, separated from ovary-base by a short perigynous zone composed of stamens and disc consolidated with an apparent corolla-tube. Stamens 10, fertile ; anthers each with a caducous apical gland. Ovary glabrous outside. Pods straight or somewhat curved, flattened, at maturity dehiscing along one of the sutures, the valves remaining attached along the other, neither splitting transversely nor into layers. Seeds flattened, ± oblong, brown, surrounded by a broad membranous wing ; the body of the seed somewhat elongate in the direction of the length of the pod ; cotyledons elongate transversely to the radicle ; funicle attached at or near the middle of the seed.

A genus of single species in the forest regions of tropical Africa.
Although the seed superficially resembles in shape that of *Newtonia*, in fact it lies transversely in the pod, the embryo being at right angles to the sutures ; the seed is thus much wider than long. In *Newtonia* the seed lies longitudinally in the pod, the embryo being parallel with the sutures, and the seed is thus much longer than wide.

P. africanum (*Hook. f.*) *Brenan* in K.B. 1955 : 179 (1955) ; Consp. Fl. Angol. 2 : 262 (1956) ; F.W.T.A., ed. 2, 1 : 489 (1958). Types : Nigeria, *Ansell* (K, syn. !) & *T. Vogel* (K, syn. !)

Tree up to 50 m. high, with buttressed base and smooth, usually grey bark. Young branchlets shortly and ± densely rusty (when dry)-pubescent, then glabrescent. Pinnae 10–19 pairs (to 23 pairs on juvenile leaves). Leaflets 30–58 pairs (to 61 on juvenile foliage), linear, ± falcate, 3–8.5 mm. long (to 10 mm. on juvenile leaves), 0.8–1.25 mm. wide. Flowers yellowish-white, in spiciform racemes 4–11 cm. long. Pod 17–36 cm. long, 2–3.2 cm. wide. Seeds (3–)5.3–9.5 cm. long, 1.8–2.5 cm. wide. Fig. 5, p. 22.

UGANDA. Bunyoro District : Budongo, near km. 88 on Masindi–Butiaba road, *Harris* 126 & B25 *in F.H.* 844 ! ; Mengo District : Entebbe, near km. 21 on road to Kampala, Feb. 1931, *Snowden* 1957 ! & Bumpenje Hill near Entebbe, 1 Mar. 1950, *Dawkins* 530 !
DISTR. U2, 4 ; from Senegal and the Sudan to Angola and the Belgian Congo
HAB. Lowland rain-forest and riverine forest ; 1100–1220 m.

SYN. *Piptadenia africana* Hook. f. in Niger Fl.: 330 (1849) ; F.T.A. 2 : 328 (1871) ; F.W.T.A. 1 : 354 (1928) ; L.T.A. : 794 (1930) ; Chalk, Burtt-Davy & Desch, Some E. Afr. Conif. & Legum. : 54 (1932) ; I.T.U., ed. 2 : 228, fig. 51 (1952) ; Gilb. & Bout. in F.C.B. 3 : 226 (1952)

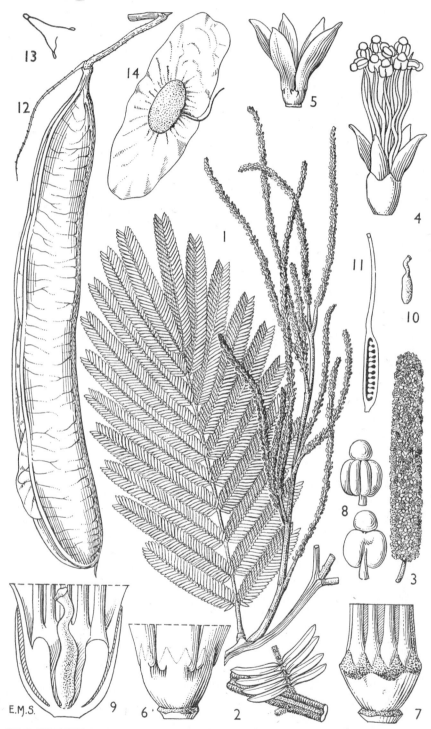

FIG. 5. *PIPTADENIASTRUM AFRICANUM*—**1**, part of branch with inflorescences in bud, × 1; **2**, part of rhachis of leaf with pinna-base, × 3; **3**, part of inflorescence, × 1; **4**, flower, × 10; **5**, petals, × 10; **6**, basal part of petals, × 30; **7**, bases of stamen-filaments, petals removed, × 30; **8**, anther, front and back views, × 30; **9**, longitudinal section through base of flower to expose short ovary, × 30; **10**, short ovary, × 10; **11**, mature ovary and ovules, × 10; **12**, dehiscing pod, × ⅔; **13**, cross-section of pod, diagrammatic, × ⅔; **14**, seed, × 1. 1, from *Harris* 844; 2, 12–14, from *Louis* 6902; 3–11, from *Snowden* 1957.

5. NEWTONIA

Baill., in Bull. Soc. Linn. Par. 1 : 721 (1888)

Trees, often tall, unarmed. Leaves bipinnate ; pinnae each with one to many pairs of leaflets ; rhachis of leaf usually (always in our species) with a gland between each pair of opposite pinnae. Flowers sessile or nearly so, in spikes or spiciform racemes. Calyx and petals pubescent or puberulous outside. Anthers with or without an apical gland. Ovary densely pilose outside. Flowers and pods otherwise as in *Piptadeniastrum* (p. 21). Seeds flattened, oblong, brown, surrounded by a membranous wing ; the body of the seed much elongate in the direction of the length of the pod ; cotyledons elongate in the same direction as the radicle ; funicle slender, attached at or near one end of the seed.

A genus of 14 or more species, 11 of them over much of tropical Africa and just reaching South Africa, the rest in tropical South America.

Leaflets(24–)38–67 pairs per pinna ; pinnae (7–)12–23
    pairs ; anthers with an apical gland soon falling
    off, but readily seen in bud   .   .   .   .   1. *N. buchananii*
Leaflets (1–)2–19(–27) pairs per pinna ; pinnae 1–7
    pairs ; anthers eglandular at apex even in bud :
    Pinnae 1–2 pairs ; leaflets (1–)2–3 pairs, obovate to
      elliptic, 5–38 mm. wide ; pinna-rhachis with
      glands between leaflet-pairs   .   .   .   2. *N. paucijuga*
    Pinnae 4–7 pairs ; leaflets 6–19 pairs per pinna, ±
      oblong or linear-oblong, 1–3 mm. wide ; pinna-
      rhachis without glands between leaflet-pairs   .   3. *N. hildebrandtii*

1. **N. buchananii** (*Baker*) *Gilb. & Bout.* in F.C.B. 3 : 213 (1952) ; Consp. Fl. Angol. 2 : 261 (1956). Type : Nyasaland, *Buchanan* 192 (K, holo. !)

Tree 10–40 m. high, somewhat buttressed at base, with smooth bark. Branchlets densely pubescent when young. Leaf-rhachis with a stipitate gland between each pinna-pair ; no glands between leaflet-pairs ; pinnae (7–)12–23 pairs ; leaflets (24–)38–67 pairs, linear, ± falcate, 2–6(–9) mm. long, 0·5–1·5(–2) mm. wide ; lateral nerves invisible beneath. Flowers yellowish, in spikes 3·5–19 cm. long. Anthers with an apical gland that soon falls off. Pod 10–32 cm. long, 1·3–2·5 cm. wide. Seeds 4·1–7·5 cm. long, 0·9–2·1 cm. wide. Fig. 6, p. 24.

UGANDA. Ankole District : Ruizi R., 18 Oct. 1950, *Jarrett* 386 ! ; Masaka District : Bugala Is., Sozi Point, Nov. 1931, *Eggeling* 256 ! ; Mengo District : Entebbe, Bumpenje Hill, 1 Mar. 1950, *Dawkins* 531 !
KENYA. Fort Hall/Kiambu Districts : Thika, by Chania R., 25 June 1947, *Bogdan* 781 ! ; Nairobi, French Mission, Dec. 1940, *Bally* 1384 ! ; Teita District : Taveta, 22 Jan. 1936, *Greenway* 4477 !
TANGANYIKA. Moshi, Nov. 1935, *R. M. Davies* 1134 ! & 10 Nov. 1949, *Wigg in F.H.* 3024 ! ; Lushoto District : Amani, 11 Feb. 1932, *Toms* H4/32/1 ! ; Songea District : top of escarpment leading to Mbamba Bay, 25 May 1956, *Milne-Redhead & Taylor* 10445 !
DISTR. U2, 4 ; K4, 7 ; T2–4, 6, 8 ; Portuguese East Africa, Nyasaland, Northern & Southern Rhodesia, Belgian Congo and Angola
HAB. Lowland and upland rain-forest (usually by streams), ground-water, riverine and swamp forest ; 600–2130 m.

2. **N. paucijuga** (*Harms*) *Brenan* in K.B. 1955 : 181 (1955). Type : Tanganyika, ? Uzaramo District, " Pandeberg," *Leopold in Holtz* 3207 (B, holo. †, BM, drawing !)

Tree up to 30 m. high, with smooth, grey to greyish-green and brown bark. Young branchlets (when dry) densely rusty-puberulous, slowly glabrescent.

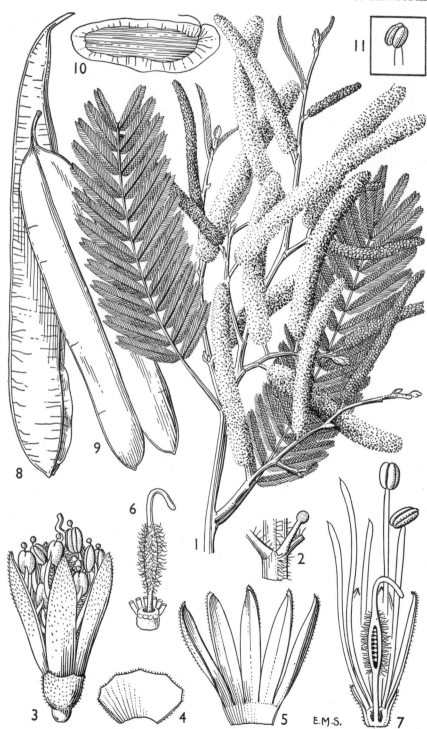

FIG. 6. *NEWTONIA BUCHANANII*—1, part of flowering branch, × ⅔; 2, part of leaf-rhachis showing gland, × 6; 3, flower, × 12; 4, calyx, × 12; 5, corolla, × 12; 6, ovary, × 12; 7, longitudinal section of mature flower, × 12; 8, pod, × ⅔; 9, pod showing unilateral dehiscence, × ⅔; 10, seed, × 1; *NEWTONIA HILDEBRANDTII*—11, anther showing absence of gland, × 12. 1–7, from *Eggeling* 3227; 8, 10, from *Wigg* 3008; 9, from *van Someren* 6771; 11, from *Lewis* 2.

Leaf-rhachis with a sessile ± depressed-hemispherical gland between each pinna-pair ; glands also present between leaflet-pairs ; pinnae 1–2 pairs ; leaflets (1–)2–3 pairs, obovate to elliptic, 0·8–7 cm. long, 0·5–3·9 cm. wide (on saplings to 11 × 5·8 cm.), rounded or emarginate at apex. Spikes about 3–10 cm. long, often paniculate. Anthers without an apical gland. Pod 23–60 cm. long, (1·8–)2·3–3·1 cm. wide. Seeds about 7–8·8 cm. long, 1·7–2·3 cm. wide.

Kenya. Without precise locality, *Battiscombe* 93 ! ; " Coast forests," *Webber* 607 *in C.M.* 16977 ; Kilifi, 17 May 1937, *Moggridge* 384 !
Tanganyika. Lushoto District : Longuza, 4 Oct. 1936, *Greenway* 4649 ! ; Tanga District : 6 km. SE. of Ngomeni, 2 Aug. 1953, *Drummond & Hemsley* 3604 ! ; Morogoro District : Turiani, June 1944, *Wigg in F.H.* 1262 !
Distr. K7 ; T3, 6, ? 8 ; not known elsewhere
Hab. Lowland rain-forest, riverine forest ; 75–300 m.

Syn. *Piptadenia paucijuga* Harms in E.J. 51 : 368 (1914) ; L.T.A. : 792 (1930) ; T.T.C.L. : 347 (1949)
    Cylicodiscus battiscombei Bak. f. in J.B. 67 : 198 (1929) ; L.T.A. : 796 (1930) ; T.S.K. : 66 (1936) ; Bogdan in Nature in E. Afr., No. 4 : 12 (1947) ; T.T.C.L. : 343 (1949). Type : Kenya, *Battiscombe* 93 (K, holo. !)
    Cylicodiscus paucijugus (Harms) Verdc. in K.B. 1950 : 364 (1950)

Note. Harms (l.c.) mentions a specimen, *Braun* 1277 (Tanganyika, Lindi, Nondora), which is said to be very near *N. paucijuga* but with somewhat narrower, longer, lanceolate, acute bracts, which are decidedly sharp ; this specimen has been identified by Dr. P. J. Greenway as *Pseudoprosopis euryphylla* Harms. *Gillman* 1558, Lindi District, may be *N. paucijuga*, but the material is totally inadequate.
    Another specimen from Lindi District, collected by the Chief Conservator of Forests at Mtene on the Rondo Plateau, is very close to *N. paucijuga* but the leaflets are mostly in 3–5 pairs per pinna and only about 1–2 cm. long and 0·5–1·3 cm. wide. This does not exactly match any Kenya material, and it may well be a distinct variety or subspecies. More material is desirable, however, before a verdict is given.

3. **N. hildebrandtii** (*Vatke*) *Torre* in Mendonça, Contrib. Conh. Fl. Moçamb. 2 : 89 (1954) ; Brenan in K.B. 1955 : 181 (1955). Type : Kenya, Teita District, Ndi, *Hildebrandt* 2492 (B, holo, †, BM, K, iso. !)

Tree up to 25 m. high, with rough or sometimes smooth bark. Branchlets puberulous or shortly pubescent when young. Leaf-rhachis with a usually barrel-shaped or cylindrical gland between each pinna-pair ; no glands between leaflet-pairs ; pinnae 4–7 pairs ; leaflets 6–19(–27) pairs, ± oblong or linear-oblong, 3–9 mm. long, 1–3 mm. wide ; lateral nerves often ± raised beneath. Flowers white or creamy, in spikes 4–9 cm. long. Anthers without an apical gland (Fig. 24/11). Pod 9–30 cm. long, 2·2–2·6 cm. wide. Seeds 3·6–5·7 cm. long, 1·5–2·1 cm. wide.

var. **hildebrandtii** ; Brenan in K.B. 1955 : 181 (1955)

Leaflets, except for ciliation, glabrous or sparingly pubescent. Inflorescence-axis puberulous with very small arcuate hairs appearing nearly appressed.

Kenya. Northern Frontier Province : Mathews Range, Ngeng, 20 June 1944, *Mrs. J. Bally* 1 *in Bally* 3601 ! ; Machakos District : Kibwezi, 7 Mar. 1906, *Scheffler* 1171 ! ; Teita District : N. foot of Teita Hills, 7 Feb. 1953, *Bally* 8777 !
Tanganyika. Moshi District : Rau Forest, *H. A. Lewis* 2 ! ; Pare District : Gonja, 4 Feb. 1930, *Greenway* 2120 ! ; Lushoto District : without locality, July 1893, *Holst* 8818 !
Distr. K1, 4, 7 ; T1–3, 5 ; Portuguese East Africa, Southern Rhodesia and Zululand
Hab. In riverine forests and dry areas with high water-table, associated with *Acacia albida*, *A. polyacantha* subsp. *campylacantha* and *Dobera*, and also in bushland ; 120–1100 m.

Syn. *Piptadenia hildebrandtii* Vatke in Oesterr. Bot. Zeitschr. 30 : 273 (1880) ; P.O.A. B : 304 (1895) ; L.T.A. : 793 (1930) ; T.S.K. : 66 (1936) ; Bogdan in Nature in E. Afr., No. 4 : 11 (1947) ; T.T.C.L. : 346 (1949).

Note. Some specimens, e.g. *B. D. Burtt* 3318 (Tanganyika, Shinyanga District, Ningwa R. near Mwantine village, 23 July 1931) and *Savile* 12 (Tanganyika, Dodoma

District, Msakiri, 17 Aug. 1938) have somewhat more pubescence than usual on the leaflets, and may be looked upon as transitional to the following variety.

var. **pubescens** *Brenan* in K.B. 1955 : 131 (1955). Type : Tanganyika, Shinyanga *Koritschoner* 2009 (K, holo. !, EA, iso. !)

Leaflets ± densely pubescent or puberulous on both surfaces. Inflorescence-axis puberulous with small spreading straight or ± flexuous hairs.

Tanganyika. Singida District : Iwumbu R., 10 Mar. 1928, *B. D. Burtt* 1397 ! ; Dodoma District : Manyoni Boma, 18 Feb. 1957, *F. G. Smith* 1388 ! ; Ruaha R. 6 Nov. 1941, *Greenway* 6417 !
Distr. T1, 4, 5, ? 7 ; Portuguese East Africa
Hab. Similar to that of var. *hildebrandtii*, apparently

Doubtful species

N. erlangeri (*Harms*) *Brenan* in K.B. 1955 : 180 (1955). Type : Somalia, Juba Province, *Ellenbeck* 2317 (B, holo.†)

Tree. Leaf-rhachis with a ± cylindrical or ellipsoid gland between each pinna-pair ; no glands between leaflet-pairs ; pinnae 1–4 pairs ; leaflets 5–12 pairs, elliptic or oblong, 7–20 mm. long, 3–8 mm. wide. Spikes up to 10 cm. long, axillary and racemose or paniculate towards branchlet-ends. Anthers without an apical gland. Pods 10–25 cm. long, 1·5–2·2 cm. wide. Seeds 5–5·5 cm. long, 1·6–1·8 cm. wide.

Kenya. Lamu District : Iwezo, 23 Feb. 1955, *Power in E.A.H.* 10971 ! & Boni Forest, about 85 km. from Mkoe, 26 Oct. 1957, *Greenway & Rawlins* 9435 !
Tanganyika. Uzaramo District : Msua, 1 Nov. 1925, *Peter* 31682 !
Hab. Deciduous woodland, *Euphorbia* bushland and thickets
Distr. K7 ; T6 ; Somalia

Syn. *Piptadenia erlangeri* Harms in E.J. 33 : 151 (1902) & in V.E. 3 (1) : 411, fig. 233 (p. 410) (1915) ; L.T.A. : 793 (1930)

The above description is taken from the cited references by Harms and from Somalia specimens kindly sent on loan from the University of Florence. Peter's specimen consists of leafy shoots only, differing from the other material of *P. erlangeri* only in having branchlets puberulous not glabrous or nearly so, also in the hairier leaf-rhachides. Power's specimen likewise lacks flowers and pods, but is less hairy than Peter's and with the leaflets mostly smaller. *Greenway & Rawlins* 9435 is similar but has pods.

The identification of these specimens is thus uncertain and more material is required before the species can be admitted with confidence to our flora.

6. PSEUDOPROSOPIS

Harms in E.J. 33 : 152 (1902)

Shrubs or lianes or sometimes small trees, unarmed. Leaves bipinnate ; petiole and rhachis eglandular ; pinnae each with few to many pairs of opposite leaflets. Inflorescences of racemes which are axillary and solitary or up to three together, and often aggregated on ± shortened shoots. Flowers ☿. Calyx gamosepalous with 5 teeth. Petals 5, free or almost so, valvate in bud (or very slightly overlapping in their lower part in *P. fischeri*). Stamens 10, fertile ; anthers each with a caducous apical gland. Ovary hairy. Pods straight or somewhat curved, compressed, woody, oblong, dehiscing from apex downwards into 2 recurving valves. Seeds lying ± obliquely in the pod, each sunk in a depression in the valve, compressed, brown, glossy, unwinged, elliptic to ± rhombic, without endosperm.

A genus of 4 species, confined to tropical Africa.

Leaflets normally persistently puberulous on both sides,
    oblong-elliptic, 3–5·5(–8) mm. wide, in 7–15 pairs ;
    bracts enlarged and trilobed or obtriangular at apex   1. *P. fischeri*

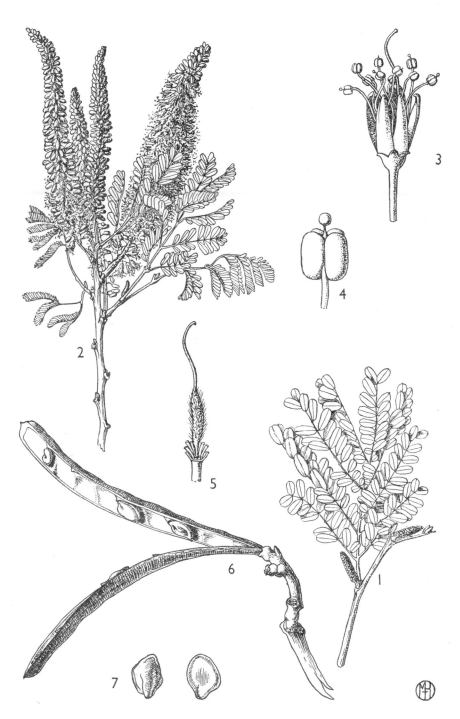

FIG. 7. *PSEUDOPROSOPIS FISCHERI*—1, part of branch showing mature leaf and inflorescence-buds, × 1; 2, part of flowering branch with young leaves, × 1; 3, flower, × 6; 4, anther, × 30; 5, ovary, sepals, petals and stamens removed, × 6; 6, dehisced pod, × 1; 7, seed, two views, × 1½. 1, from *B. D. Burtt* 3504; 2–5, from *B. D. Burtt* 5444; 6–7, from *B. D. Burtt* 1789.

Leaflets glabrous or almost so except on midrib beneath,
    rhombic-ovate or -obovate, (3–)6–14 mm. wide, in
    4–9 pairs ; bracts linear-lanceolate, acute, not
    enlarged at apex .   .   .   . 2. *P. euryphylla*

1. **P. fischeri** (*Taub.*) *Harms* in E.J. 33 : 152 (1902) ; L.T.A. : 806 (1930) ;
T.T.C.L. : 347 (1949) ; Gilb. & Bout. in F.C.B. 3 : 223 (1952). Type :
Tanganyika, Dodoma District, Saranda, *Fischer* 158 (B, holo.†)

Shrub or small spreading tree 3–6 m. high, forming coppice, said sometimes to scramble ; bark silvery-grey, smooth. Young branchlets, petioles
and rhachides densely puberulous ; older branchlets longitudinally striate.
Leaves : petiole 0·5–2·5 cm. long ; rhachis 2·7–6·5 cm. long ; pinnae opposite, 3–5(–6) pairs ; leaflets 7–15 pairs, 0·5–1·4(–1·9) cm. long, 0·25–0·55
(–0·8) cm. wide, oblong-elliptic, rounded or slightly emarginate and mucronate at apex, normally puberulous on both surfaces. Racemes 4–13 cm.
long, on peduncles 0·3–1 cm. long ; pedicels 1·5–5(–7) mm. long ; axes and
pedicels densely puberulous or shortly pubescent ; bracts enlarged and
trilobed or obtriangular at apex. Flowers creamy, honey-scented. Calyx
1–1·5 mm. long, densely puberulous or shortly pubescent. Petals 3·5–5 mm.
long, ± puberulous to almost glabrous outside. Stamen-filaments 6–8 mm.
long. Ovary densely hairy. Pods 8·5–16 cm. long, 1·3–2·2 cm. wide,
attenuate below, blackish when dry, obliquely longitudinally ± striate.
Seeds 10–12 mm. long, 7–9 mm. wide. Fig. 7, p. 27.

TANGANYIKA. Ufipa District : Kasanga, 19 June 1957, *Richards* 10168 ! ; Dodoma
    District : Hika, 10 Dec. 1931, *B. D. Burtt* 3536 ! & Manyoni, Kilimatinde road,
    12 Dec. 1935, *B. D. Burtt* 5385 ! ; Dodoma/Singida Districts : about 32 km. N. of
    Manyoni, 15 Aug. 1948, *van Rensburg* 462 !
DISTR. **T**4, 5 ; Belgian Congo and Northern Rhodesia
HAB. Tall deciduous thickets (thornless) ; dominant in the E. part of the great Itigi
    deciduous thicket in Dodoma District, and extending up the rift wall probably into
    Singida District ; about 900–1220 m.

SYN. *Prosopis fischeri* Taub. in P.O.A. C : 196 (1895)

2. **P. euryphylla** *Harms* in E.J. 49 : 419 (1913) ; L.T.A. : 806 (1930) ;
T.T.C.L. : 347 (1949). Types : Tanganyika, Kihohu, *Koerner* 2258 (B,
syn. †, EA, isosyn.!) & Lindi District, E. slope of Rondo Plateau, *Busse*
2555 (B, syn. †, EA, isosyn.!) & Ruaha–Mtua, *Braun* 1225 (B, syn. †, EA,
K, isosyn.!)

Scandent shrub or small tree 3–6 m. high. Young branchlets, petioles
and rhachides ± puberulous ; older branchlets longitudinally striate.
Leaves : petiole 0·5–2·9 cm. long ; rhachis 1·7–7 cm. long ; pinnae opposite,
2–4 pairs ; leaflets 4–9 pairs, (0·5–)0·7–2·1 cm. long, (0·3–)0·6–1·5 cm. wide,
rhombic-ovate or -obovate, rounded or slightly emarginate and mucronate
at apex, glabrous or almost so except on midrib beneath. Racemes 2–6·5 cm.
long, on peduncles 0·7–2·8 cm. long ; pedicels 1·5–4 mm. long ; axes and
pedicels densely and shortly pubescent ; bracts linear-lanceolate, acute, not
enlarged at apex. Flowers white. Calyx 0·75–1 mm. long, densely and
shortly pubescent. Petals 2·5–3 mm. long, densely puberulous outside.
Stamen-filaments 4–5·5 mm. long. Ovary densely hairy. Pods 7–11 cm.
long, 1·3–1·8 cm. wide, attenuate below, blackish when dry, obliquely ±
longitudinally striate. Seeds about 8 mm. long and 5–7 mm. wide.

TANGANYIKA. Lindi District : Mingoyo, *Gillman* 1366 ! & Rondo Plateau, S. side,
    Dec. 1951, *Eggeling* 6411 ! ; Newala District : Makonde Plateau, *Gillman* 1015 ! &
    Mahuta, 12 Dec. 1942, *Gillman* 1066 !
DISTR. **T**8 ; Portuguese East Africa
HAB. Evergreen thickets on sand or sandy loam ; 450–790 m.

7. **XYLIA**

Benth. in Hook., Journ. Bot. 4 : 417 (1842)

Trees, unarmed. Leaves bipinnate ; petiole bearing a gland at its apex at the junction of the solitary pair of pinnae ; pinnae each with few to many pairs of leaflets. Inflorescences of round heads, pedunculate, axillary, solitary or paired, or sometimes in threes, sometimes ± racemosely aggregated on short shoots. Flowers ♂ or ☿, sessile or pedicellate. Calyx gamosepalous with 5 lobes. Corolla with 5 lobes ± united below, ± pubescent or puberulous outside. Stamens 10, fertile ; anthers each with a caducous apical gland (rarely and only in extra-African species absent). Ovary pubescent. Pods normally obliquely obovate to oblanceolate or dolabriform, compressed, woody, dehiscing from apex downwards into 2 recurving valves. Seeds lying transversely in the pod, each sunk in a depression in the valve, compressed, usually brown, glossy, unwinged, without endosperm.

A genus of about 13 species in the tropics of the Old World, mostly in Africa and Madagascar.
 X. xylocarpa (Roxb.) Taub. from India is recorded in cultivation in our area (T.T.C.L. : 348 (1949) ; Dale, Introd. Trees Uganda : 70 (1953)). It is near *X. africana*, with 4–6 pairs of leaflets, 8–12 pairs of lateral nerves, and a subacute or rather pointed pod 3·5–5·3 cm. wide.

Pedicels distinct even in bud, 2–4 mm. long, densely
  puberulous all over as is the calyx . . . . 1. *X. schliebenii*
Pedicels very short and inconspicuous or absent, 1–
  1·5 mm. long or less ; calyx often glabrescent except
  on lobes . . . . 2. *X. africana*

1. **X. schliebenii** *Harms* in N.B.G.B. 13 : 413 (1936) ; T.T.C.L. : 348 (1949). Type : Tanganyika (see below), *Schlieben* 5752 (B, holo. †, BM, iso. !)

Tree 12–18 m. high. Branchlets puberulous at first. Leaves : petiole 1·2–7 cm. long ; pinnae 6–12 cm. long ; leaflets 4–6 pairs, 2–8 cm. long, 1·3–6 cm. wide, elliptic to obovate or ovate, obtuse, obtusely subacuminate or sometimes rounded at apex, rounded to obtuse or subobtuse at base, glabrescent above, puberulous especially on midrib and lateral nerves beneath. Flowers yellow, in heads on 2–4 cm. long peduncles. Pedicels distinct even in bud, 2–4 mm. long, densely puberulous. Calyx densely puberulous all over, 2 mm. long. Corolla 3·5–4 mm. long ; petals lanceolate, silky-pubescent outside. Pods unknown.

Tanganyika. Lindi District : Mlinguru, 19 Dec. 1934, *Schlieben* 5752 !
Distr. **T8** ; known only from this gathering
Hab. Bushland ; 260 m.

Note. This is imperfectly known, and more material is required to confirm its status and to clarify its relationship to *X. africana*. According to Schlieben's field-notes, *X. schliebenii* has yellow and *X. africana* white flowers ; this too requires confirmation.

2. **X. africana** *Harms* in E.J. 40 : 20, 21 (fig.) (1907) ; L.T.A. : 810 (1930) ; T.T.C.L. : 348 (1949). Types : Tanganyika, Uzaramo District, Pugu Hills, *Holtz* 1065 (B, syn. †) & *Busse* 57 (B, syn. †) & *Engler* 3622 (B, syn. †) ; Lindi District, Netibi, *Busse* 2477 (B, syn. †, EA, isosyn. !)

Small or medium tree 5–15(–20) m. high. Young branchlets puberulous or shortly pubescent. Leaves : petiole 3–10 cm. long ; pinnae 8–18 cm. long ; leaflets 4–9 pairs, 3·5–10 cm. long, 1·5–4·5 cm. wide, obovate-elliptic to oblong or ovate, narrowed or shortly acuminate at apex, obtuse rounded or subacute at base and usually very shortly narrowed into the petiole, glabrous above, very shortly and sparsely puberulous or subglabrous beneath, or the midrib pubescent. Flowers white, in heads on 2–4 cm. long peduncles ;

pedicels thick, inconspicuous and extremely short, 1–1·5 mm. long or less. Calyx silky-puberulous on the lobes, otherwise sparingly hairy or almost glabrous, 2·5 mm. long. Corolla 3–3·5 mm. long ; lobes silky-pubescent outside. Pods oblong or more usually obliquely obovate-oblong or obovate, narrowed at base, obtuse or rounded at apex, 9–13 cm. long, 5·5–6 cm. wide. Seeds 16–17 mm. long, 10 mm. wide, glossy.

Tanganyika. Kilosa, Nov.–Dec. 1920, *Swynnerton* 704 ! ; Masasi District : Nyengedi, 23 Mar. 1943, *Gillman* 1230 ! ; Lindi District : Lake Lutamba, 3 Oct. 1934, *Schlieben* 5428 !
Distr. T6, 8 ; ? Portuguese East Africa. Most material from Portuguese East Africa, as well as Southern Rhodesian and Transvaal specimens named *X. africana*, appears to belong to another species (*X. torreana* Brenan).
Hab. Uncertain, probably woodland ; about 150–300 m.

8. ADENANTHERA
L., Sp. Pl. : 384 (1753) & Gen. Pl., ed. 5 : 181 (1754)

Trees,* unarmed. Leaves bipinnate, pinnae with several to many pairs of alternate* leaflets. Inflorescences of spiciform racemes which are axillary, solitary or sometimes paired, often aggregated at shoot-ends. Flowers normally ⚥. Calyx gamosepalous, ± 5-toothed. Petals 5, free except near base. Stamens 10, fertile, free except at extreme base ; anthers with a caducous apical gland. Pods linear, curved or spirally twisted, longitudinally dehiscing at maturity into 2 coriaceous or subcoriaceous valves which do not separate from the sutures nor lose their outer layer (exocarp). Seeds thick, hard, red or red and black.

A genus of about 8 species in tropical Asia and the Pacific region, one of them widely cultivated in the tropics ; 2 other African species described under *Adenanthera* are probably not congeneric*.
A. microsperma Teijsm. & Binn., from tropical Asia, is cultivated at Amani in Tanganyika (*Braun* 7008 !, *Greenway* 2300 !). It differs from *A. pavonina*, though not very satisfactorily or constantly, in its puberulous calyx and pedicels, pods twisted spirally before dehiscence, and the seeds 5–7 mm. in diameter.

A. pavonina *L.*, Sp. Pl. : 384 (1753) ; L.T.A. : 797 (1930) ; T.T.C.L. : 339 (1949) ; U.O.P.Z. : 106 (1949) ; Gilb. & Bout. in F.C.B. 3 : 219 (1952) ; Dale, Introd. Trees Uganda : 4 (1953). Type : Ceylon, *Hermann* (BM, lecto. !)

Trees 4–20 m. high. Young branchlets usually glabrous. Leaves up to 40 cm. long ; pinnae 3–5 pairs ; leaflets 5–9 on each side of the pinna-rhachis, elliptic to ovate- or obovate-elliptic, 1·5–4·5 cm. long, 1·2–2·3 cm. wide, shortly petiolulate, rounded at apex, minutely puberulous especially beneath (use × 20 lens). Racemes 9–26 cm. long, glabrous or slightly puberulous ; pedicels 2–3·5 cm. long. Flowers yellowish. Calyx 0·75–1 mm. long, usually glabrous. Petals 3–4·5 mm. long. Stamen-filaments 2·5–4 mm. long. Pods 18–22 cm. long, 1·3–1·7 cm. wide, brown outside, after dehiscence the valves reflexing spirally to show the satiny-yellow inner surface and the rather persistent, scarlet, elliptic-lenticular, glossy seeds which are 8–10 × 7–9 mm.

Uganda. Mengo District : Mukono Hill, Dec. 1915, *Dummer* 2694 !
Distr. U4 (not indigenous) ; native of tropical Asia, but introduced into other parts of the tropics
Hab. " Thicket " ; 1310 m.

Note. This species, with attractive seeds, is cultivated in Uganda (Entebbe) and Tanganyika (Moshi, Amani, Dar es Salaam), and is said to be a quite common roadside tree in Zanzibar. It may well become naturalized more often than the evidence suggests.

* Certain African species with climbing habit and opposite leaflets, described under *Adenanthera* but not so far known from our area, are very possibly in the wrong genus.

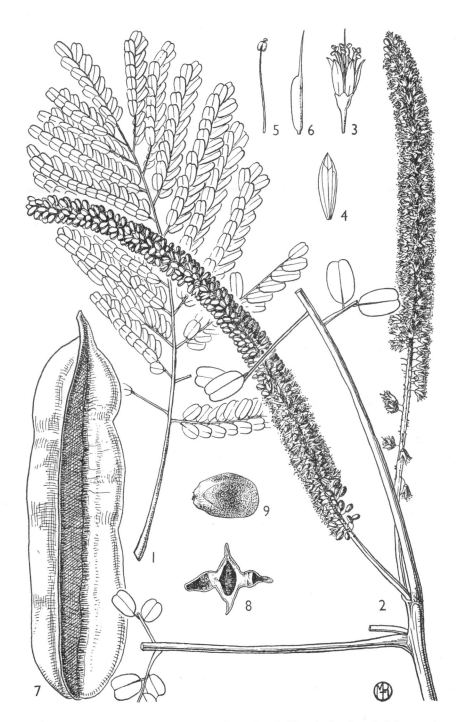

FIG. 8. *TETRAPLEURA TETRAPTERA*—1, leaf, × ½; **2**, part of flowering branch, × 1; **3**, flower, × 4; **4**, petal, × 6; **5**, stamen, × 6; **6**, ovary, × 6; **7**, pod, × ⅓; **8**, cross-section of pod, × ⅓; **9**, seed, × 2. 1–6, from *Chandler* 1190; 7–9, from *Drummond & Hemsley* 3960.

9. TETRAPLEURA

Benth. in Hook., Journ. Bot. 4 : 345 (1841)

Unarmed trees. Leaves bipinnate, eglandular ; pinnae each with few to many alternate leaflets. Inflorescences of axillary, solitary or rarely paired racemes. Flowers ☿. Calyx gamosepalous, with 5 teeth. Petals 5, free. Stamens 10, fertile ; anthers with a caducous apical gland. Pods straight or slightly curved, oblong, woody, indehiscent, the valves with a thick wing-like projection arising from the middle of the valve and running longitudinally for the whole length of the latter, the pod in section thus ± cruciform, internally septate between the seeds. Seeds hard, dark brown, unwinged.

A genus of 2 tropical African species.

T. tetraptera (*Schumach. & Thonn.*) *Taub.* in Bot. Centralbl. 47 : 395 (1891) ; L.T.A. : 803 (1930) ; T.T.C.L. : 348 (1949) ; Gilb. & Bout. in F.C.B. 3 : 218 (1952) ; I.T.U., ed. 2 : 231 (1952) ; Consp. Fl. Angol. 2 : 264 (1956) ; F.W.T.A., ed. 2, 1 : 493, fig. 157 (1958). Type : Ghana, Akwapim, *Thonning* (C, holo.)

Tree 6–30 m. high. Bark smooth to rather rough, grey or brown. Young branchlets glabrous or almost so. Leaves : petiole 4–11 cm. long, glabrous to puberulous ; rhachis 7·5–23 cm. long, puberulous ; pinnae 5–7(–10) pairs, opposite to alternate ; leaflets 6–11(–13) on each side of a pinna, oblong-elliptic to elliptic, (0·4–)0·65–2·1 cm. long, (0·3–)0·55–1·3 cm. wide, rounded to emarginate at apex, appressed-puberulous beneath, glabrous or almost so above, on petiolules 0·5–0·7(–1) mm. long. Racemes 4–14 cm. long, on peduncles 1·5–4 cm. long ; pedicels 1–3 mm. long, puberulous. Flowers yellowish to pinkish. Calyx ± puberulous, 0·5–1 mm. long. Petals 2–3·5 mm. long, 0·5–0·9 mm. wide, puberulous outside towards apex, or rarely glabrous. Stamen-filaments 2·5–4·5 mm. long. Pods 12–23(–25, *fide* F.C.B.) cm. long, 3·5–5·7(–6·5) cm. wide, dark brown, glossy, rounded and sometimes emarginate at apex. Seeds 9–9·5 mm. long, 7–8 mm. wide and 3·5–4 mm. thick. Fig. 8, p. 31.

UGANDA. Bunyoro District : Budongo Forest, Mar. 1933, *Eggeling* 1137 *in F.H.* 1271 ! ; Mengo District : Lwamkima Forest, S. of Mabira, 17 Mar. 1950, *Dawkins* 541 !
KENYA. Kwale District : Buda Mafisini Forest, 13 km. WSW. of Gazi, 22 Aug. 1953, *Drummond & Hemsley* 3960 !
TANGANYIKA. Bukoba District : Minziro F.R., 5 Nov. 1947, *Wye in F.H.* 2268 ! ; Kilosa District : Vigude, Kidodi, Nov. 1952, *Semsei* 1026 ! ; Morogoro District : Nguru Mts., Liwale Valley, 2 Apr. 1953, *Drummond & Hemsley* 2020 ! ; Lindi District : Rondo escarpment, Mchinjiri, Dec. 1951, *Eggeling* 6429 !
DISTR. U1 (*fide* I.T.U.), 2, 4 ; K7 ; T1, 6, 8 ; from Portuguese Guinea and the Sudan in the north to the Gaboon, Belgian Congo and Tanganyika in the south.
HAB. Lowland rain-forest ; 80–1220 m.

SYN. *Adenanthera tetraptera* Schumach. & Thonn., Beskr. Guin. Pl. : 213 (1827)
      Tetrapleura thonningii Benth. in Hook., Journ. Bot. 4 : 345 (1841) ; Oliv., F.T.A. 2 : 330 (1871), *nom. illegit.* Type as for *T. tetraptera*

10. AMBLYGONOCARPUS

Harms in E.J. 26 : 255 (1899)

Unarmed tree. Leaves bipinnate, eglandular ; pinnae each with several pairs of alternate or sometimes subopposite leaflets. Inflorescences of solitary or paired, axillary racemes. Flowers ☿. Calyx gamosepalous, with 5 (rarely and casually 6) teeth. Petals 5 (rarely and casually 6), free. Stamens 10 (rarely and casually 12), fertile ; anthers eglandular at apex, even in bud. Pod straight or nearly so, oblong, woody, indehiscent, in section bluntly tetragonal or even subcylindrical, internally septate between the seeds. Seeds hard, brown, unwinged.

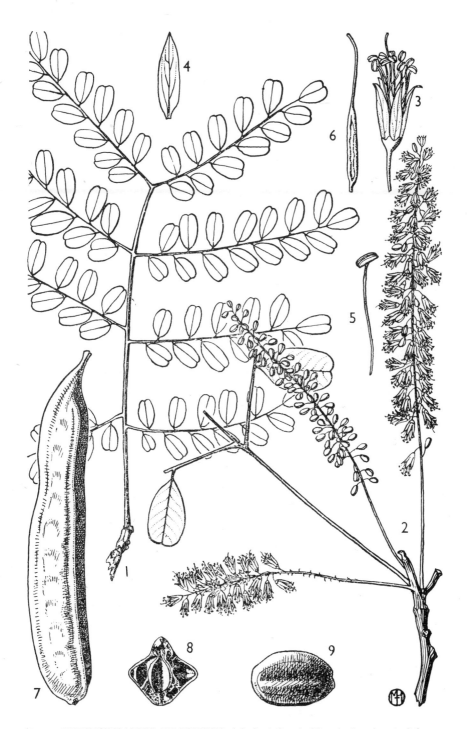

FIG. 9. *AMBLYGONOCARPUS ANDONGENSIS*—**1**, leaf, × ½; **2**, part of flowering branch, × 1; **3**, flower, × 4; **4**, petal, × 6; **5**, stamen, × 6; **6**, ovary, × 6; **7**, pod, × ⅔; **8**, cross-section of pod, × ⅔; **9**, seed × 2. 1, from *Eggeling* 6409; 2–6, from *Eggeling* 3421; 7–9, from *Dalziel* 26.

A genus of a single tropical African species.

Immature pods may have four rather prominent ribs, simulating the shape in section of those of *Tetrapleura*. In *Amblygonocarpus* the whole width of the valve is thickened, while in *Tetrapleura* the thickening is restricted to a narrow band running along the face of each valve. The eglandular anthers of *Amblygonocarpus* also distinguish this genus from *Tetrapleura*.

A. andongensis ([*Welw. ex*] *Oliv.*) *Exell & Torre* in Bol. Soc. Brot., ser. 2, 29 : 42 (1955) ; Consp. Fl. Angol. 2 : 264 (1956) ; F.W.T.A., ed. 2, 1 : 492 (1958). Type : Angola, Cuanza Norte, Pungo Andongo, *Welwitsch* 618 (LISU, ? holo., BM & K, iso.!)

Tree 6–25 m. high, altogether glabrous. Bark reticulate or scaly, grey or brown. Leaves : petiole 4–9 cm. long ; rhachis 2–18 cm. long ; pinnae 2–5(–6) pairs, opposite or subopposite ; leaflets 5–9(–10) on each side of a pinna, elliptic to obovate-elliptic, usually emarginate at apex, 1·2–3 cm. long, 0·7–1·9 cm. wide, on petiolules 1·5–3 mm. long. Racemes (3–)6–17 cm. long, on peduncles 1·5–4·5 cm. long. Flowers yellow, on pedicels 1·5–3·5 (–5) mm. long. Calyx 0·5–1 mm. long. Petals 3–4·75 mm. long, 0·8–1·5 mm. wide. Stamen-filaments 5–6 mm. long. Pods 9–17(–19) cm. long, 2–3·3 cm. wide, brown, glossy, blunt or ± pointed at apex. Seeds 10–13 mm. long, 7–8 mm. wide, 4–5 mm. thick. Fig. 9, p. 33.

UGANDA. West Nile District : Amua, Dec. 1931, *Brasnett* 310! & Mt. Otzi, Sept. 1937. *Eggeling* 3421!
TANGANYIKA. Rufiji District, *F. J. Ross* 96! ; Lindi District : Milolo, 12 Dec. 1942, *Gillman* 1110! & Rondo Plateau, Dec. 1951, *Eggeling* 6409!
DISTR. U1 ; T6, 8 ; Ghana, Nigeria, Ubangi-Shari, the Sudan, French Cameroons, Belgian Congo, Angola, Northern & Southern Rhodesia, Nyasaland and Portuguese East Africa
HAB. Little evidence as far as our area is concerned, but probably in deciduous woodland or wooded grassland especially on sandy soil ; from probably near sea-level to 1370 m.

SYN. *Tetrapleura andongensis* [Welw. ex] Oliv., F.T.A. 2 : 331 (1871)
      T. obtusangula [Welw. ex] Oliv., F.T.A. 2 : 331 (1871). Type : Angola, Cuanza Norte, Golungo Alto, *Welwitsch* 1751 (BM, drawing!)
      Amblygonocarpus schweinfurthii Harms in E.J. 26 : 255 (1899) ; L.T.A. : 804 (1930). Types : the Sudan, Seriba Siber Ruchama, *Schweinfurth* ser. II, 92 (B, syn. †, BM, K, isosyn.!) & Seriba Agad, *Schweinfurth* 1692 (B, syn. †, BM, K, isosyn.!) ; Angola, Malange, *Marques* 23 (B, syn.†)
      A. obtusangulus ([Welw. ex] Oliv.) Harms in E.J. 26 : 256, in obs. (1899) ; L.T.A. : 804 (1930) ; T.T.C.L. : 343 (1949) ; Gilb. & Bout. in F.C.B. 3 : 217 (1952) ; I.T.U., ed. 2 : 223, fig. 50a (1952)
      Tetrapleura andongensis [Welw. ex] Oliv. var. *schweinfurthii* (Harms) Aubrév., Fl. Forest. Soudano-Guin. : 287 (1950)

11. **PROSOPIS**

L., Mant. : 10 (1767)

Shrubs or trees, spinous, prickly, or unarmed. Leaves bipinnate, rarely absent or very reduced ; rhachis glandular at insertion of pinnae, with glands often also between leaflet-pairs ; pinnae each with one to many pairs of opposite or rarely alternate leaflets. Inflorescences of spikes, spiciform racemes or heads. Flowers ⚥. Calyx gamosepalous, with 5 teeth. Petals 5, free or connate below. Stamens 10, fertile ; anthers with an apical gland which may sometimes be sessile and inconspicuous. Pods straight, curved or spirally coiled, woody or coriaceous, subcylindrical or ± compressed, internally septate between the seeds. Seeds hard, unwinged, with endosperm.

A genus of 37 species, one in tropical Africa, two in the Middle East from Cyprus and Turkey to Arabia and western India, the remainder American.

The spiny American species *P. chilensis* (Molina) Stuntz is occasionally cultivated in our area (Uganda, Acholi District, Kitgum, *Eggeling* 1699!)

FIG. 10. *PROSOPIS AFRICANA*—**1**, part of flowering branch, × ½; **2**, gland on leaf-rhachis, × 4; **3**, flower, × 4; **4**, petal, × 6; **5**, stamen, × 6; **6**, anther, front and back views, × 30; **7**, ovary, × 6; **8**, pod, × ⅔; **9**, cross-section of pod, × ⅔; **10**, part of longitudinal section of pod, × ⅔; **11**, seed, × 2. 1–6, 11, from *Chandler* 615; 7–10, from *H. Brown* 65.

P. africana (*Guill. & Perr.*) *Taub.* in E. & P. Pf. III. 3 : 119 (1891) ;
Burkart in Darwiniana 4 : 62 (1940) ; Gilb. & Bout. in F.C.B. 3 : 212
(1952) ; I.T.U., ed. 2 : 230 (1952) ; F.W.T.A., ed. 2, 1 : 492 (1958). Type :
Senegal, Kounoun (? P, holo.)

Tree 4·5–12(–21) m. high, unarmed, with grey, rough, scaly or fissured
bark. Young branchlets shortly pubescent or puberulous. Leaves : petiole
2·5–6·6 cm. long, pubescent or puberulous as in the (0–)2·7–9·5 cm. long
rhachis ; pinnae (1–)2–4 pairs, glandular between most of the pairs of
leaflets ; leaflets opposite, in (5–)7–15 pairs, oblong-lanceolate or elliptic-
lanceolate, (1·3–)1·5–3(–4) cm. long, 0·4–1(–1·5) cm. wide, narrowed to an
usually acute or subacute apex, inconspicuously appressed-puberulous on
both sides. Flowers creamy-white or yellow-green, fragrant, sessile or
nearly so, in 3–6 cm. long spikes borne on 1–3·5 cm. long peduncles. Calyx
1·5–2 mm. long, puberulous. Petals free, 3–4·5 mm. long, glabrous or nearly
so outside. Stamen-filaments 5·5–6·5 mm. long. Ovary hairy. Pods 10–
20 cm. long, 1·5–3·3 cm. in diameter, black or brown, glossy, subcylindrical
or slightly compressed, thickened. Seeds ellipsoid, 8–10 mm. long, 4–9 mm.
wide, blackish-brown, glossy. Fig. 10, p. 35.

UGANDA. West Nile District : Payida, 21 Mar. 1945, *Greenway & Eggeling* 7236 !
 Acholi District : Adilang, Apr. 1943, *Purseglove* 1536 ! ; Teso District : Serere,
 Mar. 1932, *Chandler* 615 !
DISTR. U1, 3 ; Senegal, Gambia, Portuguese and French Guinea, Sierra Leone, Ghana,
 Nigeria, French Cameroons, Ubangi and the Sudan
HAB. Wooded grassland : 910–1220 m.

SYN. *Coulteria* ? *africana* Guill. & Perr. in Fl. Seneg. 256 (1832)
 Prosopis ? *oblonga* Benth. in Hook., Journ. Bot. 4 : 348 (1841) ; F.T.A. 2 : 331
 (1871) ; L.T.A. : 805 (1930). Type : Senegal, Combo and Cayor, *Heudelot*
 14 (K, holo. !)
 Prosopis ? *lanceolata* Benth. in Hook., Journ. Bot. 4 : 347 (1841). Type : the
 Sudan, Sennar, *Kotschy* 387 (K, holo. !)

12. DICHROSTACHYS

(DC.) Wight & Arn., Prodr. Fl. Ind. Or. : 271 (1834)

Desmanthus Willd. sect. *Dichrostachys* DC., Mém. Leg. 12 : 428 (1825)

Shrubs or small trees ; spines present or absent ; prickles absent. Leaves
bipinnate ; rhachis glandular at insertion of pinnae ; pinnae each with
several to many pairs of leaflets. Inflorescences of axillary spikes, solitary
or appearing clustered ; upper part of spike cylindrical, of ☿ flowers, lower
part broader, of differently coloured neuter flowers. Calyx shortly 5-toothed.
Corolla-lobes 5, ± united below. Stamens 10, all fertile in hermaphrodite
flowers ; anthers (in our species) with a stalked apical gland which is caducous. Staminodes of neuter flowers elongate, without anthers. Pods
clustered, coriaceous, narrowly oblong, compressed, usually irregularly
contorted or spiral, indehiscent or opening irregularly or (not in our species)
dehiscent. Seeds (in the African species at least) ± compressed, ovoid to
ellipsoid, smooth.

A genus of about 20 species in the tropics of the Old World from Africa to Australia,
most in Madagascar. The generic limits require revision, however (see p. 40).

D. cinerea (*L.*) *Wight & Arn.*, Prodr. Fl. Ind. Or. : 271 (1834). Type :
Ceylon, *Hermann* Mus. Zeyl. No. 215 (BM., syn. !)

Shrub or small tree 1–8(–12) m. high, sometimes suckering and thicket-
forming or (*fide* Greenway) even scandent, with rough bark and armed with
spines terminating short lateral spreading twigs which often bear leaves
and flowers. Young branchlets ± pubescent. Leaves with (2–)5–19(–21
fide F.C.B.) pairs of pinnae ; rhachis (with petiole) 0·5–20 cm. long, with

one stalked gland between each pair of pinnae ; leaflets 9–41 pairs, 1–11 (–14) mm. long, 0·3–4(–5·5) mm. wide, linear to oblong. Inflorescences yellow in apical hermaphrodite part, mauve, pink or sometimes white in lower neuter part, 2–5 cm. long, pendent on solitary or apparently fascicled peduncles 1–9 cm. long. Calyx 0·6–1·25 mm. long. Corolla 1·5–3 mm. long. Stamen-filaments of hermaphrodite flowers 3·25–3·5 mm. long ; staminodes 4–17 mm. long. Pods 2–10 cm. long, 0·5–2(–2·5 *fide* F.C.B.) cm. wide. Seeds 4–6 mm. long, 3–4·5 mm. wide, deep brown, glossy.

subsp. **cinerea** ; Brenan in K.B. 1957 : 358 (1958)

Leaflets 0·3–1·5 mm. wide (rarely wider). Inflorescences solitary or fascicled. Fig. 11, p. 38.

UGANDA. Karamoja District : Kakumongole, 6 Jan. 1937, *A. S. Thomas* 2193 ! ; Teso District : Kasilo, Mar. 1932, *Chandler* 613 ! ; Toro District : Mohokya, 10 Sept. 1941 *A. S. Thomas* 3952 !
KENYA. Northern Frontier Province : Moyale, 23 Apr. & Dec. 1952, *Gillett* 12910 ! ; Masai District : near Mara R., Feb. 1932, *Rammell* 2755 ! ; Kwale District : Shimba Hills, Pengo Forest, 9 Feb. 1953, *Drummond & Hemsley* 1178 !
TANGANYIKA. Shinyanga, *Koritschoner* 1738 ! ; Tabora District : near Urambo, 7 Oct. 1949, *Bally* 7535 ! ; Mpwapwa District : Kiboriani Mts., 10 Jan. 1938, *Hornby* 882 !
ZANZIBAR. Zanzibar Is., Kiwengwa–Kinyasini, 29 Jan. 1929, *Greenway* 1248 ! & Chwaka, 21 Dec. 1930, *Vaughan* 1741 ! ; Pemba, Vumba–Shengejun, 19 Feb. 1929, *Greenway* 1502 !
DISTR. U1–4 ; K1, 2, 4, 7 ; T1–8 ; Z ; P ; widespread in tropical and subtropical Africa from the Cape Verde Islands, the Gambia and the Sudan southwards to Natal and South West Africa ; also in Arabia, tropical Asia and Australia ; introduced into Florida and Cuba
HAB. Various : deciduous bushland, scrub, wooded grassland, deciduous woodland, even occurring near forest and in the more open parts of swamp-forest ; forming secondary bush in native cultivation areas, and an indicator of overgrazing in low rainfall areas (T.T.C.L.) ; a pioneer on sites previously occupied by the Masai and their cattle (Greenway) ; near sea-level to 1710 m.

SYN. *Mimosa cinerea* L., Sp. Pl. : 520, No. 25 (1753)
    M. glomerata Forsk., Fl. Aegypt. Arab. : 177 (1775). Type : Mimosa glomerata; "foliis bipinnatis ; leguminibus nigris, contorto-glomeratis." (Forsk., Fl. Aegypt. Arab. : 177 (1775)—no specimen in Forskål's herbarium)
    M. nutans Pers., Syn. 2 : 266 (1807). Type : Senegal, *Adanson* (P–JU, holo.)
    Dichrostachys nutans (Pers.) Benth. in Hook., Journ. Bot. 4 : 353 (1841) ; F.T.A. 2 : 333 (1871) ; P.O.A. C : 195 (1895)
    D. platycarpa [Welw. ex] Oliv., F.T.A. 2 : 333 (1871). Types : Angola, Cuanza Norte, Golungo Alto, *Welwitsch* 1797, 1797b (BM, isosyn. !)
    D. nutans (Pers.) Benth. var. *grandifolia* Lanza in Boll. Ort. Giard. Col. Palermo, 8 : 106 (1909) ; F.R. 9 : 413 (1911). Type : Eritrea, Hamasen, between Ghinda and Filjil, *Senni* 256 (PAL, holo. !)
    D. glomerata (Forsk.) Chiov. in Ann. Bot. Roma 13 : 409 (1915) ; Hutch. & Dalz. ex Greenway in K.B. 1928 : 204 (1928) ; L.T.A. : 807 (1930) ; T.S.K. : 66 (1936) ; Bogdan in Nature in E. Afr., No. 4 : 11 (1947) ; F.P.N.A. 1 : 382 (1948) ; T.T.C.L. : 344 (1949) ; U.O.P.Z. : 230, 231, fig. (1949) ; I.T.U., ed. 2 : 224 (1952) ; Gilb. & Bout. in F.C.B. 3 : 202 (1952) ; Consp. Fl. Angol. 2 : 265 (1956) ; F.W.T.A., ed. 2, 1 : 494, fig. 158 (1958)
    "*Acacia* sp. nr. *erubescens* Welw." sensu Battiscombe, Cat. Trees Kenya Col. : 54 (1926)
    D. glomerata (Forsk.) Chiov. var. *grandifolia* (Lanza) Bak. f., L.T.A. : 808 (1930) ; Chiov., Racc. Bot. Miss. Consol. Kenya : 40 (1935)
    D. glomerata (Forsk.) Chiov. subsp. *glomerata*; Brenan in K.B. 1956 : 187 (1956).

NOTE. Very variable in indumentum, size of leaves, number, shape and size of leaflets, etc. In E. Africa two of the most distinct forms included under subsp. *cinerea* are the following. In our area subsp. *cinerea* normally has obtuse to subacute leaflets, but in Uganda plants occur with acute leaflets. These correspond with *D. platycarpa* [Welw. ex] Oliv. and seem to prefer moister habitats than usual ; representative specimens of this are : Busoga District : Buyamtole, Mar. 1931, *Harris* 7 *in F.H.* 74 ! ; Mengo District : Kireka, Feb. 1932, *Eggeling* 203 *in F.H.* 430 ! & Mabira Forest near Najembe, 17 Mar. 1950, *Dawkins* 542 ! Secondly, there is a form with less indumentum and very small leaflets that seems characteristic of the Kenya coast which is represented by the following : Kilifi District : Mida, *R. M. Graham* A510 *in F.H.* 1907 ! & Kibarani, 21 Feb. 1945, *Jeffery* K83 ! According to Dr. P. J. Greenway a scandent

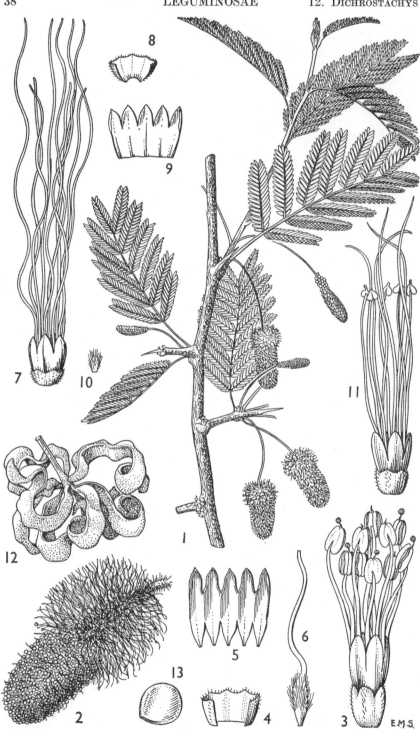

FIG. 11. *DICHROSTACHYS CINEREA* subsp. *CINEREA*—1, part of flowering branch, × ⅔; 2, inflorescence, × 2; 3, ♀ flower, × 12; 4, calyx, × 12; 5, corolla, × 12; 6, ovary, × 12; 7, neuter flower, × 12; 8, calyx of neuter flower, × 12; 9, corolla of neuter flower, × 12; 10, rudimentary ovary of neuter flower, × 12; 11, neuter flower showing intermediate stage in reduction of stamens, × 12; 12, cluster of pods, × ⅔; 13, seed, × 3. 1–11, from *Drummond & Hemsley* 1178; 12–13, from *B. D. Burtt* 1762.

form occurs in evergreen bushland on the Kenya coast, extending southwards to the Pangani River in Tanganyika. These extremes are, however, linked with the main run of subsp. *cinerea* by so many intermediates that at present it seems better to look on them as merely parts of its range of variation.

Mr. H. J. de S. Disney has made careful and interesting observations on *D. cinerea* in the region of Dodoma in Tanganyika. These indicate the presence of two forms, one of the dark brown to black soils of " mbugas " (*Disney* 57/1 !, 57/5 !), the other of dry soils (*Disney* 57/4 !, 57/6 !). The first form is more upright and less spreading than the second, with grey, fairly smooth bark, comparatively few pinnae (3–6 pairs per leaf) and larger leaflets than the second, with raised venation beneath ; the second form is more rounded and flatter in its habit, with usually blackish-brown bark, pinnae about 11–18 pairs per leaf, and very small leaflets without raised venation beneath. Near Dodoma these two forms seem distinct enough, but it is hard to link them with material from elsewhere. Further evidence is needed about their occurrence.

The diversity of habitat and altitude suggests that subsp. *cinerea* may be an aggregate of ecotypes with minor morphological distinctions. Experimental cultivation in East Africa under controlled conditions of plants or seed from various habitats should help to prove or disprove this theory.

Even the following subsp. *nyassana* is not clear-cut, and a number of specimens can only be looked upon as intermediate between it and subsp. *cinerea*. Such are :—

UGANDA. Teso District : Serere, Apr. 1932, *Chandler* 644 !
KENYA. Machakos, 2 Jan. 1931, *van Someren* 1639 !
TANGANYIKA. Mwanza District : Ngasama, Nassa, 24 Nov. 1951, *Tanner* 495 ! ; Handeni District : Kangata, Nov. 1949, *Semsei in F.H.* 2894 ! ; Morogoro District : Uluguru Mts., Nov. 1934, *E. M. Bruce* 209 !

The specimen from Kenya, ? S. Nyeri District : Moranga, *Balbo* 215 (TOM !) recorded as *D. glomerata* var. *grandifolia* (Lanza) Bak. f. by Chiovenda (Racc. Bot. Miss. Consol. Kenya: 40 (1935)), is also one of these intermediates.

subsp. **nyassana** (*Taub.*) *Brenan* in K.B. 1957 : 358 (1958). Type : Nyasaland, Buchanan 195 (B, holo.†, K, iso. !)

Leaflets 2–7 mm. wide. Inflorescences usually fascicled, sometimes solitary.

UGANDA. Ankole District : Ruborogoto, Oct. 1932, *Eggeling* 684 *in F.H.* 1058 !
KENYA. Kiambu/Fort Hall Districts : 10 km. beyond Thika, 31 Oct. 1938, *Bally in C.M.* 8447 !
TANGANYIKA. Buha District : Kasulu, Nov. 1930, *Rounce* B5 ! ; Dodoma District : Manyoni, 21 Nov. 1931, *B. D. Burtt* 3420 ! ; Mpwapwa, 24 Dec. 1933, *Hornby* 569 ! ; Rungwe District : Kyimbila, 26 Nov. 1910, *Stolz* 431 !
DISTR. U2 ; K4 ; T1, 4–8 ; Belgian Congo, ? Ethiopia, Portuguese East Africa, Nyasaland, Northern and Southern Rhodesia and the Transvaal ; ? Angola
HAB. Deciduous woodland (*Brachystegia-Julbernardia* and *Combretum*) and wooded grassland ; 300–1625 m.

SYN. *Dichrostachys nyassana* Taub. in P.O.A. C : 195 (1895) ; L.T.A. : 807 (1930) ; T.T.C.L. : 344 (1949) ; I.T.U., ed. 2 : 225 (1952) ; Gilb. & Bout. in F.C.B. 3 : 199 (1952) ; Consp. Fl. Angol. 2 : 265 (1956)
D. glomerata (Forsk.) Chiov. subsp. *nyassana* (Taub.) Brenan in K.B. 1956 : 188 (1956)

NOTE. When extreme, distinct and easily recognized, but the diagnostic characters are insufficient and too inconstant for it to be maintained as a species (see note above under subsp. *cinerea*). At low altitudes subsp. *nyassana* seems less common than subsp. *cinerea* or completely absent.

Doubtful species

D. sp. A

Bush to 2 m., similar in appearance to *D. cinerea*, but leaves all with only 2–3 pairs of pinnae ; rhachis (with petiole) 0·7–3 cm. long ; leaflets 6–11 pairs, 4–8 mm. long, 1·5–3·5 mm. wide, obovate-oblong. Flowers and pods similar to those of *D. cinerea*.

KENYA. Northern Frontier Province : Wajir, Jan. 1955, *Hemming* 450 ! & 483 !
HAB. Bushland with *Acacia tortilis*, *Delonix elata*, mixed *Commiphora* etc. on red sandy soil.

NOTE. This may perhaps be no more than a very unusual extreme of *D. cinerea* with fewer pinnae and fewer, broader, and more obovate leaflets than usual, but it seems better to keep it provisionally separate, to await further collections and observations. It has been suggested that this plant might be *D. benadirensis* Chiov., but that must, from the description, have a very different pod, similar to that of Species B, below.

D. sp. B

Shrub 2 m. high. Young stems pubescent, older glabrescent and grey to grey-brown with rather few lateral branches diverging at wide angles and spinescent at ends. Internodes very short (5 mm. or less). Leaves very small, up to 9 mm. long in all, with a single pair of short pinnae, each pinna with 4–6 pairs of leaflets ; leaflets tiny, pubescent, up to 2·5 mm. long and 1 mm. wide, lateral nerves raised beneath. Flowers unknown. Pods brown, straight, oblanceolate, about 3–5 cm. long and 0·8 cm. wide, the sutures much thickened, dehiscing from apex downwards, the valves recurving.

KENYA. Northern Frontier Province : 22 km. SSW. of El Wak, 26 May 1952, *Gillett* 13351 !

HAB. In rich *Commiphora-Acacia* open scrub on red sandy soil ; about 400 m.

NOTE. The exact generic position of this plant can only be settled when the flowers are known. It does not fit any described species of *Dichrostachys*, and more material is thus very much wanted.

The straight, dehiscent pods are very different from those of typical *Dichrostachys* (*D. cinerea*), but similar to those of *D. dehiscens* Balf. f. and, probably, *D. kirkii* Benth. and also to those of certain Madagascar species described either under *Dichrostachys* or *Desmanthus* Willd., a genus which seems to be at present artificially separated from *Dichrostachys* by the absence of glands on the anthers.

These genera may well require recasting, so that *Dichrostachys* would be characterized by contorted indehiscent pods and *Desmanthus* by straight dehiscent ones. The further problem of whether, after such recasting, species such as *D. dehiscens* could be maintained in the same genus as *Desmanthus virgatus* (L.) Willd., whose pods are considerably different, would also require investigation.

13. **NEPTUNIA**

Lour., Fl. Coch. : 653 (1790)

Herbs, aquatic or terrestrial, unarmed. Leaves bipinnate ; pinnae each with several to numerous pairs of leaflets. Inflorescences of globose to ellipsoid heads which are solitary and axillary. Flowers in upper part of head ♀, in lower part of head ♂ or neuter with ± elongate staminodes. Calyx 5-toothed. Corolla-lobes 5, free or ± united. Stamens 5 or 10, free, all fertile in ♀ flowers ; anthers glandular or not at apex. Pods clustered, membranaceous to subcoriaceous, oblong to suborbicular, compressed, not contorted or spiral, dehiscent. Seeds ± compressed, oblong-ellipsoid to obovoid, smooth.

A genus of 14 or more species, widely distributed and mostly tropical. In Africa, however, only one species occurs.

N. oleracea *Lour.*, Fl. Coch.: 654 (1790) ; Oliv., F.T.A. 2 : 334 (1871) ; Benth. in Trans. Linn. Soc. 30 : 383 (1875) ; Consp. Fl. Angol. 2 : 267 (1956) ; F.W.T.A., ed. 2, 1 : 496 (1958). Type : Cochinchina, *Loureiro* (BM, ? holo. !)

Aquatic herb with creeping stems usually floating, swollen, and rooting especially at nodes, glabrous or rarely puberulous when young. Leaves very sensitive ; stipules obliquely ovate, 5–9 mm. long, 3–5 mm. wide, thin ; petiole 2·5–9 cm. long ; rhachis 1·1–4·2(–6·5) cm. long ; pinnae 2–4 pairs ; leaflets 7–22 pairs, oblong, 5–20 mm. long, 1·5–4 mm. wide, basal ones smaller, glabrous or with a few hairs on margins. Flowers yellow, in heads 1·5–2·5 cm. long ; peduncles 6·5–23(–30) cm. long. Calyx 1–3 mm. long. Corolla about 3–4 mm. long. Stamens 10 ; anthers eglandular at apex, even in bud ; staminodes up to 17–21 mm. long. Pods bent at an angle to the short basal stipe, shortly oblong, 1·3–2·7(–3·8) cm. long, 1–1·2 cm. wide. Seeds 5–5·5 mm. long and 3–3·5 mm. wide. Fig. 12.

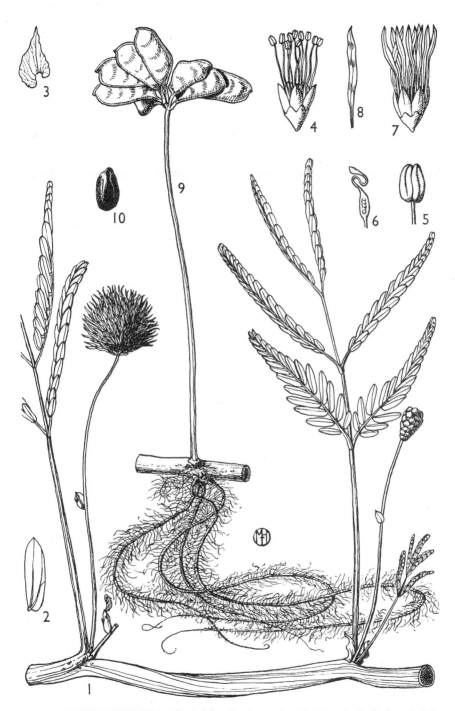

FIG. 12. *NEPTUNIA OLERACEA*—**1,** part of flowering stem, × 2; **2,** leaflet, × 2; **3,** stipule, × 2; **4,** ♀ flower, × 2; **5,** anther, × 10; **6,** ovary, × 5; **7,** neuter flower, × 2; **8,** staminode from neuter flower, × 2; **9,** part of fruiting stem, × 2; **10,** seed, × 2. 1–8, from *Bally* 6133; 9–10, from *Peter* 44973.

KENYA. Northern Frontier Province : Kolbio, *Ritchie in C.M.* 1910 ! ; Lamu District : Kiunga, 23 Oct. 1947, *Mrs. J. Adamson* 434 *in Bally* 6133 !
TANGANYIKA. Dodoma District : Ndachi, May 1938, *Savile* 41 ! ; " Rufiji," 3 Feb. 1931, *Musk* 40 ! ; Mafia Is., Kiwawe, 1 Sept. 1937, *Greenway* 5202 !
ZANZIBAR. Pemba, Vitongoge, 15 Feb. 1929, *Greenway* 1455 !
DISTR. K1, 7 ; T5, 6 ; P ; tropics of Old and New Worlds
HAB. In and by fresh water of pools, lakes and swamps ; 10–1220 m.

SYN. *Mimosa prostrata* Lam., Encycl. 1 :   10 (1783), excl. β *M. natans* L.f., *nom. illegit.* Types : *Niti-todda-vaddi* Rheede Hort. Malabar. 9 : 35, t. 20 (1689) (syn. !) ; *Mimosa orientalis non spinosa* . . . Pluk., Almagest. Bot. : 252, t. 307, fig. 4 (1696) (syn. !)
      Neptunia prostrata (Lam.) Baill. in Bull. Soc. Linn. Par. 1 : 356 (1883) ; L.T.A. : 809 (1930) ; Bogdan in Nature in E. Afr., No. 4 : 11 (1947) ; U.O.P.Z. : 379, fig. (1949) ; Gilb. & Bout. in F.C.B. 3 : 198 (1952)

14. MIMOSA

L., Sp. Pl. : 516 (1753) & Gen. Pl., ed. 5 : 233 (1754)

Mostly herbs or shrubs, rarely trees, sometimes scrambling or climbing ; prickles usually present. Leaves bipinnate, or the pinnae seeming almost digitate on account of the very short rhachis, rarely (not in our species) absent or modified to phyllodes ; pinnae each with few to many pairs of leaflets. Inflorescences of ovoid or subglobose heads or (not in our species) spikes, which are axillary, solitary or more usually clustered and often ± aggregated. Flowers ⚥ or ♂, small, sessile. Calyx very small, irregularly laciniate or denticulate in our species. Corolla gamopetalous, 4- or sometimes 3-, 5- or 6-lobed. Stamens as many as or twice as many as the corolla-lobes, fertile. Anthers without any apical gland. Pods straight to circinate, flat, in our species ± bristly or prickly ; at maturity the valves between the sutures splitting ± transversely into 1-seeded segments or rarely (not in our species) remaining entire ; exocarp (at least in our species) not separating from the endocarp ; sutures persistent.

A genus of about 450–500 species, widely distributed through the tropics, but the vast majority of the species South American.
The two following species from South America are cultivated in our area:—*M. elliptica* Benth., similar to *M. pigra*, but with flowers in oblong-elliptic heads, and usually sparsely and shortly bristly pods (Tanganyika, Lushoto District, Amani, 2 Nov. 1928, *Greenway* 966 !) ; *M. scabrella* Benth., an unarmed tree with dense minute stellate or branched indumentum, and yellow flowers (Uganda, Bunyoro District, Budongo, Sept. 1941, *Eggeling* 4490 ! ; Tanganyika, Iringa District, Mufindi, 29 July 1955, *Benedicto* 51 !), referred to under the synonymous name *M. bracaatinga* Hoehne by Dale, Introd. Trees Uganda : 51 (1953).

Leaves ± prickly on petiole and/or rhachis ; pinnae in
      3–20 pairs, not subdigitate, arranged along the
      rhachis, which is usually as long as or longer than
      the petiole ; stamens (7–)8 :
Rhachis of the leaf with a short or long, straight, ±
      erect or forward-pointing, slender prickle at the
      junction of each of the pairs of pinnae, often with
      other prickles also ; leaflets often setulose on
      margins, in 20–42 pairs ; pods densely bristly all
      over, 0·9–1·4 cm. wide   .   .   .   .   .   1. *M. pigra*
Rhachis of the leaf without prickles at junctions of
      pinna-pairs, though often prickly otherwise ;
      leaflets not setulose ; if pods with bristles all
      over, then not more than 0·45 cm. wide :
Prickles on stems and leaves upcurved towards
      tip ; stems glabrous or very nearly so ; calyx
      about 1·4 mm. long ; prickles on leaf-rhachis
      conspicuously broad-based   .   .   .   .   3. *M. suffruticosa*

Prickles on stems and leaves downwardly hooked or
bent ; stems ± pubescent ; calyx 0·4–0·8 mm.
long ; prickles on leaf-rhachis not especially
broad-based :
> Leaflets in 3–6 pairs, (5–)8–18 mm. long, (2–)
> 5–10 mm. wide, venation ± visible beneath ;
> pods prickly on margins only or ± unarmed,
> 6–10 cm. long, 1·5–2·5 cm. wide . .   2. *M. busseana*
> Leaflets in 11–30 pairs, 2–6 mm. long, 0·7–1·5 mm.
> wide, venation not or scarcely visible ; pods
> shortly bristly on margins and sides, 1·5–
> 3·5 cm. long, 0·4–0·45 cm. wide. . . 4. *M. invisa*

Leaves without prickles on petiole or rhachis (though
bristly hairs may be present) ; pinnae in 1–2 pairs,
subdigitately arranged on the very short rhachis
which is much exceeded by the petiole ; stamens
4 ; pods setose-prickly on margins only . . 5. *M. pudica*

1. **M. pigra** *L.*, Cent. Pl. 1 : 13 (1755) ; Fawc. & Rendle, Fl. Jam. 4 :
135 (1920) ; T.T.C.L. : 346 (1949) ; Gilb. & Bout. in F.C.B. 3 : 230 (1952) ;
Consp. Fl. Angol. 2 : 268 (1956) ; F.W.T.A., ed. 2, 1 : 495 (1958). Type :
Aeschynomene spinosa quinta Commelin, Rar. Pl. Amst. : 59, t. 30 (1697)
(lecto. !)

Shrub 0·6–3(–4·5) m. high, sometimes scandent or rambling. Stems
armed with broad-based prickles up to 7 mm. long, also usually ± appressed-
setose. Leaves sensitive ; petiole 0·3–1·5 cm. long ; rhachis 3·5–12(–18)
cm. long, with a straight ± erect or forward-pointing slender prickle (sometimes short) at the junction of each of the 6–14(–16) pairs of pinnae, sometimes with other stouter spreading or deflexed prickles between the pairs ;
leaflets 20–42 pairs, linear-oblong, 3–8(–12·5) mm. long, 0·5–1·25(–2) mm.
wide ; venation nearly parallel with midrib, margins often setulose. Flowers
mauve or pink, in subglobose pedunculate heads about 1 cm. in diameter,
1–2(–3) together in upper axils. Calyx minute, laciniate, 0·75–1 mm. long.
Corolla about 2·25–3 mm. long. Stamens 8. Pods clustered, brown, densely
bristly all over, 3–8 cm. long, 0·9–1·4 cm. wide, breaking up transversely
into segments 3–4 mm. long, the sutures persisting as an empty frame.
Fig. 13, p. 44.

UGANDA. Acholi District : Madi, Dec. 1862, *Speke & Grant* 575 ! ; Ankole District :
Kifunfu near Mbarara, Dec. 1930, *Harris in F.H.* 45 ! ; Busoga District, Mar. 1931,
Harris 12 *in F.H.* 78 !
KENYA. Northern Frontier Province : Balambala, Sept. 1947, *Mrs. J. Adamson* 404
in Bally 5890 ! ; Kisumu-Londiani District : Kisumu, 14 Mar. 1951, *Bogdan* 2922 ! ;
Kilifi District : Sokoke, 16 Apr. 1945, *Jeffery* K165 !
TANGANYIKA. Mwanza, Aug. 1932, *Rounce* 196 ! ; Tanga District : Kange Estate,
2 Nov. 1951, *Faulkner* 816 ! ; " Rufiji," 6 Dec. 1930, *Musk* 7 !
ZANZIBAR. Zanzibar Is., Zingwe-Zingwe, 21 Jan. 1929, *Greenway* 1106 !
DISTR. U1–4 ; K1, 4–7 ; T1, 3–8 ; Z ; widespread in tropical Africa and America,
also in Madagascar and Mauritius ; in Asia apparently only a rare introduction ;
not in Australia
HAB. Swamps, especially along rivers and lake-shores ; 2–1520 m.

SYN. *Mimosa asperata* L., Syst. Nat., ed. 10 : 1312 (1759) ; L.T.A. : 812 (1930) ;
T.S.K. : 66 (1936) ; Bogdan in Nature in E. Afr., No. 4 : 11 (1947) ; F.P.N.A.
1 : 381 (1948). Type : *Herb. Linnaeus* 1228.32 (LINN, holo. !)

VARIATION. The setose hairs clothing the stems and the rhachides of leaves and pinnae
are normally appressed, but are ± spreading in e.g. three Uganda specimens, *Harris*
45 (K !, see above), *Jarrett* 486 (Ruizi R., EA !) and *Chandler* 927 (Teso District, K !) ;
also in *Culwick* 54 and *Schlieben* 1089 (both from Tanganyika, Ulanga District, K !).
The armature also varies somewhat : some plants appear to lack the prickles on the
leaf-rhachis between the pairs of pinnae, while in others they may be well developed ;
however, every intermediate condition can also be found.

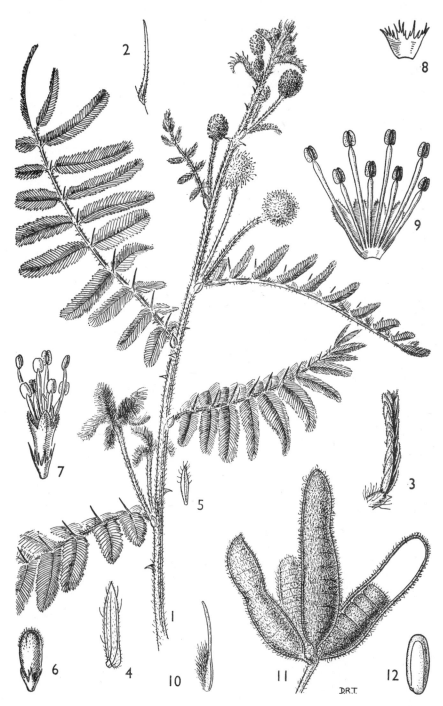

FIG. 13. *MIMOSA PIGRA*—**1,** part of flowering stem, × 1; **2,** setiform hair from peduncle, × 6; **3,** part of pinna showing leaflets closed up in " sleeping " condition, × 4; **4,** leaflet, × 4; **5,** bract subtending flower, × 6; **6,** flower-bud, × 6; **7,** flower, × 6; **8,** calyx, opened out, × 6; **9,** corolla and stamens, opened out × 6; **10,** ovary, × 6; **11,** pods, × 1; **12,** seed, × 3. From *Harris* 45.

2. **M. busseana** *Harms* in E.J. 49 : 419 (1913) ; L.T.A. : 813 (1930) ; T.T.C.L. : 345 (1949). Types : Tanganyika, Masasi/Newala Districts, Makonde Plateau, near Mkomadatchi, *Busse* 1086 (B, syn. †, EA, isosyn. !) ; Lindi District, Muera [Rondo] Plateau near Mpunga, *Busse* 2621 (B, syn. †, EA, isosyn. !)

Scandent shrub 1·5–2 m. high. Stems shortly pubescent and densely armed with downwardly hooked prickles up to 2·5 mm. long. Leaves with petiole (1·9–)3·3–4·5 cm. long and prickly ; rhachis about 3·5–5·5 cm. long, prickly ; pinnae (2–)3–7 pairs ; leaflets 3–6 pairs, ± obovate-elliptic, (5–)8–18 mm. long, (2–)5–10(–12) mm. wide, ± appressed-pubescent or puberulous on both surfaces, not setulose, venation pinnate, ± visible beneath. Flowers pink or mauve, in subglobose heads about 0·7–1·5 cm. in diameter, on pubescent unarmed or sparingly prickly peduncles 1–3 cm. long, 1–6 together from axils, usually aggregated into a panicle. Calyx 0·5–0·8 mm. long. Corolla (2–)2·5–3 mm. long. Stamens 7–8. Pods ± curved, with scattered recurved prickles on margins only (sometimes prickles very sparse), 5–10 cm. long, 1·2–2·5 cm. wide.

TANGANYIKA. Lindi District : Mingoyo, 25 May 1949, *Gillman* 1453 ! & Mlinguru, 3 Jan. 1935, *Schlieben* 5827 ! & Rondo Plateau, Feb. 1951, *Eggeling* 6052 ! & 9 Dec. 1955, *Milne-Redhead & Taylor* 7594 !
DISTR. T8 ; Portuguese East Africa
HAB. Secondary deciduous bushland ; said by Gillman to grow " on pale yellow sands " ; 280–340 m.

3. **M. suffruticosa** (*Vatke*) *Drake*, Hist. Pl. Madag. 1 : 30, 45 (1902). Type : Madagascar, *Hildebrandt* 3046 (B, holo. †, K, iso. !)

Shrub. Stems glabrous or very nearly so, very sparingly armed with upcurved prickles about 2·5 mm. long. Leaves : petiole 0·7–2·1 cm. long, unarmed ; rhachis 6·5–18(–20 *fide* Drake) cm. long, armed with many brown upcurved prickles whose bases are compressed and broadly triangular ; no prickles on rhachis at junction of each of the 4–20 pairs of pinnae ; leaflets 4–7 pairs, shortly oblong to obovate-oblong, 5–11 mm. long, 2·5–5·5 mm. wide, rounded at apex, appressed, puberulous both sides, not setulose, venation (other than midrib and basal nerves) not visible beneath. Flower-heads subglobose, about 0·7–0·9 cm. in diameter, ± racemosely arranged, 1–2 together from axils, on unarmed almost glabrous peduncles 4–7 mm. long. Calyx 1·4 mm. long. Corolla 2·5 mm. long. Stamens 8. Pods slightly curved, 4–9 cm. long, 0·9–1·4 cm. wide, puberulous over surface, not bristly, unarmed or (*fide* Drake) prickly on margins.

ZANZIBAR. Zanzibar Is., without locality, 1833, *Bojer* !
DISTR. Z ; Madagascar
HAB. Unknown

SYN. *M. decurrens* [Boj. ex] Benth. in Trans. Linn. Soc. 30 : 420 (1875) ; Drake, Hist. Pl. Madag. 1 : 45 (1902) ; L.T.A. : 813 (1930), *non* Wendl. (1798) ; *nom. illegit.* Type : Zanzibar, *Bojer* (G, holo. !, K, photo. !)
Entada suffruticosa Vatke in Linnaea 43 : 108 (1880–82)

NOTE. Careful search is desirable to ascertain if it is still on Zanzibar. Bojer, Hort. Maurit. : 113 (1837) and Drake (*l.c.*) give *M. decurrens* as native of Madagascar only, and it is thus possible that the label of the holotype bears a wrong locality, although it seems authentic.
      The holotype of *M. decurrens* is in flower, and the only material of *E. suffruticosa* that I have seen is a fruiting isotype. I can see no evidence for maintaining them as separate, but allowance must be made for the specimens not being entirely comparable. Drake (*l.c.*, p. 45) suspected that *M. suffruticosa* might be only a variety of *M. decurrens*, but I do not see that even that separation is justified.

4. **M. invisa** [*Mart. ex*] *Colla*, Herb. Pedem. 2 : 255 (1834) ; Mart. in Flora 20, Beibl. : 121 (1837) ; Gilb. & Bout. in F.C.B. 3 : 231 (1952). Type : Brazil, *Martius, Herb. Fl. Bras.* 172 (K, iso. !)

Shrub up to about 1 m. high, often scandent or prostrate, with long whip-like stems which are densely armed with very downwards-bent prickles 1·5–5 mm. long and ± pubescent as well. Leaves sensitive ; petiole 2–6 cm. long, prickly ; rhachis 1·3–9 cm. long, prickly or not ; pinnae 3–10 pairs ; leaflets 11–30 pairs, linear-oblong, 2–6 mm. long, 0·7–1·5 mm. wide, pubescent at least on margins but not setulose, venation not or scarcely visible. Flowers pink, in subglobose or shortly ovoid heads about 0·5–1 cm. in diameter, on prickly and pubescent peduncles 0·5–1·3(–1·6) cm. long, 1–3 together from axils. Calyx minute, about 0·3–0·4 mm. long. Corolla 1·5–2 mm. long. Stamens 8. Pods clustered, oblong, slightly curved, with short prickly bristles on margins and surface of valves, 1·5–3·5 cm. long, 0·4–0·45 cm. wide.

TANGANYIKA. Moshi District : Lyamungu, 27 Dec. 1943, *Wallace* 1172 !
DISTR. **T2** ; native of tropical America, introduced here and there in the Old World
HAB. Unknown, probably an introduced weed ; 1310 m.

5. **M. pudica** *L.*, Sp. Pl. : 518 (1753) ; L.T.A. : 812 (1930) ; T.T.C.L. : 346 (1949) ; U.O.P.Z. : 353, 354, fig. (1949) ; Gilb. & Bout. in F.C.B. 3 : 229 (1952) ; Brenan in K.B. 1955 : 184 (1955) ; F.W.T.A., ed. 2, 1 : 495 (1958). Type : cult. in Hort. Cliffort., *Linnaeus* (BM, lecto. !)

Annual or perennial herb, sometimes woody below, up to about 1 m. high, often prostrate or straggling. Stems ± sparsely armed with prickles about 2·5–5 mm. long, in addition varying from densely hispid to subglabrous. Leaves sensitive, unarmed ; petiole 1·5–5·5 cm. long ; rhachis very short, so that the 2 (rarely only 1) pairs of pinnae are subdigitate ; leaflets 10–26 pairs, linear-oblong, 6–12·5(–15) mm. long, 1·2–2·75(–3) mm. wide ; venation diverging from and not nearly parallel with midrib ; margins setulose. Flowers lilac or pink, in shortly ovoid pedunculate heads about 1–1·3 cm. long and 0·6–1 cm. wide, 1–4(–5) together from axils. Calyx minute, about 0·2 mm. long. Corolla 2–2·25 mm. long. Stamens 4. Pods clustered, densely setose-prickly on margins only, 1·0–1·8 cm. long, 0·3–0·5 cm. wide (excluding the prickles).

KEY TO INTRASPECIFIC VARIANTS

Corolla, at least in bud, ± densely grey-puberulous outside
  in the upper part :
  Heads in bud appearing ± densely bristly owing to seti-
    form hairs projecting beyond the corolla for 1–
    1·5 mm. ; stipules 8–14 mm. long .   .   .   . var. **hispida**
  Heads in bud with few and short (to *c.* 0·75 mm.) or no
    projecting setiform hairs ; stipules 4–8 (rarely to 10)
    mm. long .   .   .   . var. **tetrandra**
Corolla, even in bud, glabrous or almost so outside ; pro-
  jecting setiform hairs absent or almost so from the
  heads in bud ; stipules 4–7(–8) mm. long .   .   . var. **unijuga**

var. **hispida** *Brenan* in K.B. 1955 : 186 (1955). Type : Java, *Junghuhn* 719 (K, holo. !)

Stems densely hispid. Stipules 8–14 mm. long. Bracteoles 1·8–2·2 mm. long, longer than the corollas in bud, their margins ciliate with setiform hairs which project for 1–1·5 mm. beyond the buds. Corolla grey-puberulous above on outside.

UGANDA. Mengo District : Kampala, 1930, *Mettam* 154 !
TANGANYIKA. Lushoto District : Amani, on the way to Bulwa, 12 Apr. 1926, *B. D. Burtt* 434 !
ZANZIBAR. Zanzibar Is., without locality, *Last* ! & Kizimbani, 15 Feb. 1951, *Oxtoby* K14 ! ; Pemba, Mkoani, 8 Aug. 1929, *Vaughan* 457 !
DISTR. **U4** ; **T3** ; **Z** ; **P** ; pantropical

var. **tetrandra** ([*Humb. & Bonpl. ex*] *Willd.*) *DC.*, Prodr. 2 : 426 (1825) ; Brenan in K.B. 1955 : 187 (1955). Type : South America, *Humboldt* (B, holo., K, photo. !)

Stems densely hispid or almost glabrous. Stipules 4–8 (rarely to 9–10) mm. long. Bracteoles about 1–1·5 mm. long, usually shorter than or as long as the corolla in bud or sometimes longer ; setiform hairs on margins few and short (to about 0·75 mm.) or absent. Outside of corolla grey-puberulous above.

TANGANYIKA. Lushoto District : Amani, 14 Aug. 1916, *Peter* K297 ! & 16 Apr. 1956, *Tanner* 2749 !
ZANZIBAR. Zanzibar Is., Zingwe-Zingwe, 21 Jan. 1929, *Greenway* 1104 !
DISTR. **T3** ; **Z** ; pantropical

SYN. (of var.). *M. tetrandra* [Humb. & Bonpl. ex] Willd., Sp. Pl. 4 (2) : 1032 (1806)

var. **unijuga** (*Duchass. & Walp.*) *Griseb.* in Abh. K. Gesellsch. Wissensch. Göttingen 7 : 211 (1857) & in Fl. Brit. W. Ind. : 219 (1860) ; Brenan in K.B. 1955 : 188 (1955). Type : Guadeloupe, *Duchassaing* (GOET, isolecto. !, K, photo. !)

Stems almost glabrous or sometimes densely hispid. Stipules 4–7(–8) mm. long. Bracteoles 0·7–1·1 mm. long, shorter than the corolla in bud ; setiform hairs on margins few or absent. Outside of corolla glabrous or almost so above.

KENYA. Nairobi Hill district, June 1943, *Mrs. Copley in C.M.* 11658 ! ; Kilifi District : Kibarani, 16 June 1945, *Jeffery* K228 !
TANGANYIKA. Lushoto District, Magrotto, 10 Aug. 1932, *Geilinger* 1278 ! & 6 Oct. 1957, *Mrs. Faulkner* 2071 !
DISTR. **K4**, 7 ; **T3** ; pantropical

SYN. (of var.). *M. unijuga* Duchass. & Walp. in Linnaea 23 : 744 (1850)

HAB. (of species as whole). Insufficiently noted, but recorded from banks of drainage canals in a rice valley and on margins of rain-forest ; probably always originally an escape ; 30–1220 m.

NOTE. The original home of this plant is probably South America.

Doubtful species

M. latispinosa *Lam.*, Encycl. 1 : 22 (1783) ; V.E. 3 (1) : 391 (1925) ; L.T.A. : 813 (1930) ; T.T.C.L. : 345 (1949). Type : Madagascar, *Commerson* (P, holo.)

Shrub up to about 3(–5–6, *fide Braun* 706) m. high. Stems when young with dense yellowish (when dry) pubescence, unarmed or with scattered straight or nearly so prickles up to 4 mm. long. Leaves with petioles 0·4–3·5 cm. long ; rhachis much exceeding petiole, (2·5–)8–22 cm. long, ± armed with straight or slightly upcurved prickles whose bases are conspicuously broad and laterally compressed, sometimes unarmed or almost so ; prickles not at junction of pinna-pairs as in *M. pigra;* pinnae (5–)7–27 pairs ; leaflets 7–21 pairs, oblong-elliptic or oblong, 3–5·5 mm. long, 1–2·5 mm. wide, pubescent, not setulose. Flower-heads about 0·6–0·9 cm. in diameter, clustered and shortly stalked along the branches of a ± ample terminal panicle. Calyx 0·75–1 mm. long. Corolla 1·5–2 mm. long. Stamens 8. Pods 4–8·5 cm. long, 0·9–1 cm. wide, ± pubescent, not bristly or prickly, breaking up transversely into segments 6–12 mm. long.

TANGANYIKA. Lushoto District : Amani, 23 May 1905, *Braun* 706 ! & 10 Feb. 1906, *Braun* 1035 !

NOTE. The two cited specimens, which are in the East African Herbarium are evidently correctly identified, although belonging to a form with few or no prickles along the leaf-rhachides. As *M. latispinosa* is otherwise known only from Madagascar and, possibly Mauritius, and has never been re-collected at Amani, its status must be suspect, especially as so many species have been introduced there. The specimens give no evidence for or against its being planted, although *Braun* 706 gives a native name and use.

15. **LEUCAENA**

Benth. in Hook., Journ. Bot. 4 : 416 (1842)

Trees or shrubs, unarmed. Leaves bipinnate ; a gland often present at junction of lowest pair of pinnae, petiole and rhachis otherwise eglandular, or rarely with glands between other pairs of pinnae ; pinnae each with one to several or many pairs of leaflets. Inflorescences of round heads, pedunculate, axillary, 1–3 together, often racemosely aggregated. Flowers hermaphrodite, sessile. Calyx gamosepalous with 5 teeth. Petals 5, free, pubescent to glabrous outside. Stamens 10, fertile. Anthers eglandular at apex (except in the extra-African *L. forsteri*). Ovary pubescent or sometimes glabrous. Pods oblong or linear-oblong, compressed, usually thinly subcoriaceous, splitting into 2 non-recurring valves. Seeds lying ± transversely in the pod, compressed, brown, glossy, unwinged, with endosperm.

A genus of about 50 species, one (*L. glauca*) widespread in the tropics, one in the Pacific islands, the rest tropical American.

L. pulverulenta (Schlechtend.) Benth. from Mexico, with more numerous pinnae than in *L. glauca* and small very numerous leaflets, and *L. glabrata* Rose, also from Mexico, very near *L. glauca* but more glabrous, have been tried in cultivation at Entebbe (Dale, Introd. Trees Uganda : 48 (1953))

L. glauca (*L.*) *Benth.* in Hook., Journ. Bot. 4 : 416 (1842) ; L.T.A. : 814 (1930) ; T.T.C.L. : 345 (1949) ; U.O.P.Z. : 329, 330 (fig.) (1949) ; Gilb. & Bout. in F.C.B. 3 : 231 (1952) ; Dale, Introd. Trees Uganda : 48 (1953) ; Consp. Fl. Angol. 2 : 268 (1956) ; F.W.T.A., ed. 2, 1 : 495 (1958). Type : uncertain.

Shrub or small tree 0·6–9 m. high. Young branchlets densely grey-puberulous. Leaves : petiole 2–4·7 cm. long, often with a gland at junction of lowest pair of pinnae, glands otherwise absent ; rhachis 7–14 cm. long ; pinnae (2–)3–7 pairs, opposite ; leaflets (5–)7–17(–21) pairs, obliquely oblong-lanceolate, 7–18 mm. long, 1·5–5 mm. wide, acute at apex, puberulous on margins and sometimes also on midrib beneath. Peduncles 2–5 cm. long. Calyx 2–3·5 mm. long, puberulous above outside. Petals 4–5·25 mm. long, puberulous above outside, pale green. Stamen-filaments 6·5–7·5 mm. long, white ; anthers hairy. Pods 8–18 cm. long, (1·4–)1·8–2·1 cm. wide, with a stipe up to 3 cm. long. Seeds elliptic to obovate, 7·5–9 mm. long, 4–5 mm. wide.

UGANDA. Ankole District : Ruizi R., 16 May 1951, *Jarrett* 478 ! ; Mengo District : Namaniyama, July 1916, *Dummer* 2933 ! & clearing near Mabira Forest, 12 Nov. 1938, *Loveridge* 53 !
KENYA. Mombasa Is., Fort Jesus, 19 Aug. 1948, *Bally* 6363 !
TANGANYIKA. Pangani District : Bushiri Estate, 14 May 1950, *Faulkner* 576 !
ZANZIBAR. Zanzibar Is., Mangapwani, 23 Jan. 1929, *Greenway* 1129 !
DISTR. U2, 4 ; K7 ; T3 ; Z ; widespread in the tropics and subtropics, probably native only in the New World
HAB. A cultivated plant, escaping, seeding freely, and becoming naturalized here and there, in suitable places forming dense thickets ; used by the Germans in Ceara rubber plantations as a contour plant (Greenway) ; 0–1300 m.

SYN. *Mimosa glauca* L., Sp. Pl. : 520 (1753)

NOTE. The specimens quoted above are from localities where there is some evidence for the possible naturalization of *L. glauca*. It may, however, certainly be expected to become established elsewhere.

The hairy anthers (to see which a lens is necessary) are a most useful diagnostic character of *L. glauca*, and indeed distinguish it from all other East African *Mimosoïdeae*.

16. ACACIA

Mill., Gard. Dict., abridg. ed. : 4 (1754)

Trees or shrubs, sometimes climbing ; the native species in our area almost invariably armed with prickles or spines, the introduced ones usually unarmed. Leaves bipinnate or (in introduced species) often modified to phyllodes (entire, leaf-like often flattened organs, without pinnae or leaflets) ; pinnae each with one to many pairs of leaflets ; gland on upper side of petiole usually present ; glands often also present at insertion of pinnae. Flowers in spikes, spiciform racemes or round heads, hermaphrodite or ♂ and ♀ ; if in heads then central flowers not enlarged and modified ; inflorescences usually axillary, racemose or paniculate. Calyx (in our species) gamosepalous, subtruncate or usually with 4–5 teeth or lobes. Corolla 4–5(–7)-lobed. Stamens many (from 35–40 in *A. lahai* to about 215 in *A. thomasii*), fertile, their filaments free or (in *A. albida*) connate into a tube at their extreme base only ; anthers (at least some) glandular at apex, or all eglandular (in all native species glandular except in *A. albida*, in introduced species mostly eglandular). Ovary stipitate to sessile, glabrous to puberulous. Pods very variable, dehiscent or sometimes indehiscent, flat, ± compressed, or sometimes cylindrical, straight, curved, spiral or contorted, continuous or moniliform. Seeds unwinged, often with a hard smooth testa, without endosperm.

A genus of about 750–800 species, mostly tropical or subtropical ; more than half in Australia, many in Africa and America, fewer in Asia.

NOTES. In some of our species the remarkable structures derived from the stipules and commonly known as ant-galls occur. There is some evidence that these may not be galls at all, but natural outgrowths of the plant itself. Certainly their presence is taxonomically important. In the text I have compromised by keeping the familiar term " ant-gall," but enclosing it by inverted commas. A new and more accurate term may have to be devised.

Our acacias are taxonomically difficult. However, the statement by Fr. Marie-Victorin (in Amer. Midl. Nat. 19 : 498–9 (1938)), based apparently on personal experience, that the East African acacias have produced an enormous number of species or forms under ecological conditions (e.g. in the Ngong Hills) similar to those that have given rise to the multitude of North American hawthorns (*Crataegus*), goes, I think, too far. The taxonomic complexity of the hawthorns seems much greater than that of our acacias ; and it would be hard to prove that the rate of speciation in acacia is especially increased by savannah or scrub conditions.

KEY TO EXOTIC SPECIES

Several exotic species of *Acacia* are met with in our area. The following short key to the planted species of which I have seen material will, I hope, be useful. All, unless otherwise stated, are natives of Australia.

Leaves bipinnate :
 Plants spiny :
  Flowers creamy-yellow, in heads in much-branched panicles ; pods 0·5–1(–1·1) cm. wide . . . . . *A. leucophloea* (Roxb.) Willd. (Native of tropical Asia)

  Flowers bright or deep yellow, in axillary heads ; pods 0·9–2·2 cm. wide :
  Lateral nerves of leaflets invisible beneath ; involucel in lower half of peduncle ; pods clearly compressed . . . . . *A. nilotica* (L.) [Willd. ex] Del. (Native of tropical Africa and Asia. See p. 109)

Lateral nerves of leaflets visible and
somewhat raised beneath ; involucels apical ; pods almost round in
section   .   .   .   .   .   *A. farnesiana* (L.) Willd.
(Native of tropical Ameri-
ca, ? also of Africa and
Australia. See p. 111)

Plants unarmed ; flower-heads racemose or
panicled :
Pinnae 2–5 pairs per leaf :
Leaflets 2·5–5 cm. long ; young stems
puberulous or pubescent   .   .   *A. elata* [A. Cunn. ex]
Benth.

Leaflets 0·15–2·3 cm. long :
Petiole 2·5–7·5 cm. long ; pinnae
mostly 5–14 cm. long, not
crowded   .   .   .   .   *A. schinoïdes* Benth. (*A.
pruinosa* auct. p.p., *non*
[A. Cunn. ex] Benth.)

Petiole about 0·2 cm. long ; pinnae
1–2·5 cm. long, crowded .   .   *A. baileyana* F. v. Muell.

Pinnae 6–21 pairs per leaf :
Leaflets 5–12 mm. long, linear ; young
stems inconspicuously puberulous,
very angular   .   .   .   .   *A. decurrens* (Wendl.) Willd.

Leaflets 1·5–5 mm. long ; young stems
obviously puberulous, pubescent
or tomentellous, less angular :
Young shoots grey-pubescent or
puberulous, not golden ; glands
at junction of all or most pairs
of pinnae, but not on rhachis between insertions of pairs ; pod
not or slightly constricted be-
tween seeds   .   .   .   .   *A. dealbata* Link

Young shoots golden-yellow-tomen-
tellous ; pod constricted between seeds :
Leaf-rhachis with a gland at the
insertion of each pair of pinnae
and usually also with addi-
tional ones between insertions   *A. mearnsii* De Wild. (see
p. 95)

Leaf-rhachis with a gland at the
insertion of only the upper
pairs (usually 1–2, rarely to 7)
of pinnae ; no additional ones
between   .   .   .   .   *A. irrorata* [Sieb. ex] Spreng

Leaves apparently simple, modified to phyllodes by dilation of the petiole and
rhachis :
Flowers in spikes :
Phyllodes and inflorescence-axes puberulous or minutely grey-hoary (use × 10
lens) :

Phyllodes distinctly falcate, up to about
2 cm. wide ; 3–5 longitudinal
nerves stronger than the rest   .   *A. binervia* (Wendl.) Macbr.
(*A. glaucescens* Willd.)

Phyllodes nor or slightly falcate, up to
0·8 cm. wide ; longitudinal nerves
all of about the same strength   .   *A. aneura* F. v. Muell.*

Phyllodes and usually inflorescence-axes
glabrous :
Phyllodes straight :
Phyllodes mostly 1·3–3 cm. wide,
5–10 cm. long, mostly obovate-
oblong, oblanceolate or oblong-
elliptic, coriaceous   .   .   *A. longifolia* (Andr.) Willd.
var. *sophorae* (Labill.) [F.
v. Muell. ex] Benth.

Phyllodes mostly 0·3–1·0 cm. wide,
7·5–13 cm. long, linear-lanceolate or linear, less coriaceous than
the last   .   .   .   *A. longifolia* (Andr.) Willd.
var. *floribunda* (Vent.)
[F. v. Muell. ex] Benth.

Phyllodes strongly falcate .   .   *A. auriculiformis* [A. Cunn.
ex] Benth.

Flowers in round heads :
Phyllodes each with one main longitudinal
nerve :
Stems and phyllodes pubescent, the
latter elliptic or ovate-elliptic   .   *A. podalyriifolia* [A. Cunn.
ex] G. Don

Stems and phyllodes glabrous :
Phyllodes falcate, mostly ± obtuse at
apex, obovate-lanceolate to
oblanceolate, mostly 1·3–3 cm.
wide   .   .   .   .   *A. pycnantha* Benth.

Phyllodes straight or if slightly falcate
then narrow :
Flower-heads 7–9 mm. in diameter,
on peduncles 7–18 mm. long ;
funicle not coiled round the
seed   .   .   .   .   *A. saligna* (Labill.) Wendl.

Flower-heads 3–5 mm. in diameter,
on peduncles 3–6 mm. long ;
funicle encircling the seed in a
double fold .   .   .   *A. retinodes* Schlechtend.

Phyllodes each with 2–7 or more main
longitudinal nerves :
Phyllodes oblong-linear to linear ; veins
between longitudinal nerves very
fine, parallel, not reticulate ; heads
solitary or fascicled   .   .   *A. homalophylla* [A. Cunn.
ex] Benth.

* The form cultivated in East Africa has phyllodes 4–8 mm. wide and unusually broad
for the species ; it may be referable to var. *latifolia* J. M. Black.

4—2

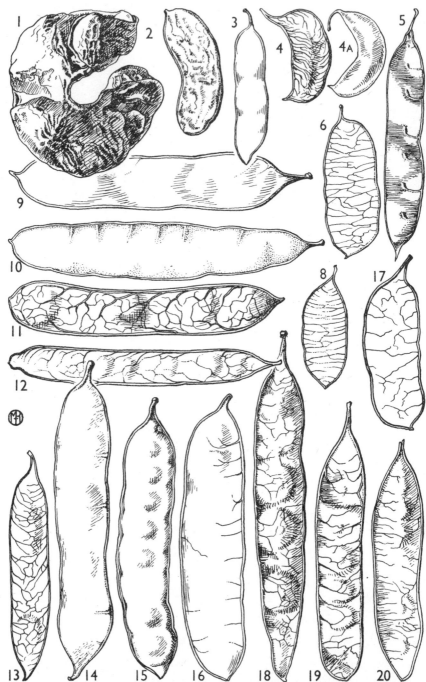

FIG. 14. *ACACIA*—Pods of spicate-flowered species, × ⅔. Species numbered as in text. **1**, *A. albida;* **2**, *A. lahai;* **3**, *A. bussei;* **4**, *A. horrida* subsp. *benadirensis;* **4A**, *A.* near *horrida* (see p. 82); **5**, *A. ataxacantha;* **6**, *A. laeta;* **8**, *A. mellifera;* **9**, *A. nigrescens;* **10**, *A. persiciflora;* **11**, *A. hecatophylla;* **12**, *A. polyacantha* subsp. *campylacantha;* **13**, *A. erubescens;* **14**, *A. tanganyikensis;* **15**, *A. rovumae;* **16**, *A. goetzei;* **17** *A. senegal;* **18**, *A. circummarginata;* **19**, *A. condyloclada;* **20**, *A. thomasii*. 1, from *B. D. Burtt* 697; 2, from *Greenway* 9054; 3, from *Trapnell* 2210; 4, from *Dale* K732; 4A, from *Gillett* 13305; 5, from *Deighton* 3459; 6, from *Trapnell* 2203; 8, from *Bally* 2168; 9, from *B. D. Burtt* 3440; 10, from *Dale* 3434; 11, from *Schweinfurth* 254; 12, from *Semsei* 2211; 13, from *Michelmore* 623; 14, from *Milne-Redhead & Taylor* 11263; 15, from *Greenway* 8859; 16, from *Semsei* 69 *in F.H.* 2117; 17, from *Faulkne* 1526; 18, from *Hornby* 140; 19, from *Gillett* 13279; 20, from *Vesey-FitzGerald* 29.

Phyllodes oblanceolate to narrowly
obovate ; veins between longitudinal nerves ± reticulate ; heads
usually racemose   .   .   .   *A. melanoxylon* R. Br.

KEY TO NATIVE AND NATURALIZED SPECIES

A. FLOWERS IN SPIKES OR SPICIFORM RACEMES * (B on p. 58)

Stipules spinescent, straight, rarely somewhat
curved, often conspicuously enlarged and
inflated below and ashen or whitish (to p. 54):
Leaf-rhachis with a conspicuous gland between
every pair of pinnae ; stipules not enlarged and inflated below ; calyx 1–1·7
(–2·5) mm. long ; corolla-lobes 5, 1·5–
2·5 mm. long ; anthers 0·2–0·4 mm. wide,
eglandular even in bud ; stamen-filaments
connate and tubular for about 1 mm. at
base ; branchlets pale grey to whitish ;
pods indehiscent, falcate or coiled, orange   1. *A. albida*

Leaf-rhachis without conspicuous glands ** ;
stipules often enlarged and inflated below ;
calyx 0·7–1 mm. long ; corolla-lobes 4,
sometimes 5, 0·5 mm. long ; anthers
0·1 mm. wide, at least in bud with a
caducous apical gland ; stamen-filaments
free to base ; branchlets dull grey, brown
or purplish ; pods dehiscent, commonly
straight or nearly so :
Stipular spines neither inflated nor fusiform ;
inflorescence-axis (of *A. lahai*) pubescent
and with many reddish sessile or subsessile glands ; pinnae up to 14 pairs per
leaf ; leaflets up to 35 pairs per pinna ;
at comparatively high altitudes (1150–
2440 m.) :
Spikes 2·5–7 cm. long, on peduncles 0·7–
2·2 cm. long ; spinescent stipules up
to 7 cm. long .   .   .   .   .   2. *A. lahai*

Spikes very short, 1–1·5 cm. long, on
peduncles 4–7·5 cm. long ; spinescent
stipules on mature shoots up to about
4 mm. long .   .   .   .   .   2a. *A. dolichocephala*

Stipular spines (or some of them at least)
characteristically inflated or fusiform
(" ant-galls ") ; inflorescence-axis with
few or no glands ; pinnae up to 8 pairs
per leaf ; leaflets up to 18 pairs per
pinna ; at comparatively low altitudes
(180–970 m.) :

* Spikes very short, almost headlike, in 2a, *A. dolichocephala.*
** 2, *A. lahai* has clusters of tiny red bodies on the leaf-rhachis between the pinna-
pairs ; these are, however, altogether different from the large single glands to be seen
in this position in *A. albida.*

Inflated stipular spines constricted at
  base ; tree 3–10 m. high, usually with
  a well-defined trunk ; corolla 3–5
  times as long as the 0·7–0·8 mm. long
  calyx ; pods straight, puberulous,
  0·8–1·5 cm. wide   .   .   .   .   3. *A. bussei*
Inflated stipular spines not or only slightly
  constricted at base ; bushes 1·3–3·6 m.
  high (very rarely taller), branching
  from base ; pods falcate or subreniform :
  Pods glabrous or nearly so, rarely slightly
    puberulous, 1·5–2·5 cm. wide,
    venose ; young branchlets glabrous ;
    leaflets subglabrous or inconspi-
    cuously ciliate ; corolla 2–2·5 times
    as long as the 0·8–1 mm. long
    calyx .   .   .   .   .   4. *A. horrida* subsp.
                                                          benadirensis

  Pods densely and shortly pubescent all
    over, 1·1–1·2 cm. wide, not venose ;
    young branchlets pubescent ; leaf-
    lets pubescent ; flowers unknown .   *A.* sp. near *horrida**
Stipules not spinescent ; prickles (usually pre-
  sent) borne below the stipules, short, up to
  7(–12) mm. long, usually ± hooked, deflexed or curved, never inflated below, usually
  brown, blackish or dull grey :
  Prickles absent :
    Calyx short, 1–1·4 mm. long, red or purplish,
      as is the corolla :
      Inflorescences 1·5–3 cm. long ; pinnae (on
        well-developed leaves of flowering
        shoots) 4–8 pairs per leaf ; calyx and
        corolla glabrous   .   .   .   10. *A. persiciflora*
      Inflorescences 4–11 cm. long ; pinnae (on
        well-developed leaves of flowering
        shoots) 9–14 pairs per leaf ; calyx
        and corolla puberulous (often sparsely
        and minutely so)   .   .   .   *A. galpinii***
    Calyx 1·7–2·7 mm. long, not red or purplish :
      Leaflets 1·25–3·5 mm. wide, conspicuously
        pale and glaucous beneath ; lateral
        nerves somewhat prominent beneath.   11. *A. hecatophylla*
      Leaflets 0·4–0·75(–1.25) mm. wide, not
        especially pale or glaucous beneath ;
        normally midrib alone and sometimes
        some faint basal nerves visible be-
        neath   .   .   .   .   .   12. *A. polyacantha*‡
  Prickles present :
    The prickles irregularly scattered along the
      internodes :
      Calyx glabrous ; leaflets 0·5–1(–1·2) mm.
        wide ; ovary pubescent, shorter than

---

* See note under 4, *A. horrida,* on p. 82.
** See note under 10, *A. persiciflora* on p. 87.
‡ The unarmed variant of *A. polyacantha* is very rare.

its supporting stipe ; in our area a
scandent shrub ; prickles normally
occurring along the internodes .   .   5. *A. ataxacantha*
Calyx densely pubescent ; leaflets 1·25–
2·5(–3·5) mm. wide ; ovary glabrous,
much exceeding its supporting stipe ;
a tree 4·5–7·5 m. high ; prickles
occurring along the internodes only
rarely, and probably abnormally   . 11. *A. hecatophylla*
The prickles either solitary or in pairs or
threes, grouped at or just below the
nodes ; erect trees or shrubs ; ovary
glabrous (at least at first), much longer
than its supporting stipe :
Flowers distinctly but shortly pedicellate
(pedicels 0·5 mm. or more) :
Calyx 1·25–2 mm. long ; pedicels mostly
less than 0·5 mm. long, rarely as
much as 0·75 or even 1 mm. ; leaflets
mostly in 3–4 pairs (occasionally
only 2 pairs on reduced pinnae)   .   6. *A. laeta*
Calyx 0·6–1 mm. long ; pedicels mostly
0·75–1·5 mm. long, rarely as short
as 0·5 mm. ; leaflets in 1–2, very
rarely 3 pairs   .   .   .   .   8. *A. mellifera*
Flowers sessile or subsessile (pedicels 0–
0–0·3 mm.) :
Prickles in pairs near nodes (to p. 57) :
Leaflets in one, or sometimes two pairs
on each pinna ; trunk usually
with knobby prickles.   .   .   9. *A. nigrescens*
Leaflets of normal, well-developed
pinnae in three to four or more
pairs on each pinna (in 6, *A. laeta*
and 16, *A. goetzei* occasional
reduced pinnae may have but two
pairs) :
Pinnae of all leaves 2–3 pairs ;
leaflets (2–)3–4(–5) pairs,
obovate-elliptic or oblong ; pod
(where known) pale brown : *
Prickles hooked downwards ;
calyx 1·25–2 mm. long   .   6. *A. laeta*
Prickles straight or almost so,
pointing somewhat upwards ;
calyx 0·5–1 mm. long   .   7. *A.* sp. A
Pinnae of all or most leaves four
or more pairs : only occasional
reduced leaves with as few as
three :
Calyx short, 1–1·4 mm. long, red
or purplish :
Inflorescences 1·5–3 cm. long ;
pinnae (on well-developed

_____

* It is possible that certain extreme forms of *A. goetzei* subsp. *goetzei* may key out
here. They are of rare occurrence, and may be separated from *A. laeta* by their thicker
and usually red- to purplish-brown pods.

leaves of flowering shoots)
4–8 pairs per leaf; calyx
and corolla glabrous . 10. *A. persiciflora*

Inflorescences 4–11 cm. long ;
pinnae (on well-developed
leaves of flowering shoots)
9–14 pairs per leaf ; calyx
and corolla ± puberulous
(often sparsely and
minutely so) . . . *A. galpinii* *

Calyx 1·5–4·5 mm. long, normally
not purplish or red, perhaps
never so :

Pinnae of normal, well-developed leaves 12–20 or more
pairs per leaf :

Leaflets 1·25–2·5(–3·5) mm.
wide, conspicuously pale
and greyish beneath,
obliquely oblong ; midrib and lateral nerves
somewhat prominent beneath ; pods rounded to
subacute at apex . . 11. *A. hecatophylla*

Leaflets 0·4–0·8(–1·25) mm.
wide, not conspicuously
pale beneath, linear to
linear-oblong or -triangular ; midrib alone (except sometimes for small
basal nerves) visible beneath ; pods usually ±
acuminate :

Gland on petiole large, 2–4
×1·75–3 mm. ; corolla
1⅓ times or more the
length of the pubescent
or puberulous calyx ;
spikes produced with
the new leaves ; pods
1·2–2·0 cm. wide . 12. *A. polyacantha*

Gland on petiole small or
medium, 0·8–1·5 mm.
in diameter ; corolla
equalling or only
slightly exceeding the
glabrous or slightly
hairy, rarely more
densely hairy, calyx ;
spikes produced
usually before the new
leaves, on leafless
abbreviated lateral
shoots; pods 2–2·6 cm.
wide . . . 14. *A. tanganyikensis*

* See note under 10, *A. persiciflora* on p. 87.

Pinnae of normal well-developed
leaves up to 8(–10) pairs
per leaf :

Corolla-lobes ± appressed-
pubescent outside :

Inflorescences 2–4·5 cm.
long ; calyx densely
pubescent, 2·25–4·5
mm. long ; prickles
hooked ; pods dehi-
scent   .   .   13. *A. erubescens*

Inflorescences 6–10 cm.
long ; calyx puberu-
lous, 1·5–2 mm. long ;
prickles usually point-
ing a little upwards ;
pods probably inde-
hiscent   .   .   15. *A. rovumae*

Corolla-lobes glabrous outside ; calyx glabrous or
sometimes sparingly
pubescent or puberulous:

Leaflets 0·6–1·0 mm. wide,
in 19–29 pairs per
pinna ; flowers
usually precocious and
produced before the
new leaves on short
leafless lateral shoots ;
corolla equalling or
only slightly exceeding
the calyx.   .   .   14. *A. tanganyikensis*

Leaflets (0·75–)1–7(–12·5)
mm. wide, in (2–)5–31
pairs per pinna ;
flowers produced with
the leaves ; corolla
distinctly exceeding
the calyx :

Prickles usually spread-
ing or pointing a
little upwards; calyx
puberulous ; leaflets
not more than 3·5
wide ; pods pro-
bably indehiscent .   15. *A. rovumae*

Prickles hooked down-
wards ; calyx glab-
rous ; leaflets up to
7(–12·5) mm. wide ;
pods dehiscent   .   16. *A. goetzei*

Prickles in threes near nodes, the central
one hooked downwards, the laterals
± curved upwards, or else the
prickles solitary, the laterals being
absent :

Corolla about 2·75–4 mm. long ; sta-

men-filaments (where known) 4·5–
  7 mm. long ; calyx 2–2·75(–3·5)
  mm. long ; pinnae 3–7 pairs :
Leaflets 0·5–2 (–3) mm. wide ; leaf-
  rhachis glandular between the
  top 1–5 pairs of pinnae :
Rhachides of pinnae 0·5–1·5(–2·4,
  very rarely to 4) cm. long ;
  bark grey, scaly, rough ;
  branchlets dull grey to grey-
  brown or purplish-grey   .   17. *A. senegal*
Rhachides of pinnae 2–6·5 cm.
  long ; bark yellow, flaking ;
  branchlets purplish-brown to
  purplish-black   .   .   .   18. *A. circummarginata*
Leaflets 3·5–9 mm. wide ; leaf-
  rhachis eglandular ; bark white,
  peeling ; pinnae 3–4 pairs   .   19. *A. condyloclada*
Corolla 6·5–7 mm. long ; stamen-fila-
  ments 13–15 mm. long ; calyx
  3·5–4·5 mm. long ; pinnae 1–2
  (–3) pairs   .   .   .   .   20. *A. thomasii*

B. Flowers in round heads

Two alternative keys are given, one (I) based mainly on floral and vegetative characters, for use with flowering specimens, the other (II) using mainly
characters taken from the pods and vegetative parts, for use with fruiting
specimens. For the sake of convenience and to save space the keys are
split into parts, but these parts do not necessarily correspond with natural
groupings.

I. Key based mainly on floral and vegetative characters

Plant unarmed ; leaflets narrow, 0·5–0·75 mm. wide ;
  flowers pale yellow, in paniculate heads ; anthers
  eglandular   .   .   .   .   21. *A. mearnsii*
Plant armed with spines or prickles ; anthers (at least
  some) glandular at apex when young :
  Prickles scattered along the internodes of the stem,
    not grouped at or near nodes   .   .   .   **Ia** (p. 59)
  Prickles or spines in pairs at or near nodes :
    Heads of flowers in panicles ; leaves large ;
      flowers bright or orange-yellow   .   .   30. *A. macrothyrsa*
    Heads of flowers on peduncles which are axillary
      and clustered or borne singly, sometimes
      racemosely aggregated, not paniculate :
      " Ant-galls " present (i.e. some pairs of stipules
        fused at base into an inflated, ± rounded
        or bilobed, ultimately woody structure ;
        rarely the stipules are inflated but not
        fused)   .   .   .   .   .   **Ib** (p. 59)
      " Ant-galls " absent :
        Pinnae at least 15 pairs (and usually more
          than 20 pairs) per leaf on the well-
          developed leaves of flowering shoots
          (reduced leaves with fewer pairs of
          pinnae usually also present) ; flowers
          white or cream .   .   .   .   **Ic** (p. 61)

Pinnae 1–14 pairs : *
 Flowers bright or golden yellow . . **Id** (p. 62)
 Flowers cream, white, pink or greenish . **Ie** (p. 63)

Ia. Prickles scattered along the internodes of the stem

Petiole 0·5–1·5 cm. long :
 Midrib of leaflets excentric at base ; calyx eglan-
  dular . . . . . . 22. *A. brevispica*
 Midrib of leaflets almost central at base ; calyx
  glandular outside . . . . 23. *A. adenocalyx*
Petiole 1·5–6 cm. or more long :
 Stipules ** broadly ovate, 5–9 × 3–4·5 mm., sub-
  cordate at base . . . . . 24. *A. latistipulata*
 Stipules ** narrower, linear to lanceolate, oblanceo-
  late or falcate, 0·3–2 mm. wide, not subcordate
  at base :
  Midrib of leaflets markedly excentric at base :
   Leaflets 0·3–0·8 mm. wide, glabrous or sparsely
    and very inconspicuously ciliolate :
    Young branchlets subglabrous or sparsely
     puberulous, older ones going grey or
     yellowish-grey . . . . . 25. *A. kamerunensis*
    Young branchlets ± densely pubescent and
     with ± numerous dark purple glands,
     dark brown, older branchlets blackish . 26. *A. monticola*
   Leaflets 0·8–2 mm. wide :
    Young branchlets with numerous red-purple
     glands :
     Branchlets dark brown to blackish, ±
      densely pubescent when young ;
      leaflets glabrous or sparsely and in-
      conspicuously ciliolate, especially at
      first ; in upland forest . . . 26. *A. monticola*
     Branchlets olive-green to olivaceous-
      brown, puberulous when young ;
      leaflets persistently and usually
      clearly ciliolate ; in woodland and
      thickets . . . . . 27. *A. schweinfurthii*
    Young branchlets usually eglandular,
     glabrous to sparsely puberulous ;
     usually in lowland forest . . . 28. *A. pentagona*
  Midrib of leaflets subcentral at base ; young
   branchlets densely pubescent, almost eglandular . . . . 29. *A. taylorii*

Ib. Spines paired; " ant-galls " present

" Ant-galls " composed of pairs of stipules, each
 stipule fusiform-inflated, but not fused with the
 other one of the pair, and free almost or quite to
 base . . . . 41. *A. elatior*

* 2a, *A. dolichocephala*, placed among the spicate-flowered species although the inflorescences are very short and almost head-like, is liable to key out here. Its flower-colour is uncertain, but it may be distinguished from all our other capitate-flowered species by the flower heads being longer than wide and thus not truly round.
** The stipules are early caducous and on mature shoots are usually to be found only in connection with the very young leaves subtending the peduncles.

Ant-galls '' composed of pairs of stipules fused
below into an inflated, ± rounded or bilobed
structure :
Flowers bright yellow ; " ant-galls " ± bilobed :
  Calyx 0·8–1·5 mm. long ; leaflets usually more
    than 2 and up to 10 mm. wide, usually with
    lateral nerves raised especially beneath   .   31. *A. zanzibarica* *
  Calyx 2–2·5 mm. long ; leaflets at most 3 mm.
    and usually not more than 1·5 mm. wide,
    with lateral nerves invisible beneath .   .   32. *A. seyal* var.
                                              fistula

Flowers white or cream ; " ant-galls " ± rounded
or ovoid, not bilobed :
  Leaflets 0·7–1·75 mm. wide ; pinnae 3–28 pairs
    per leaf ; plant glabrous to pubescent ;
    bark various ; pods linear-falcate, usually
    several-seeded ; seeds ± elliptic or quadrate:
    Epidermis of branchlets falling away to expose
      a powdery bark-layer brick-red to red-
      brown in colour (very rarely creamy-
      buff) ; old bark similarly coloured ; larger
      leaves of mature shoots with more than
      15 and up to 28 or more pairs of pinnae :
      Spines on mature branchlets 1·5–5 (–9) cm.
        long :
        Shrub or small tree 1·8–6(–9) m. high with
          horizontal branches all the way up
          the main stems ; pinnae 1–3 cm.
          long ; leaflets 1·5–4·5 mm. long   .   49. *A. pseudofistula*
        Small tree 1–3 m. high with simple or
          scarcely branched stems   .   .   50. *A. bullockii*
      Spines on mature branchlets 0·3–1·5 cm.
        long (to 2 cm. on juvenile shoots) ;
        bark intensely brick-red   .   .   51. *A. erythrophloea*
    Epidermis of branchlets not falling and expos-
      ing any powdery layer ; old bark grey,
      black or brown ; pinnae 3–13 pairs
      (except in 53, *A. mbuluënsis*) :
      Corolla glabrous outside, or inconspicuously
        puberulous on lobes .   .   .   .   48. *A. drepanolobium*
      Corolla densely pubescent or tomentellous
        outside on the part projecting beyond
        the calyx :
        Young branchlets puberulous with hairs
          about 0·1 mm. or less long ; spines
          mostly long (1·5–5·5 cm.) ; pinnae
          3–10 pairs ; leaflets 2·5–6 mm. long,
          0·8–1·5 mm. wide   .   .   .   52. *A. malacocephala*
        Young branchlets pubescent with hairs
          0·3–0·75 mm. long ; spines mostly
          short (0·4–1 cm.), a few longer, up to
          4 cm. long ; pinnae 10–20 pairs
          (more than 15 on the best-developed
          leaves) ; leaflets 2–3 mm. long,
          0·75 mm. wide   .   .   .   53. *A. mbuluënsis*

* For more detailed discussion of the differences between *A. zanzibarica* and *A. seyal*
var. *fistula*, see notes under the variants of the former (pp. 102–3).

Leaflets (4–)5–13(–17) mm. wide ; pinnae 1–4
  pairs ; plant altogether glabrous, with buff
  or fawn bark ; pods half-moon-shaped, 1-
  seeded ; seeds suborbicular or broadly
  elliptic . . . . 54. *A. burttii*

Ic. **Spines paired; no " ant-galls "; pinnae at least 15 pairs on well-developed**
mature leaves

Leaflets all exceedingly narrow, about 0·25 mm. wide
  (at most to 0·5 mm.) ; buds, calyx and corolla
  red ; pinnae 0·4–1·5 cm. long ; flat-crowned tree,
  usually large, 6–20 m. high . . . . 39. *A. abyssinica*
Leaflets 0·5 mm. or more wide :
  Involucel apical or in upper half of peduncle :
    Calyx shorter than the projecting part of the
      corolla (corolla 2–4 × calyx) ; peduncles
      pubescent and glandular . . . . 55. *A. arenaria*
    Calyx longer than the projecting part of the
      corolla (corolla 1–1¾ × calyx) ; peduncles
      variable in indumentum but eglandular . 57. *A. sieberiana*
  Involucel basal or in lower half of peduncle :
    Corolla glabrous outside :
      Older branchlets with powdery yellow or
        sometimes greenish bark ; leaf-rhachis to
        about 1 mm. wide ; young branchlets
        densely clothed with yellowish to grey
        hairs mostly 0·5–2 mm. long . . . 40. *A. pilispina*
      Older branchlets grey, brown or blackish, not
        as above ; leaf-rhachis 1·25–2·25 mm.
        wide ; young branchlets puberulous or
        pubescent . . . . . . 56. *A. fischeri*
    Corolla ± densely pubescent or puberulous
      outside :
      Older branchlets with vividly rusty-red bark ;
        young foliage with gleaming, silky, pale
        golden indumentum . . . . 47. *A. lasiopetala*
      Older branchlets olive, grey, brown or blackish:
        Young shoots villous with rather long
          spreading hairs, some of them up to
          1·5–3 mm. long ; leaves only occa-
          sionally with 15 or more pairs of pinnae,
          and then perhaps only on juvenile shoots 61. *A. stuhlmannii*
        Young shoots with hairs up to 0·5 (–0·75)
          mm. long, or sometimes glabrous :
          Calyx shorter than the projecting part of
            the corolla (corolla 2–4 × calyx) ;
            young branchlets often with reddish
            glands ; leaf-rhachis 1·25–2·25 mm.
            wide . . . . . . 56. *A. fischeri*
          Calyx usually longer than the projecting
            part of the corolla (corolla 1–2 ×
            calyx) ; young branchlets eglan-
            dular ; leaf-rhachis up to 1·25 mm.
            wide ; usually flat-crowned, widely
            spreading shrubs to 2 m. high :

Spines on mature flowering shoots 0·4–
3·5 cm. long ; peduncles 0·4–1·5
cm. long ; calyx 1–2 mm. long ;
corolla 2·5–3 mm. long ; leaves
rarely 15-jugate or more   .   62. *A. edgeworthii*

Spines on mature flowering shoots
(2·5–)3·5–6 cm. long ; peduncles
1·5–3 cm. long ; calyx 3·75 mm.
long ; corolla 5·5 mm. long .   .   63. *A. turnbulliana*

Id. Spines paired; no " ant-galls "; pinnae 1–14 pairs; flowers bright or golden yellow

Lateral nerves of leaflets invisible beneath ; involucel
basal or up to two-thirds way up peduncle :
Spines all straight or almost so :
Stems powdery, even the branchlets, whose
epidermis conspicuously flakes off to expose
a yellow, reddish or greyish powdery undersurface ; calyx 1–2·5 mm. long ; young
branchlets almost glabrous :
Peduncles (usually at least) on ± elongate
lateral or terminal shoots of the current
season, whose leaves are persistent or
undeveloped ; bark red to yellow or
white ; involucel firmer and more opaque
than that of *A. xanthophloea* ; calyx
2–2·5 mm. long   .   .   .   .   32. *A. seyal* var.
seyal

Peduncles (usually at least) on abbreviated
lateral shoots whose axes do not elongate
and are represented by clustered scales ;
the capitula thus appear to be in lateral
fascicles on older often yellow-barked
twigs whose leaves have fallen ; bark
yellow or greenish-yellow ; involucel
thinner and more transparent-looking
towards margins than that of *A. seyal* ;
calyx 1–1·5 mm. long ; peduncles more
slender than those of *A. seyal* at com-
parable age .   .   .   .   .   36. *A. xanthophloea*
Stems not powdery ; epidermis of branchlets
not conspicuously flaking or peeling ; calyx
1–2 mm. long ; young branchlets glabrous
to subtomentose :
Spines all suberect to spreading, mostly short,
up to 2 (rarely to 4) cm. long ; branchlets
with ± numerous reddish sessile glands,
± densely puberulous in our area ; pods
dehiscent, 0·3–0·8 cm. wide   .   .   33. *A. hockii*
Spines, some at least, characteristically deflexed (at least in East African specimens),
often 3–4 cm. long or more ; branchlets
varying from glabrous to (most commonly) densely pubescent or tomentose,
with glands inconspicuous or absent ;
pods indehiscent, 1·3–2·2 cm. wide   .   37. *A. nilotica*

Spines mostly strongly hooked downwards, a few
    only longer and straight ; twigs grey to grey-
    brown, their epidermis not peeling or flaking ;
    bark of main stem peeling off in large pieces ;
    branches to within a few feet of ground   .   34. *A. ancistroclada*
Lateral nerves of leaflets visible and somewhat raised
    beneath ; involucel at apex of peduncle .   .   38. *A. farnesiana*

Ie. Spines paired; no " ant-galls "; pinnae 1–14 pairs; flowers cream, white, pink or greenish

Spines all or many short (to about 7 mm.), ± hooked
    or curved, often with some long straight ones
    intermixed ; rhachis of leaves (except in 42, *A.*
    etbaica and 46, *A. gerrardii*) normally short, to
    about 2 cm. long ; peduncles 0·4–2·5 cm. long
    (sometimes longer in 46, *A. gerrardii*) :
Habit shrubby, obconical, branching from base ;
    young branchlets puberulous to pulverulent ;
    spines all short, hooked ; pinnae 1–3 pairs per
    leaf .   .   .   .   43. *A. reficiens*
Habit normally tree-like with a well-marked
    trunk ; young branchlets usually pubescent
    or glabrous, sometimes puberulous ; spines
    various, very often long and short mixed :
Leaf-rhachis up to about 2·5 cm. long :
    Pods * straight ; young branchlets glabrous
        or nearly so, or puberulous ; ultimate
        branches (at least in subsp. *platycarpa*)
        ascending or erect   .   .   .   42. *A. etbaica*
    Pods * contorted or spirally twisted ; young
        branchlets usually ± densely pubescent,
        very rarely (in our area) glabrous or sub-
        glabrous ; ultimate branches horizontal
        or spreading   .   .   .   .   44. *A. tortilis*
Leaf-rhachis 2·5–3 cm. or more long :
    Branchlets ± densely pubescent .   .   46. *A. gerrardii* var.
                                         latisiliqua

    Branchlets glabrous or puberulous :
        Pinnae (in our area) short, up to about 2·2
            cm. long .   .   .   .   .   42. *A. etbaica*
        Pinnae 2·5–3 cm. or more long .   .   46. *A. gerrardii* var.
                                         calvescens

Spines all straight or nearly so, varying in length ;
    rhachis of leaves usually 3 cm. or more long ;
    peduncles variable in length, but often more
    than 2·5 cm. long :
Peduncles (at least below the involucel) with ±
    numerous, very small, reddish, sessile, appa-
    rently sticky glands (use lens of × 10 or more) ;
    other hairs often sparse : similar glands on
    young branchlets and often elsewhere ; in-
    volucel mostly at or below middle of peduncle,
    usually 2–3·5 mm. long, conspicuous :
Young branchlets shortly and thinly pubescent,
    or glabrous :

* These two variable species may at times be hard to separate except by the pods,
which should be obtained if possible when certain identification is required.

Bark of twigs grey-brown to plum-coloured, not yellow, of trunk grey to brown or greenish ; pinnae of leaves of flowering shoots 6–14 pairs, but some leaves almost always with 8–9 or more pairs ; peduncles rather densely (rarely sparsely) pubescent and glandular throughout . . . 35. *A. kirkii*

Bark of twigs soon becoming pale yellow, of trunk lemon-coloured or greenish-yellow ; pinnae of leaves of flowering shoots 3–6 (–8) pairs (only on juvenile shoots as many as 10 pairs) ; peduncles sparingly pubescent to subglabrous (very rarely rather densely pubescent), glandular below and sometimes also above the involucel . . . . . . 36. *A. xanthophloea*

Young branchlets ± densely and coarsely pubescent, often showing a rusty-red colour . 46. *A. gerrardii*

Peduncles eglandular or with very small inconspicuous glands ; involucel variable in position, mostly 1–2 mm. long :

Involucel at apex of or above middle of peduncle 57. *A. sieberiana*

Involucel at base of or below middle of peduncle (sometimes in 58, *A. nubica* at about the middle) :

Corolla-lobes glabrous or almost so outside :

Twigs normally becoming coated with yellowish or greenish powdery bark ; * young branchlets with spreading, often slightly yellowish hairs 0·75–1·5(–2) mm. long ; best-developed leaves usually each with more than 10 pairs of pinnae . . . . . 40. *A. pilispina*

Twigs without yellowish or greenish powdery bark :

Young branchlets pubescent or hairy :

Twigs grey-brown, not rusty-red ; large tree to 18–25 m. high ; branchlets finely pubescent with hairs less than 0·5 mm. long . . . 41. *A. elatior* subsp. *turkanae*

Twigs usually showing rusty-red where the epidermis has fallen away ; shrub or small tree 3–15 m. high ; branchlets often coarsely hairy with hairs more than 0·5 mm. long . 46. *A. gerrardii*

Young branchlets glabrous (in our area) :

Spines straight :

Leaflets mostly less than 1·25 mm. wide ; pods straight . . 41. *A. elatior* subsp. *elatior*

Leaflets mostly more than 1·25 mm. wide ; pods falcate . . . 45. *A. clavigera*

Spines mostly hooked or recurved . 46. *A. gerrardii* var. *calvescens*

* Rare forms of *A. xanthophloea* with the peduncles almost eglandular may key out to here. They are easily distinguished from 40, *A. pilispina* by their almost glabrous young branchlets.

Corolla-lobes conspicuously hairy outside,
sometimes very densely so :
Young branchlets villous with rather long
spreading hairs, some of which are up
to 1·5–3 mm. long :
Bracteoles and calyx glabrous or almost so   59. *A. paolii*
Bracteoles conspicuously ciliate or pube-
scent ; calyx ± pubescent outside .   61. *A. stuhlmannii*
Young branchlets with shorter hairs up to
0·5(–0·75) mm. long, or sometimes
glabrous :
Large tree up to 18–25 m. high ; spines on
flowering shoots usually small, to
about 7 mm. long ; peduncles 2–5 cm.
long .   .   .   .   .   .   41. *A. elatior* subsp.
turkanae

Shrubs 0·2–5 m. high, obconical or flat-
topped ; spines on flowering shoots
often 1–6·5 cm. long ; peduncles
(except in 63, *A. turnbulliana*) 0·4–
1·5 cm. long :
Bracteoles and calyx glabrous or almost
so ; pinnae (? always) 1–5 pairs per
leaf .   .   .   .   .   60. *A.* sp. B
Bracteoles and calyx ± pubescent outside ; pinnae 3–17 pairs per leaf :
Spines on mature flowering shoots
0·4–3·5 cm. long ; peduncles
0·4–1·5 cm. long ; calyx 1–2 mm.
long ; corolla 2·5–3 mm. long :
Shrub 1–5 m. high ; branching not
horizontal ; leaflets 2·5–6(–9)
mm. long ; pods narrowly
winged ; seeds usually grey ;
widespread .   .   .   58. *A. nubica*
Shrub 0·3–2 m. high, usually flat-
topped with horizontal widely
radiating branches ; leaflets
0·75–3·5 mm. long ; pods unwinged, thick ; seeds blackish;
N. Kenya only .   .   62. *A. edgeworthii*
Spines on mature flowering shoots
(2·5–)3·5–6 cm. long (? more) ;
peduncles 1·5–3 cm. long ; calyx
3–3·75 mm. long ; corolla 4–
5·5 mm. long; a low, flat-crowned
shrub .   .   .   .   63. *A. turnbulliana*

II. Key based mainly on pod and vegetative characters *

Plant unarmed ; pods jointed, almost moniliform,
dehiscing, 0·5–0·8 cm. wide .   .   .   21. *A. mearnsii*
Plant armed with spines or prickles ; pods very
varied : (Key continued on p. 68).

* 56, *A. fischeri*, whose pods are imperfectly known, cannot be accounted for in this
part of the key at present It probably comes either in IIc or IId.

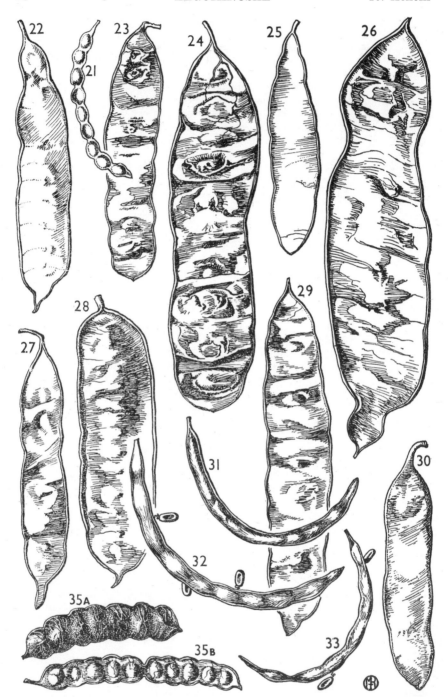

FIG. 15. *ACACIA*—Pods of capitate-flowered species, × ⅔. Species numbered as in text. **21**, *A. mearnsii*; **22**, *A. brevispica*; **23**, *A. adenocalyx*; **24**, *A. latistipulata*; **25**, *A. kamerunensis*; **26**, *A. monticola*; **27**, *A. schweinfurthii*; **28**, *A. pentagona*; **29**, *A. taylorii*; **30**, *A. macrothyrsa*; **31**, *A. zanzibarica*; **32**, *A. seyal*; **33**, *A. hockii*; **35A**, *A. kirkii* var. *intermedia*; **35B**, *A. kirkii* subsp. *mildbraedii*. 21, from *Greenway* 1612; 22, from *Gane* 16; 23, from *Faulkner* 1776; 24, from *Parry* 214; 25. from *Dummer* 2446; 26, from *Hornby* 864; 27, from *Gerstner* 6083; 28, from *Chase* 18029; 29, from *Milne-Redhead & Taylor* 7588; 30, from *Dale* 3287; 31, from *Greenway* 2049; 32, from *Trump* 38; 33, from *Tanner* 670; 35A, from *van Someren* 2700; 35B, from *Trapnell* 2184.

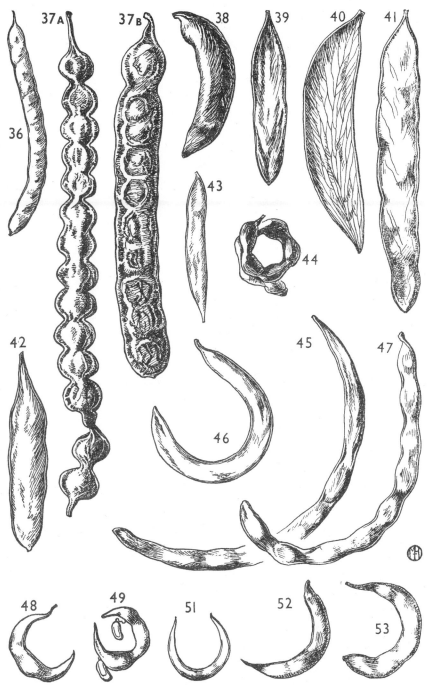

FIG. 16. *ACACIA*—Pods of capitate-flowered species, × ⅔. Species numbered as in text. **36,** *A. xanthophloea;* **37A,** *A. nilotica* subsp. *indica;* **37B,** *A. nilotica* subsp. *subalata;* **38,** *A. farnesiana;* **39,** *A. abyssinica* subsp. *calophylla;* **40,** *A. pilispina;* **41,** *A. elatior* subsp. *turkanae;* **42,** *A. etbaica* subsp. *platycarpa;* **43,** *A. reficiens* subsp. *misera;* **44,** *A. tortilis* subsp. *spirocarpa;* **45,** *A. clavigera* subsp. *usambarensis;* **46,** *A. gerrardii* var. *gerrardii;* **47,** *A. lasiopetala;* **48,** *A. drepanolobium;* **49,** *A. pseudofistula;* **51,** *A. erythrophloea;* **52,** *A. malacocephala;* **53,** *A. mbuluensis.* 36, from *Michelmore* 731; 37A. from *B. D. Burtt* 5132; 37B, from *Porter* 3742; 38. from *Foster* 179; 39, from *Greenway* 7860; 40, from *Michelmore* 1056; 41. from *Hemming* 250; 42, from *Dale* 790; 43, from *Trapnell* 2209; 44, from *Eggeling* 2512; 45, from *Faulkner* 778; 46, from *A. S. Thomas* 3076; 47, from *B. D. Burtt* 6049; 48, from *Savile* 10; 49, from *B. D. Burtt* 1943; 51, from *Jackson* 44; 52, from *Doggett* 108; 53, from *B. D. Burtt* 4271.

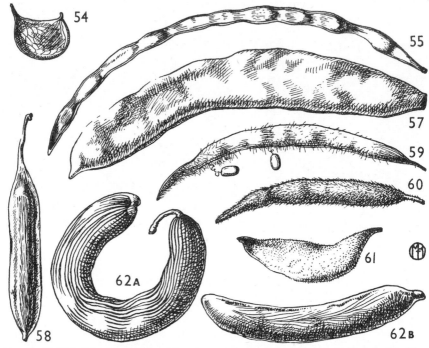

FIG. 17. *ACACIA*—Pods of capitate-flowered species, × ⅔. Species numbered as in text. **54,** *A. burttii;* **55,** *A. arenaria;* **57,** *A. sieberiana;* **58,** *A. nubica;* **59,** *A. paolii;* **60,** *A.* sp.; **61,** *A. stuhlmannii;* **62A & B,** *A. edgeworthii.* 54, from *B. D. Burtt* 6454; 55, from *Doggett* 120; 57, from *B. D. Burtt* 1643; 58, from *Dale* 3882; 59, from *Gillett* 12520; 60, from *Dale* K378; 61, from *B. D. Burtt* 1651; 62A, from *Gillett* 13333; 62B, from *Gillett* 13283.

Prickles scattered along the internodes of the stem,
    not grouped at or near nodes ; pods flattened
    or compressed, 1 cm. or more wide, usually
    brown, dehiscent (except in 28, *A. pentagona*)   **IIa** (p. 69)
Prickles or spines in pairs at or near nodes ; pods
    very varied :
    " Ant-galls " present (i.e. some pairs of stipules
        fused at base into an inflated, ± rounded or
        bilobed, ultimately woody structure ; rarely
        the stipules are inflated but not fused) ;
        pods mostly falcate and dehiscent (straight
        in 41, *A. elatior* and sometimes in 31, *A.*
        zanzibarica, half-moon-shaped or reniform
        in 54, *A. burttii*)   .   .   .   .   .   **Ib** (p. 59)
  ' Ant-galls " absent :
    Pods contorted or spirally twisted ; spines a
        mixture of some short and hooked with
        others long and straight   .   .   .   **44.** *A. tortilis*
    Pods straight, curved or falcate, but not con-
        torted or spirally twisted :
        The pods indehiscent and thin-valved
            (except often for tubercles in the centre
            of the joints), usually ± moniliform or
            jointed and breaking up transversely,
            ± transversely or net-veined, glabrous
            except usually for sessile glands ; bark

on trunk usually yellow to green, sometimes grey or brown, powdery or with
papery peel :
Bark of twigs grey-brown to plum-
coloured, not yellow, of trunk grey
to brown or greenish ; pinnae of
leaves of fruiting shoots 6–14 pairs,
but some leaves almost always with
8–9 or more pairs ; joints of pod
often tubercled in middle, mostly as
wide as or wider than long   .   .   **35. A. kirkii**
Bark of twigs soon becoming pale yellow,
of trunk lemon-coloured or greenish-
yellow ; pinnae of leaves of fruiting
shoots 3–6(–8) pairs (only on juvenile
shoots as many as 10 pairs) ; joints
of pods not tubercled, mostly longer
than wide   .   .   .   .   .   **36. A. xanthophloea**
The pods indehiscent or dehiscent ; if indehiscent, then the valves markedly
thickened, woody or pulpy in texture,
not venose and glandular as above :
Valves of the pod markedly thickened,
woody or pulpy in texture ; pods
indehiscent or slowly dehiscent (or
easily dehiscent only in 58, *A. nubica*)   **IIb** (p. 70)
Valves of the pod membranaceous to subcoriaceous   or   coriaceous,   not
markedly thickened :
Pods straight or nearly so .   .   .   **IIc** * (p. 71)
Pods ± falcate   .   .   .   .   **IId** * (p. 72)

IIa. Prickles scattered along the internodes of the stem

Petiole 0·5–1·5 cm. long :
Midrib of leaflets excentric at base ; branchlets
usually pale brown, then grey   .   .   .   **22. A. brevispica**
Midrib of leaflets almost central at base ; branch-
lets dark brown to blackish, very glandular .   **23. A. adenocalyx**
Petiole 1·5–6 cm. or more long :
Midrib of leaflets markedly excentric at base :
Pods dehiscent, with margins about 1–1·5 mm.
thick :
Leaflets 0·3–0·8 mm. wide ; seeds with small
areole 4–7 × 1·5–3·5 mm. :
Pods rather narrow, 1·7–2·8 cm. wide, pale,
flat ; branchlets usually yellow-brown,
then grey, with glands few or none,
subglabrous or sparsely and minutely
puberulous   .   .   .   .   .   **25. A. kamerunensis**
Pods about 3–4·5 cm. wide, dark brown ;
branchlets dark brown, then blackish,
with ± numerous red-purple glands
mixed with fulvous pubescence .   .   **26. A. monticola**

* 56, *A. fischeri*, whose pods are imperfectly known, cannot be accounted for in this
part of the key at present. It probably comes either in IIc or IId.

Leaflets 0·8–3 mm. wide :
 Pods 1–2·9 cm. wide ; areole of seeds largish,
  6–8 × 3–5 mm. ; branchlets glandular
  when young, olive-green to olivaceous-
  brown ; in woodland and thickets . 27. *A. schweinfurthii*
 Pods (2·5–)3·6–4·2 cm. wide ; areole of
  seeds small, 4·5–7 × 2–3·5 mm. :
  Branchlets eglandular, pale brown, later
   grey ; pods pale ; in coastal forest
   and thickets . . . . . 24. *A. latistipulata*
  Branchlets with numerous red-purple
   glands, dark brown to blackish ;
   pods dark brown ; in or near
   montane forest . . . . 26. *A. monticola*
 Pods indehiscent, thick and hard, with margins
  2–4 mm. thick, dark brown ; areole of
  seeds large, 7–10 × 4·5–6 mm. ; a tall rain-
  forest liane . . . . . 28. *A. pentagona*
Midrib of leaflets subcentral at base ; pods dehiscent, pale brown ; areole of seeds small,
 4–5 × 1·5–2·5 mm. . . . . . 29. *A. taylorii*

IIb. Spines paired; no " ant-galls "; pod-valves thick, woody or pulpy
Pods glabrous or very sparingly hairy :
 The pods very fat, almost round in section, glabrous,
  finely longitudinally striate . . . . 38. *A. farnesiana*
 The pods distinctly compressed or flattened :
  Pods moniliform or jointed or with the valves
   marked with distinct raised bumps each one
   corresponding to a seed inside ; pods indehiscent . . . . . . 37. *A. nilotica*
  Pods not moniliform or jointed, the valves with
   an irregular but ± smooth and glossy surface, without any bumps corresponding to
   the seeds inside ; pods very slowly dehiscent 57. *A. sieberiana*
Pods ± densely puberulous, pubescent, tomentellous
 or villous :
 Ripe pods distinctly compressed or flattened :
  Pods quickly and readily dehiscent, with thin
   narrow wing-like margins and a longitudinally veined central part . . . . 58. *A. nubica*
  Pods indehiscent, without thin wing-like margins
   (in 37, *A. nilotica* often with a thick border) :
   Indumentum on pod short, the hairs less than
    0·5 mm. long ; pods moniliform or
    jointed or marked on outside with bumps
    each corresponding to a seed within ;
    seeds blackish-brown, smooth . . 37. *A. nilotica*
   Indumentum on pod long, the hairs spreading,
    2–4 mm. long ; pods not in the least
    moniliform or jointed and not marked
    with bumps as above ; seeds olive,
    punctate . . . . . . 61. *A. stuhlmannii*
 Ripe pods cylindrical, round or nearly so in section,
  velvety-pubescent, longitudinally veined (at
  least in 62, *A. edgeworthii*) :

Spines on mature shoots 0·4–3·5 cm. long ;
  peduncles 0·4–1·5 cm. long   .   .   .   62. *A. edgeworthii*
Spines on mature shoots (2·5–)3·5–6 cm. long
  (? more) ; peduncles 1·5–3 cm. long .   .   63. *A. turnbulliana*

IIc. Spines paired; no " ant-galls "; pods straight or nearly so; valves membranaceous to coriaceous, not markedly thickened

Leaves very large for the genus, 10–20 cm. wide,
  rhachis together with petiole 10–37 cm. long ;
  pods coriaceous, glossy, glabrous   .   .   .   30. *A. macrothyrsa*
Leaves much smaller, less than 10 cm. long and
  wide ; pods various :
Pods clothed with whitish spreading hairs up to
  3–4 mm. long ; a shrub branching from base   59. *A. paolii*
Pods glabrous or clothed with shorter hairs at most
  (in 60, *A.* sp. B) up to 1·5 mm. long :
Seeds ± strongly compressed ; pods glabrous or
  subglabrous to grey-puberulous or pulverulent :
Leaflets all exceedingly narrow, about 0·25 mm.
  wide (at most to 0·5 mm.) ; pinnae on
  well-developed mature leaves at least 15
  and up to 36 pairs ; flat-crowned tree,
  usually large, 6–20 m. high   .   .   39. *A. abyssinica*
Leaflets 0·5 mm. or more wide :
Twigs normally becoming coated with
  yellowish or greenish powdery bark ;
  young branchlets with spreading, often
  slightly yellowish hairs 0·75–1·5(–2)
  mm. long ; pinnae 8–16(–21) pairs per
  leaf   .   .   .   .   .   .   40. *A. pilispina*
Twigs without yellowish or greenish powdery
  bark ; young branchlets glabrous to
  puberulous or shortly pubescent, with
  the hairs up to at most 0·5 mm. long ;
  pinnae 1–13 pairs per leaf :
Habit shrubby, obconical, branching from
  base ; young branchlets puberulous
  to pulverulent ; spines all short,
  hooked ; pinnae 1–3 pairs per leaf ;
  pods (in our area) 0·6–0·95 cm. wide ;
  seeds lying longitudinally in the pod   43. *A. reficiens*
Habit normally tree-like, with a well-
  marked trunk ; young branchlets
  pubescent, glabrous or puberulous ;
  spines various, very often long and
  short mixed ; pinnae often more
  than 3 pairs per leaf ; pods 0·6–
  2·2 cm. wide ; seeds lying transversely or obliquely in the pod :
Spines all straight or nearly so ; pods
  1·2–1·8 cm. wide ; tree up to 18–
  25 m. high   .   .   .   .   41. *A. elatior* *

* Fruiting specimens of 2a, *A. dolichocephala*, placed among the spicate-flowered species, although the inflorescences are very short and almost head-like, are liable to key out here. They may be separated from *A. elatior* by the usually much longer peduncles ending in a distinctly clavate scar-covered receptacle, and in the ellipsoid, less compressed seeds with a small central areole.

Spines all or many of them hooked ;
pods 0·6–2·2 cm. wide ; tree 2·4–
12 m. high ; (occasionally spines
all straight or nearly so, but then
distinguished from 41, *A. elatior* by
the pods being only 0·6–1·2 cm.
wide)   .   .   .   .   42. *A. etbaica*

Seeds not or scarcely compressed, ellipsoid to
subglobose ; pods puberulous to densely
hairy :

Pods 0·9–2·2 cm. wide, with narrow wing-like
margins 1–3·5 mm. wide, usually pale
brown to straw-coloured, puberulous to
densely and shortly pubescent   .   .   58. *A. nubica*

Pods 0·7–0·9 cm. wide, without any wing-like
margins, densely clothed with rather
matted ascending hairs mostly concealing
the surface of the valves   .   .   .   60. *A.* sp. B

**IId. Spines paired; no " ant-galls "; pods ± falcate or arcuate; valves
membranaceous to coriaceous, not markedly thickened**

Pinnae 15 or more (up to 36) pairs per leaf on well-
developed leaves of mature shoots :

Leaflets all exceedingly narrow, about 0·25 mm.
wide (at most 0·5 mm.) ; usually a large, flat-
crowned tree 6–20 m. high ; pods at most
slightly arcuate, flattened, 1·2–2·8 cm. wide,
glabrous to puberulous   .   .   .   .   39. *A. abyssinica*

Leaflets 0·5 mm. or more wide :

Older branchlets with vivid rusty-red powdery
bark ; pods ± moniliform and turgid over
the seeds, grey-tomentellous ; young foliage
with gleaming, silky, pale golden indumentum   .   .   .   .   .   47. *A. lasiopetala*

Older branchlets without rusty-red powdery
bark ; pods flattened, glabrous or subglabrous ; young foliage without gleaming,
silky indumentum :

Pods 1·4–2·9 cm. wide ; branchlets with
spreading hairs mostly 0·5–2 mm. long ;
epidermis of branchlets falling away to
expose a yellow or sometimes greenish
powdery bark layer   .   .   .   40. *A. pilispina*

Pods 0·5–0·8 cm. wide ; branchlets with short
puberulence or pubescence of hairs up to
about 0·25 mm. long ; epidermis of
branchlets not falling away so as to expose
a powdery layer   .   .   .   55. *A. arenaria*

Pinnae 1–14 pairs per leaf :

Pods glabrous, subglabrous, somewhat puberulous
or rarely tomentellous with short hairs much
less than 0·5 mm. long ; seeds smooth, ±
strongly compressed :

The pods 0·3–1·2 cm. wide :

Young branchlets glabrous, subglabrous or
densely puberulous ; pods 0·3–1·2 cm.
wide ;

Stems powdery, even the branchlets, whose
epidermis conspicuously flakes off to
expose a greyish to ± reddish powdery
undersurface ; pods glabrous except
for sessile glands, 0·5–0·9 cm. wide   .   32. *A. seyal*

Stems and twigs not powdery ; epidermis
of branchlets not conspicuously flaking
off :

Spines a mixture of short downwardly
hooked ones and a few long straight
ones ; pods glabrous, 0·3–0·5 cm.
wide .   .   .   .   .   34. *A. ancistroclada*

Spines all straight or almost so :

Pods 0·3–0·6(–0·8) cm. wide ; bark on
older stems peeling off in papery
layers when not burned ; branchlets ± densely puberulous, rarely
glabrous, with ± numerous sessile
glands   .   .   .   .   33. *A. hockii*

Pods 0·7–1·2 cm. wide ; bark without
papery peeling layers ; branchlets
glabrous, eglandular   .   .   45. *A. clavigera*
subsp. *usambarensis*

Young branchlets ± densely pubescent ; pods
0·6–1·2 cm. wide ; epidermis of branchlets
often splitting or falling away to expose a
rusty-red inner layer   .   .   .   46. *A. gerrardii* var.
gerrardii

The pods 1·2–2·9 cm. wide :

Epidermis of branchlets falling away to expose
an inner yellow or sometimes greenish
powdery bark layer ; young branchlets
densely clothed with spreading, grey to
slightly yellowish hairs mostly 0·5–2 mm.
long ; pods 1·4–2·9 cm. wide   .   40. *A. pilispina*

Epidermis of branchlets not falling away to
expose any yellow or greenish layer ;
pods up to 1·7 cm. wide :

Branchlets glabrous or nearly so :

Spines straight or nearly so   .   .   45. *A. clavigera*
subsp. *clavigera*

Spines mostly hooked or recurved .   46. *A. gerrardii* var.
calvescens

Branchlets ± densely pubescent ; spines
often hooked or recurved, sometimes
straight   .   .   .   .   .   46. *A. gerrardii* var.
latisiliqua

Pods densely hairy with hairs up to 1–4 mm. long ;
seeds punctate, scarcely compressed :

The pods clothed with spreading hairs up to
3–4 mm. long ; pods attenuate for about
2–3 cm. at base ; young branchlets villous
with rather long spreading hairs, some of
them up to 1·5–3 mm. long ; pinnae 4–9
pairs per leaf .   .   .   .   .   59. *A. paolii*

The pods clothed with rather ascending and matted hairs up to 1–1·5 mm. long ; pods attenuate for about 0·5–1 cm. at base ; young branchlets with short hairs up to about 0·5(–0·75) mm. long ; pinnae (? always) 1–5 pairs per leaf.   .   .   .   60. *A.* sp. B

The species of *Acacia* occurring in our area can be divided into groups, some of which may be natural ones. These are the basis of the sequence in which the species are arranged here. The order and principal characters of these groups are given in the key and conspectus below :

KEY TO THE GROUPS

Flowers in spikes or spiciform racemes :
  Leaf-rhachis with a conspicuous gland between each
    pair of pinnae   .   .   .   .   .   .   **A** (sp. 1)
  Leaf-rhachis without conspicuous glands :
    Stipules spinescent, straight, rarely somewhat
    curved, often conspicuously enlarged below and
    ashen or whitish .   .   .   .   .   .   **B** (spp. 2–4)
    Stipules not spinescent ; prickles below the stipules,
      short, usually ± hooked or curved, usually
      brown, blackish or dull grey, never inflated,
      very rarely absent :
      Prickles irregularly scattered along the inter-
        nodes ; a climbing shrub (in our area).   .   **C** (sp. 5)
      Prickles grouped or solitary at or just below the
        nodes (very rarely and probably abnormally
        a few along the internodes) ; erect trees or
        shrubs :
        Flowers usually distinctly but shortly pedicel-
          late (pedicels 0·5 mm. or more long).   .   **D*** (spp. 6,8)
        Flowers sessile or subsessile (pedicels 0–0·3 mm.):
          Prickles in pairs   .   .   .   .   **E** (spp. 7*, 9–16)
          Prickles in threes or solitary   .   .   **F** (spp. 17–20)
Flowers in heads :
  Plant unarmed ; naturalized, but not native   .   .   **G** (sp. 21)
  Plant armed with spines or prickles :
    Stipules not spinescent ; prickles scattered along
      the stems ; usually climbing, sometimes not .   **H** (spp. 22–29)
    Stipules spinescent, paired at nodes ; other prickles
      absent :
      Heads of flowers in panicles   .   .   .   **I** (sp. 30)
      Heads of flowers axillary, sometimes racemose,
        not panicled :
      Flowers bright or golden-yellow :
        Pods dehiscent, falcate, less than 1 cm. wide   **J** (spp. 31–34)
        Pods indehiscent :
          Involucel basal or to half-way up peduncle:
            Bark on trunk green, yellow or lemon-
              coloured ; pods 0·7–1·4 cm. wide   .   **K** (sp. 36)

---

* The relationship of group D with E and F requires further study. As explained further on, spp. 6 and 7 may perhaps be hybrids between *A. mellifera* in group D and other species in groups E or F.

Bark on trunk blackish, grey or brown ; pods 1·3–2·2 cm. wide . . . **L** (sp. 37)

Involucel at apex of peduncle ; pods subterete, dark brown or blackish, glabrous. . . . . . **M** (sp. 38)

Flowers white, pink, cream or very pale yellow :
Pods ± moniliform or jointed, and often tubercled, indehiscent, straight or slightly curved . . . . . . **K** (spp. 35–36)

Pods not jointed or tubercled, usually not moniliform, but if so then dehiscent :
Pods contorted or spirally twisted . . **O** (sp. 44)

Pods straight, curved or falcate, not as above :
Pods straight or almost so :
Pods papery to subcoriaceous, flattened . . . . . **N** (spp. 39–43)

Pods thick, ± compressed to almost round in section, woody (at least when dry) :
Involucel normally apical or in upper half of peduncle ; pods glabrous or slightly hairy, without veins . . . . **S** (sp. 57)

Involucel normally basal or in lower half of peduncle ; pods densely clothed with puberulence or longer hair, often veined. . **T** (spp. 58–63)

Pods falcate or curved :
" Ant-galls " present . . . **Q** (spp. 48–54)

" Ant-galls " absent :
Involucel normally apical or in upper half of peduncle :
Pods narrow, 0·5–0·8 cm. wide ; valves not markedly thick ; areoles on seeds 1·5–2·5 mm. wide . . . . . **R** (sp. 55)

Pods wide, 1·5–3·5 cm. wide ; valves thick and almost woody when dry ; areoles on seeds 5–6 mm. wide . **S** (sp. 57)

Involucel normally basal or in lower half of peduncle :
Corolla-lobes glabrous or almost so outside ; seeds compressed . . . . . **P** (spp. 45–47)

Corolla-lobes ± densely white-pubescent outside ; seeds not or only slightly compressed . . . . **T** (spp. 58–63)

NOTE. 56, *A. fischeri*, whose pods I have not seen, is not accounted for in the above key.

CONSPECTUS OF THE GROUPS

I. Flowers in spikes or spiciform racemes

A (sp. 1). Stipules spinescent, not inflated. Leaf-rhachis with a conspicuous gland between each pair of pinnae. Stamen-filaments shortly connate for about 1 mm. at base ; anthers eglandular even in bud. Pods indehiscent, falcate or coiled. Seeds rather large, with large areole. A tree.

B (spp. 2–4). Stipules spinescent, often inflated or fusiform. Leaf-rhachis without conspicuous glands. Stamen-filaments free ; anthers glandular, at least in bud. Pods dehiscent. Seeds rather small, with small or narrow areole. Trees and shrubs.

C (sp. 5). Stipules not spinescent. Prickles irregularly scattered along the stems. Leaf-rhachis without conspicuous glands. Stamen-filaments free ; anthers glandular, at least in bud. Pods dehiscent. Seeds with small areole. A climbing shrub (in our area).

D* (spp. 6, 8). Stipules not spinescent. Prickles in pairs just below each node. Leaf-rhachis without conspicuous glands. Flowers pedicellate (pedicels 0·5 mm. or more long). Stamen-filaments free ; anthers glandular, at least in bud. Pods dehiscent. Seeds with small areole. Small trees and shrubs.

E (spp. 7*, 9–16). Stipules not spinescent. Prickles in pairs just below each node. Leaf-rhachis without conspicuous glands. Flowers sessile or almost so. Stamen-filaments free ; anthers glandular, at least in bud. Pods dehiscent, except in sp. 15. Seeds and areole variable. Trees and shrubs.

F (spp. 17–20). Leaf-rhachis without conspicuous glands. Stipules not spinescent. Prickles in threes just below each node, or solitary. Flowers sessile or almost so. Stamen-filaments free ; anthers glandular, at least in bud. Pods dehiscent. Seeds subcircular with markedly impressed small to large areole. Trees and shrubs.

II. Flowers in heads

G (sp. 21). Unarmed tree. Flower-heads paniculate. Pods jointed, almost moniliform, narrow, dehiscent. Seeds black, compressed, with conspicuous caruncle, and areole about 2 mm. wide. An introduced, naturalized tree.

H (spp. 22–29). Stipules not spinescent. Prickles irregularly scattered along the stems. Flower-heads often paniculate. Flowers white, cream or pale yellow. Pods dehiscent (except in sp. 28), flat or compressed, straight. Seeds brown to black, usually elliptic, ± compressed, smooth ; areole variable. Small trees or shrubs, usually climbing.

I (sp. 30). Stipules spinescent. Leaves large. Flowers orange or yellow. Flower-heads in large panicles. Pods dehiscent, coriaceous, straight. Seeds deep brown, subcircular or elliptic, compressed, smooth ; areole small. A tree.

J (spp. 31–34). Bark red, yellow, green or white, powdery or with papery peel. Stipules spinescent, often modified to ± bilobed " ant-galls." Flower-heads not panicled ; involucel basal or in lower half of peduncle (rarely two-thirds of the way up). Flowers bright yellow. Pods dehiscent, falcate, less than 1 cm. wide. Seeds olive to olive-brown, compressed, usually smooth, longer than wide ; areole medium in size. Trees or shrubs.

* See footnote on p. 74.

K (spp. 35–36). Bark yellow to green, powdery or with papery peel. Stipules spinescent ; no " ant-galls." Flower-heads not panicled ; involucel half-way up peduncle or below. Flowers white, pink or bright yellow. Pods indehiscent, moniliform or jointed, often tubercled, straight or slightly curved. Seeds olive to blackish-olive, subcircular to elliptic, compressed, smooth ; areole small or large. Trees.

L (sp. 37). Bark blackish, grey or brown, neither powdery nor peeling. Stipules spinescent ; no " ant-galls." Flower-heads not panicled ; involucel in lower half of peduncle. Flowers bright yellow. Pods indehiscent, straight or slightly curved, compressed, moniliform or not. Seeds deep blackish-brown, subcircular, compressed, smooth ; areole large, 4·5–5 mm. wide. A tree.

M (sp. 38). Stipules spinescent ; no " ant-galls." Flower-heads not panicled ; involucel at apex of peduncle. Flowers bright yellow. Pods indehiscent, straight or curved, subterete, dark brown to blackish, glabrous. Seeds chestnut-brown, only slightly compressed, longer than wide, smooth ; areole large, 4 mm. wide. A shrub.

N (spp. 39–43). Bark black, grey or brown (yellow to greenish and powdery on branches in sp. 40 only). Flower-heads not panicled ; involucel usually in lower half of peduncle, rarely higher. Flowers white, pink, cream or very pale yellow. Pods dehiscent, straight or only slightly curved, flat, papery to subcoriaceous. Seeds olive-brown or brown, compressed, subcircular or elliptic, smooth ; areole variable in size. Trees or sometimes shrubs.

O (sp. 44). Bark grey to black, neither powdery nor peeling. Stipules spinescent ; no " ant-galls." Flower-heads not panicled ; involucel basal or in lower half of peduncle. Flowers cream or whitish. Pods contorted or spirally twisted. Seeds olive- to red-brown, compressed, longer than wide, smooth ; areole 3–4 mm. wide. A tree.

P (spp. 45–47). Bark grey to dark brown or black, or else rusty-red. Stipules spinescent ; no " ant-galls." Flower-heads not panicled ; involucel basal or in lower third of peduncle. Flowers white or cream. Corolla-lobes glabrous or almost so outside. Pods dehiscent, falcate. Seeds blackish-olive to olive-brown, longer than wide, compressed, smooth, lying longitudinally in the pod ; areole 2·5–4·5 mm. wide. Trees or shrubs.

Q (spp. 48–54). Bark grey to brown or black, or often red and powdery. Stipules spinescent, some always enlarged and modified to rounded, not lobed " ant-galls." Flower-heads not panicled ; involucel basal or sometimes a short way above. Flowers white or cream. Pods dehiscent, falcate to annular, or half-moon-shaped in sp. 54. Seeds grey or mottled with dark brown, compressed, longer than wide and often curved except in sp. 54, smooth, lying longitudinally in the pod ; areole small or large. Shrubs or normally small trees.

R (sp. 55). Bark dark. Stipules spinescent ; no " ant-galls." Flower-heads not panicled ; involucel apical or in upper half of peduncle. Flowers white or pink. Pods dehiscent, arcuate, not conspicuously thickened. Seeds olive-grey, compressed, longer than wide ; areole small, 1·5–2·25 mm. wide. Shrub or small tree.

S (sp. 57). Bark grey to brown or yellowish, sometimes flaking. Stipules spinescent ; no " ant-galls." Flower-heads not panicled ; involucel normally apical or in upper half of peduncle. Flowers white or very pale yellow. Pods very slow in dehiscing, straight or sometimes ± curved, flattened but thick and almost woody in texture when dry, glabrous or slightly hairy, without veins. Seeds olive-grey, compressed, smooth ; areole large, 5–6 mm. wide. A tree.

T (spp. 58–63).   Bark on branchlets usually pale, green to whitish, or grey, olive or pale brown, and marked with pale dot-like lenticels, occasionally with papery peeling on the old stems.   Stipules spinescent ; no " ant-galls."   Flower-heads not panicled ; involucel normally at or below middle of peduncle.   Flowers white, pink, cream or greenish ; corolla-lobes ± densely white-pubescent outside.   Pods dehiscent, sometimes slowly so, straight or annular, often thick hard and woody, ± densely clothed with very short to rather long hairs, sometimes wing-margined. Seeds not or only slightly compressed, their surface often minutely punctate or wrinkled ; areole 3–5 mm. wide.   Mostly shrubs, sometimes a small tree.

NOTE. *A. fischeri,* whose pods I have not seen, is provisionally placed after group R (as sp. 56). Its true position is doubtful.

1. **A. albida** *Del.,* Fl. Egypt. : 142, t. 52, fig. 3 (1813) ; Harms in N.B.G.B. 4 : 198, fig. 2 (1906) ; L.T.A. : 825 (1930) ; T.S.K. : 67 (1936) ; Bogdan in Nature in E. Afr., ser. 2, No. 1 : 13 (1949) ; T.T.C.L. : 330 (1949) ; I.T.U., ed. 2 : 205, fig. 47a (1952) ; Young in Candollea 15 : 89 (1955) ; Consp. Fl. Angol. 2 : 272 (1956) ; F.W.T.A., ed. 2, 1 : 499 (1958). Type : Egypt, above Philae, *Nectoux* (MPU, holo.)

Tree 6–30 m. high with rough, dark brown or greenish-grey bark and spreading branches. Young branchlets ashen to whitish. Stipules spinescent, up to 1·3(–2·3) cm. long, straight, never enlarged and inflated ; no prickles below the stipules. Leaves : rhachis with a single conspicuous gland at junction of each of the (2–)3–10 pairs of pinnae ; no gland on petiole ; leaflets 6–23 pairs, (2·5–)3·5–6(–12) mm. long, 0·7–2·25(–4) mm. wide, rounded to subacute and mucronate at apex. Flowers cream, sessile or to 0·5(–2.0) mm. pedicellate, in inflorescences 3·5–14 cm. long on peduncles 1·3–3·5 cm. long. Calyx 1–1·7(–2·5) mm. long. Corolla 3–3·5(–4·5) mm. long, with 5 lobes 1·5–2·5 mm. long. Stamen-filaments 4–5 mm. long, tubular for about 1 mm. at base ; anthers 0·2–0·4 mm. across, eglandular even in bud. Pods (Fig. 14/1, p. 52) bright orange, thick, indehiscent, glabrous or very rarely puberulous, falcate or curled into a circular coil, 6–25 cm. long, (1·8–)2–3·5(–5) cm. wide. Seeds elliptic-lenticular, 9–11 mm. long, 6–8 mm. wide ; central areole large, 7–9 × 4–6 mm.

UGANDA. Karamoja District : Kangole, 22 May 1940, *A. S. Thomas* 3476 ! ; Ankole   District : Ruampara, Gayaza, Oct. 1932, *Eggeling* 634 *in* F.H. 1008 ! & Ibanda,   29 Nov. 1951, *Trapnell* 2189 !
KENYA. Naivasha, Sept. 1933, *van Someren* 2721/5142 ! ; Teita Hills, *Grenfell* !
TANGANYIKA. Pare District : Kisiwani, 5 Feb. 1930, *Greenway* 2164 ! ; Dodoma   District : Mvumi, 17 Aug. 1928, *Greenway* 787 ! ; Southern Highlands Province :   96 km. from Iringa [on Kilosa road], *Hughes* 116 !
DISTR. U1, 2, 4 ; K3, 7 ; T1–8 ; widespread in tropical and subtropical Africa from   Egypt, Senegal and the Gambia southwards to Bechuanaland, the Transvaal and   Natal ; also in Syria, Palestine and (? native) in Cyprus
HAB. Riverine and ground-water forest and woodland ; 600–1830 m.

SYN. *Faidherbia albida* (Del.) A. Chev. in Rev. Bot. Appliq. 14 : 876 (1934) ; Gilb. &   Bout. in F.C.B. 3 : 169 (1952)

NOTE. The stamen-filaments shortly connate at base, the large anthers eglandular at   apex even in bud, and the very distinctive pods, whose appearance when ripe has   inspired the popular name of " Apple-Ring Acacia," occur nowhere else among the   East African acacias with spiciform inflorescences. Similar stamen-filaments, how-   ever, occur in *A. ogadensis* Chiov. from NE. tropical Africa and in the Rhodesian   *A. eriocarpa* Brenan.
VARIATION. In our area *A. albida* can be sorted into two well-marked geographical   races :—
    A : with young branchlets glabrous or almost so, also the inflorescence-axis, calyx   and corolla ; leaflets ciliolate on margins otherwise glabrous or nearly so, usually   rather small, to 6 × 1·5 mm.

B : with young branchlets pubescent, also the inflorescence-axis, calyx (and often corolla) ; leaflets ± pubescent on surface, often larger than in A, and up to 12 × 4 mm.

B occurs in our area in Tanganyika only, from Mbulu (*Matalu* 3131, EA !) and Dodoma Districts southwards, and appears to be the exclusive race in all territories to the south.

A occurs in Kenya and Uganda, and in Tanganyika as far south as Dodoma District. An intermediate between A and B has been collected in Tanganyika, Bukoba District, Kyaka, *Proctor* 795 ! In territories to the north and in West Africa (north of the Gulf of Guinea) B is common, together with many plants with the leaflets of B but the lack of indumentum of A, and others intermediate in leaflet-size between A and B.

A. albida var. *senegalensis* Benth. in Hook., Lond. Journ. Bot. 1 : 505 (1842), based on *Robert s.n.* (K, syn. !) and *Brunner* 72 (K, syn. !), both from Senegal, is B, although the leaflets of *Brunner* 72 are very sparingly pubescent.

A. albida var. *microfoliolata* De Wild., Pl. Bequaert. 3 : 56 (1925), based on *Delevoy* 443 ! and *A. albida* var. *variofoliolata* De Wild., *op. cit.* p. 57, based on *Delevoy* 227 !, both from the Belgian Congo, and with holotypes at Brussels, are both also referable to B. The var. *microfoliolata* is recorded from Tanganyika Territory in L.T.A. : 825 (1930).

I have also no doubt that *A. mossambicensis* Bolle in Peters, Mossamb. Bot. : 5 (1861) (not, however, as interpreted in L.T.A. : 831 (1930)) refers to B.

2. **A. lahai** [*Steud. & Hochst. ex*] *Benth.* in Hook., Lond. Journ. Bot. 1 : 506 (1842) ; L.T.A. : 826 (1930) ; T.S.K. : 71 (1936) ; Bogdan in Nature in E. Afr., ser. 2, No. 1 : 13 (1949) ; T.T.C.L. : 331 (1949) ; I.T.U., ed. 2 : 210, fig. 47e (1952). Type : Ethiopia, Tigré, near Adowa, *Schimper* 119 (K, holo. !)

Flat-crowned tree 3–15 m. high with rough, brown or grey-brown bark. Young branchlets brown to blackish-purple, pubescent. Stipules spinescent, up to 7 cm. long, straight (very rarely somewhat curved), subulate but not enlarged or fusiform, grey to grey-brown ; no prickles below the stipules. Leaves : rhachis without conspicuous single glands (only clusters of tiny red bodies) between the (3–)6–15 pairs of pinnae ; often a conspicuous gland on the petiole ; leaflets 10–28 pairs, 1·5–4·5 mm. long, 0·3–0·75 (–1·0) mm. wide, glabrous or ciliolate on the margins especially near the rounded to subacute apex, lateral nerves invisible. Flowers cream or white, sessile, in spikes 2·5–7 cm. long on peduncles 0·7–2·2 cm. long ; axis ± pubescent and with many reddish, sessile or subsessile glands. Calyx 0·5–1·25 mm. long. Corolla 2–3 mm. long, glabrous, with 4–5 lobes 0·5 mm. long. Stamen-filaments 4·5–5 mm. long, free ; anthers 0·1 mm. across, with a caducous gland. Pods (Fig. 14/2, p. 52) brown, dehiscent, glabrous, or puberulous on the stipe, elliptic-oblong or oblong, straight or ± falcate, mostly 4–7 cm. long, 1·5–3 cm. wide. Seeds ± obliquely obovate-flattened, 6–7 mm. long, 5 mm. wide ; central areole small, 1·5–2·5 × 1·5 mm.

UGANDA. Karamoja District : Karakau, Karasuk, July 1956, *Philip* 785 ! ; Mbale District : Kaburon, *Eggeling* 2477 !
KENYA. West Suk District : Kapenguria, May 1932, *Napier* 1952 *in C.M.* 5675 ! & 6 May 1953, *Padwa* 56 ! ; Elgon, 24 Jan. 1931, *Lugard* 513 ! ; Kericho District : Sotik, 6 Feb. 1949, *Bally* 6589 !
TANGANYIKA. Masai District : Ngorongoro Crater, 20 Feb. 1953, *Wigg* 1061 ! & *Ivens* 459 ! & SW. slopes of Lemagrut, 24 Nov. 1956, *Greenway* 9054 !
DISTR. U1, 3 ; K2, 3, 5, 6 ; T2 ; Ethiopia and Eritrea
HAB. Woodland and wooded grassland ; 1830–2440 m.

NOTE. *A. lahai* is characteristic of much higher altitudes than its closest relatives, *A. bussei* and *A. horrida* subsp. *benadirensis*. Unlike them, it seems never to produce " galled " spines, and its densely glandular inflorescence-rhachis is most distinctive.

2a.* **A. dolichocephala** *Harms* in Ann. Ist. Bot. Roma 7 : 86 (1897) ; L.T.A. : 854 (1930). Type : Ethiopia, between Rogono and Gobo Duaya [Galla Sidamo, about 5° 33′ N., 38° 8′ E.], *Riva* 599 (FI, holo. !)

---

* This species was identified only when the text of this part of the Flora was almost complete and the sequence of numbering of the species could not be changed to accommodate it.

Small tree 4–7·5 m. high. Young branchlets brown or grey-brown, longitudinally ridged and very sparsely puberulous when young, soon glabrous. Stipules spinescent, short, straight, slender, suberect, to about 4 mm. long (on mature shoots), not enlarged or fusiform ; no prickles below the stipules. Leaves : petiole (1·5–)2–3 cm. long ; rhachis sparsely puberulous, eglandular, 5–9 cm. long ; pinnae 6–14 pairs ; leaflets 12–35 pairs, 2·5–7·5 mm. long, 0·6–1·5 mm. wide, rounded to subacute at apex, glabrous unless for a few very inconspicuous appressed cilia ; lateral nerves invisible ; surface of leaflets paler beneath than above. Flowers sessile, in very short spikes or ellipsoid heads 1–1·5 cm. long, on usually long (4–7·5 cm., to 9 cm. in fruit) peduncles which are puberulous and with some sessile glands above. Calyx 0·5–1 mm. long, glabrous. Corolla 3·75–4 mm. long, glabrous, with 4 lobes about 1 mm. long. Stamen-filaments about 7 mm. long, free or irregularly connate (not tubular) near base ; anthers about 0·1 mm. across, with a caducous gland. Pods purplish-brown, dehiscent, almost or quite glabrous, oblong, straight or very slightly falcate, ± reticulate-venose, flattened, rounded or acute at apex, stipitate at base, 5·5–10 cm. long, 1·3–1·9 cm. wide. Seeds ± elliptic, somewhat compressed, olive-green, 5·5–6 mm. long, 4–5 mm. wide ; central areole small, 2·5 × 1–1·25 mm. Fig. 18.

FIG. 18. *ACACIA DOLICHOCEPHALA*—**1**, inflorescence, × 1; **2**, pod, × 1. 1, from *Kuls* 934; 2, from *Gillman* 802.

UGANDA. Karamoja District : 1·5 km. N. of Moroto, Jan. 1956, *Wilson* 219 ! ; Mbale District : NW. Elgon, Sebei flats, Dec. 1942, *Dale* U340 !
TANGANYIKA. Lushoto District ? : W. Usambara Mts., Kihitu, *Gillman* 802 !
DISTR. **U**1, 3 ; **T**3 ; the Sudan (Didinga Mts., *Myers* 11219 !) and Ethiopia
HAB. " Riverine woodland " (*Wilson* 219) ; " rocky hill " (*Dale* U340) ; 1150–1520 m.

NOTE. This remarkable species is difficult to place taxonomically, since it seems to bridge the gap between the capitate- and spicate-flowered groups of acacias. It seems, however, best considered as a relative of *A. lahai*, on account of the flower characters.

3. **A. bussei** [*Harms ex*] *Sjöstedt*, Schwed. Zool. Exped. Kilimandjaro 8 : 117–118, t. 6, fig. 4–5, t. 8, fig. 3 (1908) ; Harms in E.J. 51 : 365 (1914) ; L.T.A. : 825 (1930) ; Bogdan in Nature in E. Afr., ser. 2, No. 1 : 13 (1949) ; T.T.C.L. : 331 (1949). Types* : Tanganyika : Lushoto District : Mazinde, by Kisiwani road, *Busse* 361 (B, syn. †, BM, K, isosyn. !) ; Lushoto/Pare Districts : between Usambara Mts. and Kihurio, *Engler* 1506 (B, syn. †, K, drawing !) ; Pare District : between Kihurio and Gonja, *Zimmermann* 1758 (B, syn.†, EA, isosyn. !)

* Sjöstedt's publication cited no specimens, but mentioned a locality in Pare District (between Same and Mwembe). In the circumstances I have given as syntypes those specimens cited by Harms in 1914.

Tree 3–10 m. high, usually flat-crowned with a well-defined trunk, some-times branching from base ; bark black or brown, roughish. Young branch-lets grey-brown to purplish, glabrous or pubescent. Stipules spinescent, up to 9 cm. long, some normally with their lower part enlarged and ovoid or fusiform but much constricted at base, whitish or ashen ; no prickles below the stipules. Leaves : rhachis eglandular between the 2–8 pairs of pinnae ; usually a conspicuous gland on the petiole ; leaflets (7–)10–18 pairs, 1·5–5 mm. long, 0·5–1·5 mm. wide, ciliate or ± pubescent, apex obtuse or rounded, lateral nerves invisible. Flowers cream, sessile, in spikes 1·8–5 cm. long on peduncles usually 0·5–1·2 cm. long ; axis pubescent, rarely subglabrous, with few or no glands. Calyx 0·7–0·8 mm. long. Corolla 2·5–3·5 mm. long, glabrous, with 4 lobes 0·5 mm. long. Stamen-filaments about 6 mm. long, free ; anthers 0·1 mm. across, with a caducous gland. Pods (Fig. 14/3, p. 52) brown, dehiscent, puberulous, narrowly oblong, straight, 2–6·5 cm. long, 0·8–1·5 cm. wide. Seeds ± obovate-compressed, 5 mm. long, 4–4·5 mm. wide ; central areole small, 2 × 1·5 mm.

KENYA. Northern Frontier Province : Moyale, 27 Aug. 1952, *Gillett* 13758 ! ; Kitui District : Mutha Plains, 24 Jan. 1942, *Bally* 1661 ! ; Teita District : Voi, 2 Feb. 1952, *Trapnell* 2210 ! & near Taru, 9 Sept. 1953, *Drummond & Hemsley* 4224 !
TANGANYIKA. Pare District : Kisiwani, 1 Feb. 1936, *Greenway* 4558 ! ; Lushoto District : Mkomazi, 23 Apr. 1934, *Greenway* 3973 !
DISTR. **K**1, 4, 7 ; **T**3 ; Ethiopia, Somaliland Protectorate and Somalia
HAB. Deciduous bushland and dry scrub with trees ; 300–970 m.

SYN. [*A. benadirensis* sensu Chiov., Fl. Somala 2 : 183 (1932), pro parte, saltem quoad spec. cit. e Somalia *Senni* 798 (FI !), *non* Chiov.]

NOTE. *A. bussei* is most closely related to *A. horrida* subsp. *benadirensis* and *A. lahai*, under which the distinctions are discussed.

4. **A. horrida** (*L.*) *Willd.*, Sp. Pl. 4 : 1082 (1806), *non* sensu auct. mult. ; Hillcoat & Brenan in K.B. 1958 : 39 (1958). Type : Plukenet, Phyto-graphia, t. 121, fig. 4 (1692) (holo. !) backed by the specimen drawn by Plukenet in Herb. Sloane, vol. 95, fol. 3 (BM !)

Shrub 1·3–3·6 m. high, normally flat-crowned, obconical and branching from base, very rarely taller (to 10 m.) but still with fastigiate branching from base. Young branchlets grey-brown to brown or blackish-purple, glabrous. Stipules spinescent, up to 9·5 cm. long, some normally with their lower part enlarged and ovoid or fusiform, not or only slightly constricted at base ; no prickles below the stipules. Leaves : rhachis eglandular between the 2–6 pairs of pinnae ; often a conspicuous gland on the petiole ; leaflets 5–11 pairs, 2–6 mm. long, 0·75–1·8 mm. wide, subglabrous or incon-spicuously ciliate, rounded to subacute or acute at apex, lateral nerves invisible. Flowers cream, sessile, in spikes 1–4·5 cm. long on peduncles 0·5–1 cm. long. Calyx 0·3–1 mm. long. Corolla 2–2·5 mm. long, glabrous, with 4 lobes 0·5–0·7 mm. long. Stamen-filaments about 5 mm. long ; anthers 0·1–0·15 mm. across, with a caducous gland. Pods brown, dehiscent, glabrous or slightly puberulous especially near base and along sutures, oblong, subreniform or ± shortly falcate, (2·5–)3–6 cm. long, (1·2–)1·5–2·5 cm. wide. Seeds ± obovate-compressed, 5 mm. long, 4–4·5 mm. wide ; central areole small, 2 × 1·5 mm.

SYN. *Mimosa horrida* L., Sp. Pl. : 521 (1753)
    Mimosa latronum Linn. f., Suppl. : 438 (1781). Type : India, frequent below Mt. Tripully, and very abundant between Tanschu and Tirut Schinapally, *Koenig in Herb. Linnaeus* 1228, 26, pro parte (LINN, lecto. !, BM, isolecto. !)
    A. latronum (Linn. f.) Willd., Sp. Pl. 4 : 1077 (1806) ; Brenan in K.B. 1956 : 188 (1956)

subsp. **benadirensis** (*Chiov.*) *Hillcoat & Brenan* in K.B. 1958 : 40 (1958). Types : Somalia, Mogadiscio [Mogadishu], *Paoli* 94 (FI, lecto. !) 131 bis (FI, syn. !)

Inflorescence-axis 1–1·5(–4·5) cm. long.   Calyx 0·8–1 mm. long.   Corolla 2 mm. long.
Pods (Fig. 14/4, p. 52) usually glabrous or almost so.

UGANDA.   Karamoja District :   between Loyoro and Timu Forest, June 1946, *Eggeling*
   5699!
KENYA.   Northern Frontier Province :   Dandu, 21 June 1952, *Gillett* 12743! & 27 km.
   NE. of Wajir, 18 Jan. 1955, *Hemming* 489! ;   Northern Frontier Province/Meru
   District :   Isiolo–Garba Tula road, 19 July 1952, *Bally* 8227! ;   Teita District :   Voi,
   2 Feb. 1952, *Trapnell* 2212!
DISTR. (of subsp.).   U1 ; K1, 2, 4, 7 ;   Somaliland Protectorate and Somalia, Ethiopia
   and the Sudan
HAB.   Deciduous bushland and dry scrub with trees ;   180–910 m.

SYN. (of subsp.).   *A. bussei* [Harms ex] Sjöstedt var. *benadirensis* Chiov. in Miss.
   Stefanini-Paoli, Bot. :   72 (1916) ;   I.T.U., ed. 2 :   205 (1952) (but flowers not
   in round heads, as there stated).
   A. benadirensis Chiov., Fl. Somala 2 :   183 (1932), pro parte, saltem quoad
   spec. cit. e Somalia *Senni* 196 (FI !) & 691 (FI !)
   A. latronum (Linn. f.) Willd. subsp. *benadirensis* (Chiov.) Brenan in K.B. 1956 :
   191 (1956)

DISTR. (of species as a whole).   As for the subsp., but also in India

NOTES.   This is closely related to *A. bussei*, differing in habit, the usually more
   glabrescent leaflets and inflorescence-axes, which (in subsp. *benadirensis*) are usually
   shorter, the enlarged spines not or only slightly constricted at base, the corolla (of
   subsp. *benadirensis*) 2–2·5 times as long as the calyx, not 3–5 times as in *A. bussei*,
   and in the wider curved pods.   My colleague, Mr. J. B. Gillett, states that where
   the two species occur in the same neighbourhood *A. horrida* subsp. *benadirensis*
   occurs usually on the clayey alluvial soils, while *A. bussei* is on the more sandy
   eluvial soils, and that the latter is generally in higher-rainfall regions.
   Gillett 13305 (Kenya, Northern Frontier Province, Mandera, 300 m., in bush-
   Acacia-Commiphora open scrub, 24 May 1952) is perhaps a very abnormal *A. horrida*
   or, more probably, a related and apparently undescribed species.   It differs in having
   pubescent young branchlets, pubescent leaflets 0·3–0·75 mm. wide, and falcate pods
   3·5 cm. long and 1·1–1·2 cm. wide pubescent all over (not venose as in *A. horrida*)
   and with longer (10–12 mm.) stipes (Fig. 14/4A, p. 52).   Unfortunately it is in fruit
   only, and the leaves are fragmentary.   More material is desired.
   Gillett 13357 (Kenya, Northern Frontier Province, 85 km. NE. of Wajir) is the
   evidence for *A. horrida* attaining a height of 10 m. ;   except for the size it seems
   altogether normal.

5. **A. ataxacantha** *DC.*, Prodr. 2 :   459 (1825) ;   L.T.A. :   834 (1930) ;
Bogdan in Nature in E. Afr., ser. 2, No. 1 :   11 (1949) ;   T.T.C.L. :   332
(1949) ;   Gilb. & Bout. in F.C.B. 3 :   153 (1952) ;   Consp. Fl. Angol. 2 :   278
(1956) ;   F.W.T.A., ed. 2, 1 :   499 (1958).   Types :   Senegal, *Bacle* (G–DC,
syn.) & *Perrottet* (G–DC, syn.)

Scandent shrub up to 15 m. high or (but apparently not in our area) a
non-climbing shrub or small tree 3–6 m. high.   Young branchlets ± pubescent.   Stipules not spinescent, obliquely ovate.   Prickles scattered along
the internodes, ± hooked or deflexed, up to 6 mm. long.   Leaves :   rhachis
5–12 cm. long, prickly or unarmed ;   usually a gland on the petiole and
between the uppermost 1–3(–5) pairs of pinnae ;   pinnae 6–17(–25) pairs ;
leaflets 14–50 pairs, 2–5(–7) mm. long, 0·5–1(–1·2) mm. wide, ± ciliate
otherwise glabrous or (but not in our area) ± appressed-hairy on surface
beneath, apex obtuse to subacute, lateral nerves usually invisible or faintly
apparent.   Flowers cream to white, 0·25–0·4 mm. pedicellate, or appearing
sessile, in spiciform racemes 4–8 cm. long on peduncles 1–2·5 cm. long ;
axis ± densely puberulous or pubescent.   Calyx 1–1·7(–2·5) mm. long,
glabrous or (but not in our area) slightly pubescent.   Corolla 2·5–3 mm.
long, with 5 lobes 0·5–0·8 mm. long.   Stamen-filaments 4·5–6 mm. long, free ;
anthers 0·15 mm. across, with a caducous gland.   Ovary pubescent, on a
stipe longer than itself.   Pods (Fig. 14/5, p. 52) purple-brown, dehiscent,
puberulous or almost glabrous, linear-oblong, straight, very acuminate at
both ends, 5–14 cm. long, 1–1·9 cm. wide.   Seeds subcircular-lenticular,
6–7 mm. diam. ;   central areole small, obscure, 2·5–3 × 2·5–3 mm.

KENYA. Northern Frontier Province : 42 km. N. of Isiolo, 22 July 1958, *Dale in Verdcourt* 2213B! ; District ? : Tana R., 915 m., 5 Apr. 1910, *Battiscombe* 246 *in C.M.* 13954!

TANGANYIKA. Lindi District : Mingoyo, 25 May 1943, *Gillman* 1341 !

DISTR. **K**1, 4 ; **T**8 ; from Senegal to Ubangi-Shari, the Sudan and Kenya in the north, southwards to Southern Rhodesia and Portuguese East Africa ; a variety in Angola, Bechuanaland, Transvaal, South West Africa. Swaziland and Natal

HAB. Riverine forest and thicket ; altitude range uncertain—at 915 m. in Kenya

SYN. " *A.* sp. nr. *ataxacantha*," Battiscombe, Cat. Trees Kenya Col. : 54 (1926) ; T.S.K. : 67 (1936)

NOTE. The two above-cited specimens have bracts 1·5–3 mm. long and narrowly elongate-acuminate at apex, and calyces glabrous ; they are referable to var. *ataxacantha*, which normally also has leaflets glabrous except for the ciliate margins, though the presence of pubescence on the lower side of some of the leaflets of *Battiscombe* 246 shows that this character is not completely constant.

The var. *australis* Burtt-Davy in K.B. 1922 : 324 (1922) (*A. eriadenia* Benth., *A. lugardae* N.E. Br.), with a more southerly distribution, may turn up in our area. It has leaflets ± appressed-hairy both on the margins and on the surface beneath especially towards apex, bracts about 1–1·5 mm. long and narrowly triangular but scarcely acuminate above, and calyces glabrous or slightly puberulous. Unlike var. *ataxacantha* it seems often not to be a climber.

6. **A. laeta** [*R. Br. ex*] *Benth.* in Hook., Lond. Journ. Bot. 1 : 508 (1842) ; L.T.A. : 830 (1930) ; T.T.C.L. : 329 (1949) ; F.W.T.A., ed. 2, 1 : 498 (1958). Type : Ethiopia, without locality, *Salt* (BM, holo. !)

Shrub or small tree up to 6 m. high. Young branchlets glabrous, grey-brown or rarely purplish. Stipules not spinescent. Prickles in pairs just below each node, purplish-black, hooked, 3–5·5 mm. long, their base about 3–5 mm. long. Leaves : petiole usually glandular ; rhachis glabrous to ± pubescent, frequently with a gland between the top pair of pinnae ; pinnae 2–3 pairs ; leaflets (2–) 3–5 pairs, 4–15 (–20) mm. long, 2–7 (–10) mm. wide, obliquely obovate-elliptic or -oblong, glabrous or ± puberulous, venose, rounded to mucronate or subacute at apex. Flowers white, subsessile or up to 0·5 (–0·75, very rarely 1) mm. pedicellate, in spiciform racemes 3·5–5 cm. long on peduncles 0·5–2 cm. long ; axis glabrous or ± pubescent. Calyx 1·25–2 mm. long, glabrous. Corolla 2·75–4 mm. long, 5-lobed. Stamen-filaments 5–7 mm. long, free ; anthers 0·1 mm. across, with a caducous gland. Ovary glabrous at first, very shortly stipitate. Pods (Fig. 14/6, p. 52) pale brown, dehiscent, glabrous or slightly puberulous towards base, oblong, straight, venose, rounded to acuminate at apex, 3·5–8 cm. long, 1·7–2·8 (–4) cm. wide. Seeds subcircular-lenticular, 9 (–10) mm. diam. ; central areole small, somewhat impressed, 2 × 2 mm.

KENYA. Masai District : 32 km. S. of Kajiado, on Namanga road, 30 Jan. 1952, *Trapnell* 2203! and near Namanga, 16 July 1953, *Trump* 69 !

TANGANYIKA. Mbulu/Masai Districts : Oldeani, 31 Aug. 1932, *B. D. Burtt* 4218! ; Masai District : Nabarera, 10 Oct. 1934, *Hornby* 9!

DISTR. **K**6 ; **T**2 ; Egypt, French Sudan, the Sudan, Somalia, Somaliland Protectorate and Ethiopia, also in Arabia and the Dead Sea region ; according to Aubréville also in French Niger Colony, Ivory Coast and Nigeria

HAB. Deciduous bushland ; said by Burtt (T.T.C.L. : 329) to be locally common with *Commiphora merkeri* and *Acacia mellifera* on hard-pan soils in desert thornbush, and by Trapnell (2203) to occur with *Commiphora* and *Balanites* on sandy loam ; *Trapnell* 2203 was collected at 1680 m.

SYN. *A. senegal* (L.) Willd. subsp. *mellifera* (Vahl) Roberty var. *laeta* ([R. Br. ex] Benth.) Roberty in Candollea 11 : 154 (1948)

NOTE. The material of *A. laeta* from our area is very sparse and mostly without flowers, and its identity is thus not free from doubt.

A. laeta is in several ways, notably the leaves and pedicels, systematically intermediate between *A. mellifera*, with pedicellate flowers, and other species, e.g. *A. goetzei*, with sessile flowers. The possibility of *A. laeta* originating through hybridization between *A. mellifera* and some other species should be considered. *Hornby* 9 (see above) has the leaflets in three pairs, unusually long pedicels (up to 1 mm.) and calyces about 1·25 mm. long ; its resemblance to *A. mellifera* is striking, and strengthens the possibility of the hybrid origin of *A. laeta*.

7. A. sp. A

Low shrub 1·5 m. high. Young branchlets glabrous except for a few short hairs. Stipules not spinescent. Prickles just below the nodes, paired, up to 6 mm. long, straight or almost so and pointing somewhat upwards, brown, thickened near base. Leaves : petiole with or without a small gland about 0·25 mm. in diameter ; pinnae 2–3 pairs ; leaflets 3 pairs, obliquely obovate or obovate-elliptic, 3–11 mm. long, 2–7 mm. wide, glabrous, lateral nerves somewhat prominent beneath, apex rounded to emarginate. Flowers white, sessile or nearly so, in spikes about 1·5–2·5 cm. long ; axis glabrous. Calyx about 0·5–1 mm. long, glabrous. Corolla about 1·5–2 mm. long, 5–6-lobed. Stamen-filaments about 3–3·5 mm. long ; anthers 0·1 mm. across, with a caducous gland. Ovary glabrous, very shortly stipitate. Pod unknown.

KENYA. Kilifi District : Kibarani, 8 Dec. 1947, *Jeffery* K584 !
HAB. Grassland on sand

NOTE. This is near *A. laeta* [R. Br. ex] Benth., differing in the shorter calyx and in the prickles not being hooked. As suggested under that name, *A. laeta* may prove to be an aggregate of hybrids rather than a true species, in which *Jeffery* K584 might then be included. At present, however, it seems preferable to keep it apart, especially in view of the straight prickles, which are most unusual among the spicate acacias.

8. A. mellifera (*Vahl*) Benth. in Hook., Lond. Journ. Bot. 1 : 507 (1842) ; Harms in N.B.G.B. 4 : 208, fig. 6 (1906) ; Battiscombe, Cat. Trees Kenya Col. : 54 (1926) ; L.T.A. : 828 (1930) ; T.S.K. : 67 (1936) ; Bogdan in Nature in E. Afr., ser. 2, No. 1 : 12 (1949) ; T.T.C.L. : 329 (1949) ; I.T.U., ed. 2 : 211, fig. 47 f. (1952) ; Consp. Fl. Angol. 2 : 273 (1956). Type : Arabia, Surdud and elsewhere, *Forskål* (C, holo.)

Shrub or small tree 1–6 (–9) m. high. Young branchlets pubescent or glabrous, grey-brown to purplish-black. Stipules not spinescent. Prickles in pairs just below each node, deep brown to blackish, hooked, 2·5–5 (–6) mm. long. Leaves : petiole usually glandular ; rhachis glabrous to pubescent, frequently with a gland between the top 1–2 pairs of pinnae ; pinnae 2–3, very rarely 4 pairs ; leaflets 1–2 (very rarely 3) pairs, 3·5–22 mm. long, 2·5–16 mm. wide, obliquely obovate to obovate-elliptic or -oblong, glabrous to pubescent, venose, rounded to emarginate or subacute and often apiculate at apex. Flowers cream to white, on pedicels (0·5–)0·75–1·5 mm. long in subglobose to ± elongate racemes ; axis 0·15–3·5 cm. long, glabrous or sometimes pubescent ; peduncle 0·4–1·3 cm. long. Calyx 0·6–1 mm. long, glabrous. Corolla 2·5–3·5 mm. long, 5-lobed. Stamen-filaments 4–6 mm. long, free ; anthers 0·15–0·25 mm. across, with a caducous gland. Ovary glabrous ; stipe very short. Pods (Fig. 14/8, p. 52) pale brown to straw-coloured, dehiscent, glabrous, oblong, straight, venose, rounded to shortly and abruptly acuminate at apex, (2·5–) 3·5–8 (–9) cm. long, 1·5–2·5 (–2·8) cm. wide. Seeds subcircular-lenticular, 9–10 mm. long, 8 mm. wide ; central areole small, slightly impressed, 2–3 mm. long, 2·5–3 mm. wide.

subsp. **mellifera**; Brenan in K.B. 1956 : 191 (1956)

Pinnae normally 2 pairs. Racemes ± elongate, their 0·4–1·3 cm. long peduncles usually shorter than the 0·5–3·5 cm. long rhachis.

UGANDA. Karamoja District : Moroto, *Brasnett* 77 ! & Kanamugit, 28 Oct. 1939, *A. S. Thomas* 3084 ! ; Mbale District : Greek River Camp, Jan. 1936, *Eggeling* 2494 !
KENYA. Northern Frontier Province : Moyale, 20 Aug. 1952, *Gillett* 13731 ! ; Masai District : Ngong Hills–Magadi, 10 Mar. 1951, *Greenway* 8502 ! ; Kilifi District : Sokoke, 12 Mar. 1946, *Jeffery* K490 !
TANGANYIKA. Masai District : Kitumbeine, 7 Jan. 1936, *Greenway* 4286 ! ; Pare District : Vudee–Mwembe, 27 Jan. 1930, *Greenway* 2071 ! ; Handeni District : about 53 km. from Korogwe on road to Handeni, 20 Nov. 1955, *Milne-Redhead & Taylor* 7339 !

DISTR. U1, 3 ; K1, 2, 4, 6, 7 ; T2-3 (but see Note below) ; Arabia, Egypt, the Sudan, Eritrea, Somaliland Protectorate, Somalia, Ethiopia and Angola
HAB. Dry scrub with trees, deciduous bushland ; 300–1680 m. T.S.K. : 67 (1936) states that this species forms about 80–90% of the dense thornbush country at an altitude of 1000–3000 ft.

SYN. *Mimosa mellifera* Vahl, Symb. 3 : 103 (1791)
     A. senegal (L.) Willd. subsp. *mellifera* (Vahl) Roberty in Candollea 11 : 153 (1948)

subsp. **detinens** (*Burch.*) *Brenan* in K.B. 1956 : 191 (1956). Type : South Africa, Prieska Division, Zand Valley, *Burchell* 1628 (K, holo. !)

Pinnae normally 3, very rarely 4 pairs. Racemes very short or subglobose, their 0·4–1·1 cm. long peduncles normally longer than the very short (1·6–6·5 mm. long) axes.

TANGANYIKA. Shinyanga District : Kizumbi, 1 Dec. 1923, *Swynnerton* 4036 ! ; Nzega District : Igunga, 23 July 1949, *Doggett* 126 ! ; Mpwapwa District : Gulwe, 7 Mar. 1933, *Hornby* 547 ! ; Iringa District : valley of Great Ruaha R., about 90 km. N. of Iringa, 16 July 1956, *Milne-Redhead & Taylor* 11226 !
DISTR. T1, 4, 5, 7 ; Northern and Southern Rhodesia, Transvaal, Bechuanaland and South West Africa
HAB. Similar to that of subsp. *mellifera*

SYN. *A. detinens* Burch., Trav. 1 : 310 (1822) ; L.T.A. : 828 (1930) ; Consp. Fl. Angol. 2 : 273 (1956)

NOTE. Subsp. *detinens* shows a greater tendency to produce two pairs of leaflets per pinna than does subsp. *mellifera;* it also more often tends to be pubescent, although this seems usually not to apply in the northern part of its area, including Tanganyika. Subsp. *mellifera* not infrequently produces larger leaflets than any seen in subsp. *detinens.*
    Over most of their ranges subsp. *detinens* and subsp. *mellifera* occur exclusively, and present no difficulty in their recognition. However, in north-central Tanganyika they meet and, judging from the available material, intermediates showing various combinations of character are frequent. Various specimens from this region, and one from southern Kenya, thus cannot be referred with certainty to either subspecies ; the following is a selection :—
KENYA. Kwale District : Mwachi, *R. M. Graham* 1622 !
TANGANYIKA. Masai District : S. of Longido on Arusha–Namanga road, 30 Jan. 1952, *Trapnell* 2207 ! ; Pare District : Kisiwani, 26 June 1942, *Greenway* 6495 ! ; Mpwapwa, 24 June 1938, *Hornby* 879 ! ; Kilosa District : Mkata, 15 Jan. 1934, *Michelmore* 929 !
    The last two specimens are from T5 and 6, whence I have not seen typical subsp. *mellifera.* Conversely, subsp. *detinens* has not so far been collected from Kenya.
    Trapnell 2158 (K !), Tanganyika, Mpwapwa/Iringa Districts, Great Ruaha valley on Iringa–Dodoma road, is apparently *A. mellifera,* but with the pods strikingly caudate-acuminate at apex.

9. **A. nigrescens** *Oliv.*, F.T.A. 2 : 340 (1871) ; L.T.A. : 829 (1930) ; T.T.C.L. : 329 (1949) ; Young in Candollea 15 : 118 (1955) ; Coates Palgrave, Trees Centr. Afr. : 250–253 (1956) ; Consp. Fl. Angol. 2 : 274 (1956). Type : Nyasaland, near Mitonde, *Kirk* (K, holo. !)

Trees 4–25 m. high ; trunk usually ± beset with knobby prickles. Young branchlets glabrous to sometimes pubescent. Stipules not spinescent. Prickles in pairs just below each node, hooked, blackish, persistent, 2·5–7 mm. long (on branchlets). Leaves : petiole glandular or not ; rhachis glabrous to pubescent, sometimes with a gland between the top 1–2 pairs of pinnae ; pinnae 2–4 pairs ; leaflets 1–2 pairs, (8–) 10–26 (–50) mm. long, (6–) 7–22 (–45) mm. wide, obliquely obovate-orbicular to broadly obovate-elliptic, glabrous to sometimes pubescent, venose, subcoriaceous, apex rounded and often emarginate. Flowers white or cream, sessile, in ± aggregate or solitary spikes 1–9·5 cm. long on peduncles 0·6–2 cm. long ; axis glabrous except for minute sessile glands, sometimes pubescent. Calyx 1·5–2 mm. long, glabrous. Corolla 2–2·5 mm. long, 5-lobed. Stamen-filaments 3·5–6 mm. long, free ; anthers 0·1 mm. across, with a caducous gland. Ovary glabrous, very shortly stipitate. Pods (Fig. 14/9, p. 52)

darkish brown, dehiscent, glabrous, oblong, straight, hardly venose, acuminate at apex, 7–14·5 cm. long, 1·5–2·4(–2·7) cm. wide. Seeds subcircular-
lenticular, 12–13 mm. diam. ; central areole large, 7–8 mm. long, 7 mm.
wide, somewhat impressed.

TANGANYIKA. Shinyanga/Nzega Districts : Manyonga R., 27 June 1931, *B. D. Burtt*
3440 ! ; Handeni District : Korogwe–Morogoro road, 15 Oct. 1951, *Hughes* 132 ! ;
Mpwapwa District : Gulwe, 21 Oct. 1936, *Hornby* 689 !
DISTR.   T1, 3, ? 4, 5–8 ; southwards to Bechuanaland, the Transvaal and Zululand
HAB.   Deciduous bushland and probably also wooded grassland, especially in black-
soil areas ; 240–1160 m.

SYN.   *A. nigrescens* Oliv. var. *pallens* Benth. in Trans. Linn. Soc. 30 : 517 (1875). Type :
Portuguese East Africa, near Sena, *Kirk* 201 (K, holo. !)
A. brosigii Harms in N.B.G.B. 2 : 194 (1898).   Type : Tanganyika, Kilosa,
Brosig (B, holo. †)
A. perrotii Warb. in N.B.G.B. 2 : 249 (1898).   Type : Tanganyika, Lindi,
Perrot (B, holo.†)
A. pallens (Benth.) Rolfe in K.B. 1907 : 361 (1907)
A. schliebenii Harms in N.B.G.B. 12 : 507 (1935) ; T.T.C.L. : 329 (1949). Type :
Tanganyika, Lindi District, Lake Lutamba, *Schlieben* 5565 (B, holo.†, BM, P,
iso. !)

VARIATION.   This easily recognized species shows comparatively little variation.   It is
generally glabrous, but is occasionally puberulous and rarely quite densely pubescent.
Schlieben 5291 (P !), from the same locality as typical *A. schliebenii* (see above),
discussed by Harms in N.B.G.B. 12 : 508 (1935), is evidently one of these pubescent
forms of *A. nigrescens* which are probably best considered as no more than part of
the range of variation within the species.
    The number of leaflets per pinna is usually only two, but some specimens show an
inconstant tendency, though quite probably genetically controlled, to produce four.
    The characteristic raised knobs on the trunk are evidently variable in their occur-
rence.   B. D. Burtt, quoted in T.T.C.L. : 329 (1949), says that trees in the coastal
regions show them but not those in the Central and Lake Provinces.   Perrot, quoted
by Warburg in N.B.G.B. 2 : 247–8 (1898), says that in Lindi District the Africans
recognize " male " trees whose trunks have few and inconspicuous knobs and
" female " trees with numerous large knobs.

10. **A. persiciflora** *Pax* in E.J. 39 : 624 (1907) ; L.T.A. : 854 (1930).
Type : Ethiopia, West Shoa, Urga Valley, *Rosen* (BRSL ?, holo.)

Tree 4·5–9(–15) m. high, sometimes flat-crowned ;   bark brownish-yellow,
scaling off in vertical strips.   Young branchlets pubescent to puberulous.
Stipules not spinescent.   Prickles few (often absent from branchlets), in
pairs just below nodes, small, recurved, up to about 3 mm. long.   Leaves :
petiole usually glandular (gland 0·3–0·5 mm. in diameter) ; rhachis pubescent, glandular between the top 1–5 pairs of pinnae ; pinnae 4–8 pairs ;
leaflets 11–17 pairs, 3–5·5(–10) mm. long, 0·75–1·5(–2·5) mm. wide, oblong-
linear, ciliate on margins or glabrous or nearly so, lateral nerves almost
invisible beneath, apex obtuse to subacute.   Flowers sessile or subsessile,
precocious, in spikes 1·5–3 cm. long on peduncles 0·3–1·3 cm. long.   Calyx
cupular, 1–1·4 mm. long, red or purplish, glabrous.   Corolla 2·5–3·5 mm.
long, red or purplish, 5-lobed, glabrous.   Stamen-filaments 6–8 mm. long,
white ; anthers 0·1–0·15 mm. across, glandular at apex.   Ovary glabrous,
shortly stipitate.   Pods (Fig. 14/10, p. 52) brown, dehiscent, straight or
slightly curved, venose, with minute dark glands, otherwise sparsely puberulous to subglabrous, 6–15 cm. long, (1·4–) 1·6–2·5 cm. wide.   Seeds subcircular-lenticular, 7–8 mm. diam. ; central areole small, 2 × 2 mm.

UGANDA.   West Nile District, Payida, Feb. 1934, *Eggeling* 1453 ! & 20 Mar. 1945,
Greenway & Eggeling 7234 ! ; Mbale District : Kaburon, Jan. 1936, *Eggeling* 2490 !
& 2497 !
KENYA.   Trans-Nzoia District ? : Suam Valley, Mar. 1953, *Tweedie* 1106 ! ; N. Kavi-
rondo District : N. Kitosh Reserve, Jan. 1931, *Honoré in F.H.* 2590 ! & *in C.M.*
13961 !
DISTR.   U1, 3 (also in 4, *fide* I.T.U.) K3, 5 ; Belgian Congo, Ethiopia and the Sudan
HAB.   Singly or gregariously in woodland and wooded grassland ; 1220–2130 m.

SYN. *A. eggelingii* Bak. f. in J.B. **73** : 263 (1935) ; Bogdan in Nature in E. Afr.,
ser. 2, No. 1 : 13 (1949) ; I.T.U., ed. 2 : 209, fig. 47c (1952) ; Gilb. & Bout.
in F.C.B. **3** : 152 (1952). Type : Uganda, West Nile District, Zeio, *Eggeling*
1905 (BM, holo. !)

NOTE. The colour of the flowers, which is unusual among the African species of *Acacia*,
makes *A. persiciflora* particularly striking.
    One of the closest relatives of *A. persiciflora* is *A. galpinii* Burtt-Davy in K.B.
1922 : 326 (1922), recorded from Portuguese East Africa, Northern and Southern
Rhodesia, Bechuanaland and the Transvaal. A poor fruiting specimen without
precise locality but probably collected near the coast of Tanganyika Territory in **T6**,
Busse 68 (K !), is quite possibly *A. galpinii*, but I hesitate to include the species
without more certain evidence, but it is accounted for in the key at the beginning
of the genus. *A. venosa* [Hochst. ex] Benth. from Ethiopia is also very closely related
to *A. persiciflora*, but has larger leaflets.
    A. galpinii should be sought for in southern Tanganyika. It has a small calyx
similar to that of *A. persiciflora* but differs in the pinnae of well-developed leaves
being in 9–14 pairs, the longer inflorescences (4–11 cm.), and the ± puberulous calyces
and corollas. The open flowers are described as yellow, although the buds appear
red or pink.

11. **A. hecatophylla** [*Steud. ex*] *A. Rich.*, Tent. Fl. Abyss. 1 : 242 (1847) ;
L.T.A. : 832 (1930) ; I.T.U., ed. 2 : 210, fig. 47d (1952). Types : Ethiopia,
Schimper 628 (P, syn.) & 884 (P, syn., K, isosyn. !)

    Tree 4·5–7·5 m. high ; bark grey, longitudinally fissured. Young branch-
lets tomentellous to puberulous, glabrescent. Stipules not spinescent.
Prickles few, or absent on some shoots, in pairs just below a node (sometimes extra prickles, up to 5 in all, present below nodes), brown to purplish,
spreading or hooked, 3–6 mm. long. Leaves : petiole glandular ; rhachis
puberulous to glabrescent, glandular between the top 1–4 pairs of pinnae ;
pinnae 3–20 pairs (some leaves always with 12 or more) ; leaflets 13–40(–50)
pairs, 4–12 mm. long, 1·25–2·5(–3·5) mm. wide, obliquely oblong, pale
beneath, glabrous except at base or sparsely pubescent on margins, lateral
nerves somewhat prominent beneath, apex rounded. Flowers white, sessile
in spikes 5–12·5 cm. long on peduncles 1–3 cm. long ; axis tomentellous.
Calyx 2–2·7 mm. long, 5-lobed, densely pubescent. Corolla 2·75–3·5 mm.
long, 5-lobed. Stamen-filaments 4–6 mm. long ; anthers 0·15 mm. across,
with a caducous gland. Ovary glabrous ; stipe very short. Pods (Fig. 14/11,
p. 52) brown, dehiscent, glabrous or nearly so, oblong, straight, somewhat
glossy and venose. thickly coriaceous or almost woody, rounded to subacute
at apex, 6–15 cm. long, 1·7–2·7 cm. wide. Seeds subcircular-lenticular,
10–12 mm. diam. ; central areole medium, 3 mm. long, 4 mm. wide.

UGANDA. West Nile District : 5 km. S. of Obongi, June 1933, *Eggeling* 1235 *in F.H.*
1337 ! & Payida, escarpment edge, 20 Mar. 1945, *Greenway & Eggeling* 7233 !
DISTR. **U1** ; Belgian Congo (*van der Ben* 1332 !), the Sudan, Eritrea, Ethiopia
HAB. Scattered-tree grassland ; " solitary or in twos and threes on stony hillsides,
never gregarious " (I.T.U., ed. 2: 210) ; 600–1370 m.

12. **A. polyacantha** *Willd.*, Sp. Pl. **4** : 1079 (1806). Type : Eastern
India, collector unknown, *Herb. Willdenow* (B, holo., K, fragments !, photo. !)

    Tree up to 21 m. high ; trunk with fissured bark and knobby persistent
prickles. Young branchlets pubescent or puberulous, rarely subglabrous,
grey to brown. Stipules not spinescent. Prickles in pairs just below each
node, straw-coloured to brown or blackish, 4–12 mm. long. Leaves :
petiole glandular (gland usually 2–4 × 1·75–3 mm.) ; rhachis pubescent or
puberulous, rarely subglabrous, glandular between the top 3–17 pairs of
pinnae ; pinnae (6–)13–40(–60) pairs ; leaflets (15–)26–64 pairs, 2–5(–6)
mm. long, 0·4–0·75(–1.25) mm. wide, linear to linear-triangular, pubescent
usually only on margins, only the midrib (except sometimes some very small
basal nerves) visible, subacute to narrowly obtuse at apex. Flowers cream
or white, sessile or nearly so, in spikes (3·5–)6–12·5 cm. long, produced with

the new leaves ; axis densely pubescent or tomentellous ; peduncle (0·5–)
1·2–2 cm. long. Calyx 1·7–2·25 mm. long, pubescent or puberulous, rarely
puberulous on lobes only or subglabrous. Corolla 2–3 mm. long, 5-lobed,
usually 1⅓ times or more as long as calyx. Stamen-filaments 4·5–6 mm.
long ; anthers 0·1 mm. across, with a caducous gland. Ovary glabrous ;
stipe very short. Pods (Fig. 14/12, p. 52) brown, dehiscent, glabrous or
nearly so, rarely ± pubescent, oblong, straight, venose, usually acuminate
at apex, 7–18 cm. long, 1–2·1 cm. wide. Seeds subcircular to elliptic-lenticular, 8–9 mm. long, 7–8 mm. wide ; central areole medium to small,
3–4 × 3–3·5 mm., not impressed.

SYN.   *Mimosa suma* Roxb., Fl. Ind. 2, ed. 2 : 563 (1832). Type : India, " very
          common . . . about Calcutta and over Bengal", *Roxburgh* (whereabouts of holo.
          uncertain. Roxburgh painting No. 1867 at Kew !)
          A. suma (Roxb.) [Buch.-Ham. ex] Voigt, Hort. Suburb. Calcutt. : 260 (1845)

   subsp. **campylacantha** ([*Hochst. ex*] *A. Rich.*) Brenan in K.B. 1956 : 195 (1956) ;
F.W.T.A., ed. 2, 1 :   499 (1958). Types :   Ethiopia, Mai Dogale, *Schimper* 639 (B,
holo. †, K, iso. !) & Dscheladscheranne, [Jelajeranne], *Schimper* 893 (P, holo., K, iso. !)

Bark whitish to yellowish or grey.   Prickles ± hooked.

UGANDA.   Acholi District :   Kitgum, 10 Nov. 1945, *A. S. Thomas* 4339 ! ;   Ankole
   District :   Ruampara, Nov. 1930, *Brasnett* 41 ! ;   Teso District :   Serere, Apr.–May
   1932, *Chandler* 645 !
KENYA.   Nairobi, Chiromo estate, Jan. 1934, *Napier in C.M.* 6772 ! ;   Central Kavi-
   rondo District :   Ugungu, July 1944, *Davidson* 213 *in Bally* 4295 ! ;   Masai District :
   Emali, 10 Mar. 1940, *van Someren* 74 !
TANGANYIKA.   Bukoba District :   Bunazi, Sept.–Oct. 1935, *Gillman* 585 ! ;   Lushoto
   District :   Mshwamba, 3 Jan. 1930, *Greenway* 2025 ! ;   Iringa, *Emson* 499 ! ;   Songea
   District :   about 8 km. W. of Songea, by R. Wuwawesi, 9 Feb. 1956, *Milne-Redhead &
   Taylor* 8660 !
DISTR.   U1–4 ;   K2, 4–7 ;   T1–8 ;   widespread in tropical Africa from Gambia to Eritrea
   in the north and to the Transvaal in the south.   (The subsp. *polyacantha* in India and
   perhaps Ceylon)
HAB.   Wooded grassland, deciduous woodland and bushland, riverine and ground-water
   forest ;   locally common to dominant ;   " usually gregarious along rivers and in rich,
   alluvial valleys with *Acacia albida, Kigelia* and *Ficus sycomorus;* an indicator of
   fertile soil for tobacco and cotton " (T.T.C.L. : 331) ;   0–1830 m.

SYN.   *A. campylacantha* [Hochst. ex] A. Rich., Tent. Fl. Abyss. 1 : 242 (1847) ;   L.T.A. :
          831 (1930) ;   T.S.K. :   71 (1936) ;   Bogdan in Nature in E. Afr., ser. 2, No. 1 :
          13 (1949) ;   T.T.C.L. : 331 (1949) ;   I.T.U., ed. 2 : 207, fig. 47b (1952) ;   Young
          in Candollea 15 :   99 (1955) ;   Coates Palgrave, Trees Centr. Afr. :   235–8
          (1956) ;   Consp. Fl. Angol. 2 :   276 (1956)
          [*A. catechu* sensu P.O.A. C :   194, t. 20 D (1895), *non* (L.f.) Willd.]
          [*A. suma* sensu Harms in N.B.G.B. 4 :   210, fig. 7 (1906), *non* sensu stricto]
          A. caffra (Thunb.) Willd. var. *campylacantha* ([Hochst. ex] A. Rich.) Aubrév.,
             Fl. Forest. Soudano-Guin. : 272 (1950) ;   Gilb. & Bout. in F.C.B. 3 : 150 (1952)
          A. catechu (Linn. f.) Willd. subsp. *suma* (Roxb.) Roberty var. *campylacantha*
             ([Hochst. ex] A. Rich.) Roberty in Candollea, 11 : 157 (1948)

NOTE.   Although the numbers of pairs of pinnae range from 6–60 per leaf, the well-
   developed leaves have 15–20 or more pairs on all specimens collected.   *A. polyacantha* subsp. *campylacantha* does not vary much in our area, except to some extent
   in indumentum and, rarely, armature ;   commonly it is densely puberulous or
   pubescent, but rarely it may be sparingly puberulous to subglabrous on the vegetative
   parts (e.g. *Busse* 139 ! from Tanganyika) or on the calyx (e.g. *Hughes* 7 ! from Tanganyika, Tanga District, Maramba near Tanga).
      Semsei 1169 from Tanganyika, Morogoro, garden of House 28 of H.Q. Plan, 2 May
   1953 is a remarkable variant apparently completely lacking prickles although otherwise normal.

   13. **A. erubescens** [*Welw. ex*] *Oliv.* in F.T.A. 2 :   343 (1871) ;   L.T.A. :
830 (1930) ;   Young in Candollea 15 :   111 (1955) ;   Consp. Fl. Angol. 2 :
276 (1956).   Type : Angola, between Bumbo and Bruco, *Welwitsch* 1826
(LISU, holo., BM, K, iso. !)

Shrub or tree 2–7·5 m. high.   Young branchlets ± pubescent.   Stipules not spinescent. Prickles in pairs just below nodes, brown or grey, hooked, up to 4 (–6) mm. long. Leaves : petiole glandular or not ; rhachis pubescent ; glands variable, either between each pair of pinnae, or absent from some, or between top pair only ; pinnae 3–7 pairs ; leaflets 10–25 pairs, 3–8(–10) mm. long, (0·75–)1–2(–2.5) mm. wide, obliquely oblong, often slightly falcate or the upper somewhat obovate, slightly pubescent especially on margins, or becoming glabrous, veins somewhat prominent at first beneath, becoming obscure as the leaves age, apex usually oblique, acute or subacute, occasionally obtuse. Flowers white or cream or with pink tinge, sessile, in spikes 2–4·5 cm. long on peduncles 0·7–2·4 cm. long ; axis pubescent. Calyx 2·25–4·5 mm. long, densely pubescent. Corolla 2·5–6·5 mm. long, 5-lobed, appressed-pubescent on lobes outside. Stamen-filaments 6–10 mm. long ; anthers 0·2–0·25 mm. across, glandular at apex. Ovary glabrous ; stipe very short. Pods (Fig. 14/13, p. 52) brown or deep brown, dehiscent, subglabrous except for pubescent margins and stipe, oblong, straight, venose, coriaceous, rounded to acute, rarely acuminate at apex, 3–11 cm. long, 1·2–1·9 cm. wide.

TANGANYIKA. Ufipa District: Kelema on Lake Tanganyika, 15 Apr. 1936, *B. D. Burtt* 6047 !; Mbeya District: Sangwa, 3 Dec. 1932, *R. M. Davies* 786 ! & Ngerenge, 26 Feb. 1934, *Michelmore* 975 ! & escarpment on Great North Road on way to Mbozi, 16 Oct. 1936, *B. D. Burtt* 6050 !
DISTR. T4, 7; Belgian Congo, Nyasaland, Northern and Southern Rhodesia, Angola, South West Africa, Bechuanaland and the Transvaal
HAB. Deciduous woodland, especially of *Brachystegia*, locally common; 850–1680 m.

SYN. *A. dulcis* Marl. & Engl. in E.J. 10: 24 (1888); Gilb. & Bout. in F.C.B. 3: 151 (1952). Type: South West Africa, Hereroland, *Marloth* 1259 (B, holo.†)
　　A. kwebensis N.E. Br. in K.B. 1909: 108 (1909). Type: Bechuanaland, Kwebe Hills, *Mrs. Lugard* 24 (K, holo. !)

NOTE. The bark has been variously described: rough, grey (*Michelmore* 977, Tanganyika), pearly white (*Lugard* 24, Bechuanaland), whitish (*Dyer & Verdoorn* 4219, Transvaal), yellow, scaly (*Codd & Dyer* 4528, Transvaal); these notes may perhaps be reconcilable, as *Michelmore* 623 (Northern Rhodesia) describes the bark as " pale, peeling, grey outside, yellowish inside "; the matter, however, requires more observation in the field. The Tanganyika specimens of *A. erubescens* show large flowers with calyx 3–4·5 mm. long, corolla 4–6·5 mm. long, and stamen-filaments 8–10 mm. long, but do not seem specifically separable from typical *A. erubescens,*, although minor geographical races may ultimately be shown to exist within this species.
　　The plant from Kenya described as " *Acacia* sp. nr. *erubescens* Welw." in Battiscombe, Cat. Trees Kenya Col.: 54 (1926) is *Dichrostachys cinerea* (L.) Wight & Arn.

14. **A. tanganyikensis** *Brenan* in K.B. 1956 : 195 (1956). Type : Tanganyika, Shinyanga District, unlocalized, *B. D. Burtt* (K, holo. !, BM, iso. !)

Tree 5·5–15 m. high ; bark dark grey, corrugated. Young branchlets densely pubescent. Stipules not spinescent. Prickles in pairs just below a node, brown or grey, hooked, up to 5–6 mm. long. Leaves : petiole glandular (gland 0·8–1·5 mm. in diameter) ; rhachis pubescent, glandular between lowest pair and top 2–6 pairs of pinnae ; pinnae 6–17 pairs ; leaflets 19–32 pairs, 2·5–3·5 mm. long, 0·6–1·0 mm. wide, linear-oblong, ciliate on margins only, lateral nerves not visible, apex obtuse, base auricled on one side. Flowers white or yellowish-white, sessile, in spikes 5–11 cm. long on peduncles 0·5–1·5 cm. long, produced usually before the new leaves on short leafless lateral shoots ; axis densely pubescent. Calyx 1·5–2 mm. long, glabrous or with a few hairs above, rarely with more numerous hairs. Corolla 1·75–2·25 mm., equalling or only slightly exceeding the calyx, 5-lobed, glabrous outside. Stamen-filaments 5–5·5 mm. long, free ; anthers about 0·2 mm. across, with a caducous gland. Ovary glabrous, very shortly stipitate. Pods (Fig. 14/14, p. 52) dehiscent, puberulous, oblong, straight or nearly so, venose, shortly acuminate at apex, 8–21 cm. long, 1·6–2·6 cm. wide. Seeds subcircular-lenticular, 10–13 mm. diam. ; central areole medium, 5–6 × 4·5–5 mm., impressed,

TANGANYIKA. Mwanza District: 16 km. SW. of Karumo, July 1951, *Eggeling* 6258 !; Singida District: SW. of Ndindini, 1 July 1933, *B. D. Burtt* 4453 !; Dodoma District: about 7 km. S. of Dodoma, 20 July 1956, *Milne-Redhead & Taylor* 11263 !

DISTR. **T**1, ? 4, 5; not known elsewhere

HAB. Woodland, deciduous bushland and perhaps in ground-water forest; 1160–1490 m.

NOTE. This has in the past been misidentified with *A. rovumae*, a coastal, lowland species having leaflets more than 1 mm. wide, broader than in *A. tanganyikensis*, the corollas projecting beyond the calyces for a length at least ½ that of the latter, spikes produced at the same time as the new leaves, and usually a yellowish curled indumentum on the young parts.

SYN. [*A. rovumae*, sensu B.D. Burtt in Journ. Ecol. 30: 90, 142 etc., t. 12 (1942), *non* Oliv.]

15. **A. rovumae** *Oliv.*, F.T.A. 2 : 353 (1871) ; L.T.A. : 831 (1930), pro parte ; T.T.C.L. : 331 (1949). Type : Tanganyika or Portuguese East Africa, Ruvuma Bay, *Kirk* (K, holo. !)

Tree 10–15 m. high, with openly branched flat crown and rough or smooth dark grey or grey-green bark. Young branchlets puberulous or very shortly pubescent with short curved hairs that are yellowish, at least when dry. Stipules not spinescent. Prickles in pairs just below nodes, deep grey to blackish, spreading or pointing a little upwards, usually straight or only slightly curved (see, however, note below), up to 4–6 mm. long. Leaves : petiole with a small gland 0·4–0·7 mm. in diameter ; rhachis puberulous, glandular between the top 1–4 pairs of pinnae ; pinnae 6–9 pairs ; leaflets (9–)13–31 pairs, 4–8 mm. long, 1·5–2(–3·5) mm. wide, oblong, oblique at base and the subacute to obtuse apex, pale glaucescent beneath, puberulous especially beneath, lateral nerves visible when young, becoming obscure. Flowers sessile or nearly so, in spikes 6–10 cm. long on 1·5–3 cm. long peduncles, produced with the leaves ; axis puberulous. Calyx 1·5–2 mm. long, puberulous. Corolla 2–3 mm. long, glabrous or slightly puberulous on lobes outside, 5-lobed, exceeding the calyx. Stamen-filaments 4–5 mm. long, free ; anthers 0·1 mm. across, with a caducous gland. Ovary glabrous, very shortly stipitate. Pods (Fig. 14/15, p. 52) probably not dehiscent, apparently irregularly breaking up, glabrous, oblong, straight, and smooth or nearly so, dark brown when dry, green when living, rather thick and turgid, rounded or acute at apex, 7–15 cm. long, 1·7–2·5 cm. wide. Seeds oblong-elliptic-lenticular, 10–13 mm. long, 7–9 mm. wide, hard-walled ; central areole large, 7–9 × 4·5–5 mm., not impressed.

KENYA. Tana River District: Tana R., Garissa, 4 Feb. 1956, *Greenway* 8859 !; Lamu District: Boni forest, 26 Oct. 1957, *Greenway & Rawlins* 9432 !

TANGANYIKA. Lushoto District: Mkundi, *Gillman* 749 !; Tanga District: Amboni Kibuguni, 25 Nov. 1936, *Greenway* 4734 !; Uzaramo District: Dar es Salaam, 26 Apr. 1933, *B. D. Burtt* 4474 !; Mafia Is., Kipandeni, 26 Sept. 1937, *Greenway* 5323!

DISTR. **K**7; **T**3, 6, 8; Portuguese East Africa and Madagascar

HAB. Riverine forest and saline-water swamp-forest; 6–700 m.

SYN. *A. chrysothrix* Taub. in P.O.A. C: 194 (1895); L.T.A.: 833 (1930); T.T.C.L.: 331 (1949). Type: Tanganyika, Lushoto District, Mashewa, *Holst* 8793 (B, holo. †, K. iso. !)

A. morondavensis Drake, Hist. Pl. Madag. 1: 62 (1902). Type: Madagascar, *Grevé* (P, holo., K, iso. !)

NOTE. The name *A. rovumae* has been wrongly applied to the inland *A. tanganyikensis* Brenan. True *A. rovumae* seems usually to occur on or not far from the coast; in T.T.C.L.: 331, it is noted as occurring on margins of *Avicennia marina* mangroves, associated with *Barringtonia racemosa*, with its roots in brackish water. It also occurs in riverine forest further from the sea, e.g. at Garissa in Kenya. The appearance of the pods of *A. rovumae* suggests that they are indehiscent and perhaps water-borne in their dispersal. If this is confirmed, and it requires further observation on the spot, then it is a very unusual feature in any *Acacia*. *A. rovumae* is also outstanding among its nearest relatives by usually having its prickles not or scarcely hooked.

Plants superficially resembling *A. rovumae* occur in Portuguese East Africa, especially towards the south, and extend into Natal. They differ in their strongly hooked prickles, the longer hairs on the calyces, the usually broader leaflets and longer bracteoles, and apparently in the habitat. Their affinity seems to be with *A. burkei* Benth. rather than with *A. rovumae*.

The direction of the prickles of *A. rovumae* may prove not quite constant: *Peter*, Excursion no. O. III. 176, from Tanganyika, Tanga Province, Umba Steppe, Kigwasi Hill, 27 Aug. 1915 (EA!) is very close to and in my view a form of *A. rovumae*, although the prickles are somewhat hooked and the leaflets up to 3·5 mm. wide. The hooking is considerably less than in those plants discussed in the previous paragraph.

16. **A. goetzei** *Harms* in E.J. 28 : 395 (1900) ; L.T.A. : 830 (1930), pro parte, excl. *Eyles* 4049 ; T.T.C.L. : 329 (1949) ; Bogdan in Nature in E. Afr., ser. 2, No. 1 : 12 (1949) ; Brenan in K.B. 1956 : 198 (1956) ; Consp. Fl. Angol. 2 : 276 (1956). Type : Tanganyika, Kilosa District, Kidodi, *Goetze* 387 (B, holo.†, K, iso.!)

Tree 4–20 m. high, with rounded crown and rough, grey or brown bark. Young branchlets glabrous to pubescent. Stipules not spinescent. Prickles in pairs just below nodes, pale then dark brown or grey, hooked downwards, up to 7 mm. long. Leaves : petiole with or rarely without a small gland ; rhachis glabrous to pubescent, usually glandular between the top 1–3(–5) pairs of pinnae (and sometimes the basal pair as well) ; pinnae 3–10 pairs ; leaflets (2–)5–20(–23) pairs, (2–)3–17(–20) mm. long, (0·75–)1–7(–12·5) mm. wide, rounded to mucronate or subacute at apex, glabrous to pubescent, venation somewhat prominent beneath. Flowers sessile or nearly so, white or slightly yellowish, in spikes (2–)3–12 cm. long on 0·4–4·5 cm. long peduncles, produced with the leaves ; axis glabrous to pubescent. Calyx glabrous, 1·5–2·75 mm. long. Corolla 2–3·75 mm. long, glabrous, 5-lobed, exceeding the calyx. Stamen-filaments 4·5–6 mm. long, free ; anthers 0·2–0·25 mm. across, with a caducous gland. Ovary glabrous, very shortly stipitate. Pods (Fig. 14/16, p. 52) dehiscent, glabrous or nearly so, oblong or irregularly constricted, straight or nearly so, venose, red- to purplish-brown, acuminate or apiculate at apex, (5–)8–17 cm. long, (1·8–)2–3·3 cm. wide. Seeds subcircular-lenticular, 9–10 mm. long, 8–9 mm. wide ; central areole medium, 5–6 × 4 mm.

subsp. **goetzei** ; Brenan in K.B. 1956 : 204 (1956)

Leaflets (of leaves on mature flowering shoots) nearly all more than 3 mm. wide (range 2·5–12·5 mm.), usually wider towards apex and thus obovate, obovate-oblong or oblanceolate-oblong, often in comparatively few ((2–)5–11(–14, very rarely to 17)) pairs; rhachis of leaf frequently (by no means always) unarmed and with a gland between the topmost pair of pinnae only.

KENYA. Embu District: Emberre/Embu border, 12 Oct. 1932, *M. D. Graham* 2273 !
TANGANYIKA. Shinyanga District: Usule, *Koritschoner* 1687 ! & Shinyanga, *Koritschoner* 3050 !; E. Mpwapwa, 14 Nov. 1938, *Hornby* 878 !; Masasi, 12 Dec. 1942, *Gillman* 1116 !
DISTR. **K**4; **T**1, 3–6, 8; Belgian Congo, Portuguese East Africa, Nyasaland, Northern and Southern Rhodesia and Angola
HAB. Woodland and perhaps wooded grassland, " widely distributed and locally common, especially on banded-ironstone ridges " (T.T.C.L.: 329); 460–1220 m.

SYN. [*A. mossambicensis* sensu Bak.f., L.T.A.:831 (1930), *non* Bolle]

NOTE. Vegetatively, especially in indumentum and leaflets, variable, but constant in the characters of the individual flowers and fruits. The armature of the leaf-rhachis is also variable, the type of subsp. *goetzei* having a prickly rhachis which is, however, less common than an unarmed one in this subspecies.

The distribution of subsp. *goetzei* is apparently not continuous in East Africa, and it is still unknown whether variation can be clearly correlated with this. In the neighbourhood of Shinyanga and in Kondoa District several gatherings have been made which show unusually few leaflets per pinna (2–8 pairs) which are also larger than usual (up to 20 × 12·5 mm.). It is much to be desired that observers should ascertain how much variability there is in local populations of this species.

subsp. **microphylla** *Brenan* in K.B. 1956: 204 (1956). Type: Nyasaland, Mombera District, Njakwa to Fort Hill, *Greenway* 6393 (K, holo.!, EA, iso.!)

Leaflets (of leaves on mature flowering shoots) nearly all narrower than 3 mm. (range 0·75–3 mm.), usually (except terminal pairs) not wider towards apex, and thus oblong or linear-oblong, often in more numerous (8–23) pairs than in subsp. *goetzei;* rhachis of leaf frequently (by no means always) ± prickly and with glands between the topmost 1–3 (–5) pairs of pinnae.

KENYA. Northern Frontier Province: Moyale, 1 Apr. 1952, *Gillett* 12668 !
TANGANYIKA. Handeni District: Kangata, Nov. 1949, *Semsei in F.H.* 2925 !; Morogoro District: Melela, 12 Jan. 1934, *Michelmore* 923 !; Masasi District: flood-plain of R. Bangala, 16 Dec. 1955, *Milne-Redhead & Taylor* 7678 !
DISTR. K1; T1, 3, 4, 6, 8; Belgian Congo, Ethiopia, Portuguese East Africa, Nyasaland, and Northern and Southern Rhodesia
HAB. Woodland, semi-evergreen bushland, and perhaps elsewhere; said sometimes to be riverine; 170–1520 m.

SYN.   *A. ulugurensis* [Taub. ex] Harms in E.J. 28: 396 (1900); L.T.A.: 831 (1930); T.T.C.L. : 332 (1949). Type : Tanganyika, Uluguru foothills near Tununguo, *Stuhlmann* 8947 (B, holo.†)
        A. kinionge De Wild., Pl. Bequaert. 3: 60 (1925); Gilb. & Bout. in F.C.B. 3: 153 (1952). Type : Belgian Congo, Bas-Katanga, Kabalo, *Delevoy* 118 (BR, holo.)
        A. joachimii Harms in N.B.G.B. 12: 507 (1935); T.T.C.L.:331 (1949). Type: Tanganyika, Lindi District, Lake Lutamba, *Schlieben* 5636 (B, holo.†, BR, P, iso !)
        A. van-meelii Gilb. & Bout. in B.J.B.B. 22: 177 (1952) & in F.C.B. 2: 149 (1952). Types: Belgian Congo, Parc Nat. Upemba, R. Kilwezi, *Van Meel in de Witte* 4103 (BR, holo. !) ; Tanganyika, Lindi District, Lake Lutamba, *Schlieben* 5636a (BR, P, co. !)

NOTE. Like subsp. *goetzei*, this is variable, although less so in our area than in other parts of its range. The solitary Kenya specimen is from the northern frontier, but it seems hard to believe that it is really absent from the rest of the territory; the specimen however is somewhat aberrant, having few spines (as in *A. persiciflora*), and rather dense indumentum. It is matched by *Gillett* 14764 from southern Ethiopia, and these may later prove to be worth distinguishing taxonomically.
    A. mandunduënsis Engl. was quoted in T.T.C.L.: 338 (1930) as a *nomen nudum* of doubtful identity. A specimen in E. A. H., *Busse* 545, without locality, is annotated as " *Acacia mandandensis* Harms n. sp." The two spellings of the epithet are probably mere variants, and *Busse* 545 is certainly *A. goetzei* subsp. *microphylla*.

17. **A. senegal** (*L.*) *Willd.*, Sp. Pl. 4 : 1077 (1806) ; L.T.A. : 827 (1930), pro majore parte ; T.S.K. : 69 (1936) ; Bogdan in Nature in E. Afr., ser. 2, No. 1 : 12 (1949) ; T.T.C.L. : 330 (1949) ; I.T.U., ed. 2 : 212, fig. 47g, t. 10 (1952) ; Gilb. & Bout. in F.C.B. 3 : 149 (1952) ; Young in Candollea 15 : 93 (1955) ; Consp. Fl. Angol. 2 : 273 (1956) ; F.W.T.A., ed. 2, 1 : 498, fig. 159 (1958). Whereabouts of type uncertain.

Shrub or tree up to 12 m. high ; bark grey, scaly, rough. Young branchlets densely to sparsely pubescent, soon glabrescent. Stipules not spinescent. Prickles just below nodes, either in threes, up to 7 mm. long, the central one hooked downwards, the laterals ± curved upwards, or else solitary, the laterals being absent. Leaves : petiole glandular or not (gland about 0·5–0·75 mm. in diameter) ; rhachis ± pubescent, glandular between the top 1–5 pairs of pinnae, prickly or not ; pinnae (2–)3–6 pairs, 0·5–1·5 (–2·4, very rarely to 4) cm. long ; leaflets 8–18 pairs, 1–4(–7) mm. long, 0·5–1·75 mm. wide ; linear- to elliptic-oblong, ciliate on margins only or ± hairy on surface, or wholly subglabrous, lateral nerves not visible or sometimes somewhat prominent beneath, apex obtuse to subacute. Flowers white or cream, fragrant, sessile, in spikes 2–10 cm. long on peduncles 0·7–2 cm. long, normally produced with the leaves ; axis pubescent to glabrous. Calyx 2–2·75(–3·5) mm. long, glabrous to somewhat pubescent. Corolla 2·75–4 mm. long, exceeding the calyx, 5-lobed, glabrous outside. Stamen-filaments 4·5–7 mm. long, free ; anthers 0·2–0·25 mm. across, with a caducous gland. Ovary glabrous, very shortly stipitate. Pods (Fig. 14/17, p. 52) usually grey-brown, sometimes pale or dark brown, dehiscent, densely to sparsely appressed-pubescent to -puberulous, oblong, straight, venose, rounded to acuminate at apex, (3–)4–14 cm. long, (1·3–)2–3·3 cm. wide. Seeds ± subcircular-lenticular, 8–12 mm. diam. ; central areole small to medium, 2·5–6 × 2·5–5 mm., markedly impressed.

var. senegal

Tree 3–12 m. high; crown apparently variable, flat and spreading, or umbrella-shaped, or lax and rounded. Twigs dull grey to grey-brown or purplish-grey, ultimately often minutely flaking, striate and lenticellate.

UGANDA. Acholi District: Chua, Kitgum, 10 Nov. 1945, A. S. Thomas 4338 !; Mbale District: N. Bugishu, Cheptui, 10 Oct. 1933, Tothill 2241 !; Bunyoro District: Butiaba escarpment, 2 July 1951, Trapnell 2147 !
KENYA. Turkana District: Oropoi valley, June–July 1930, Liebenberg 147 (Field No. 292) !; Machakos District: Mua Hills, 2 Mar. 1953, Trump 27 !; Masai District: 37 km. S. of Kajiado on the Namanga road, 30 Jan. 1952, Trapnell 2204 !
TANGANYIKA. Musoma, 1933, Emson 329 !; Moshi District (or ? in Kenya): Himo R. – Taveta road, 25 Jan. 1936, Greenway 4505 !; Mpwapwa/Kilosa Districts: Chakwale, 15 June 1938, Hornby 945 !
DISTR. U1–4; K1–7; T1–6, 8; widespread in tropical Africa, extending southwards to Zululand
HAB. Wooded grassland, deciduous bushland, dry scrub with trees; 120–1680 m.

SYN. Mimosa senegal L., Sp. Pl.: 521 (1753)
    Acacia verek Guill. & Perr. in Fl. Seneg. Tent. 1: 245, t. 56 (1832); Oliv., F.T.A. 2: 342 (1871), nom. illegit. Type : from Senegal (P, syn.)
    A. virchowiana Vatke & Hildebr. in Oesterr. Bot. Zeitschr. 30: 275 (1880), pro parte, quoad fol. et flor. tantum. Type: Kenya, Teita District, Voi R. and elsewhere, Hildebrandt 2486 (?B, holo. †, K, iso. !)
    " A. campanulata Hochst." T.S.K. : 69 (1936)
    [A. somalensis sensu T.T.C.L.: 330 (1949), non Vatke; spec. cit. Zimmermann 7006 est A. senegal ad var. leiorhachidem vergens, Zimmermann 7007 est A. senegal var. leiorhachis]
    [A. thomasii sensu T.T.C.L.: 330 (1949), non Harms; spec. cit. Greenway 4505 est A. senegal var. leiorhachis]
    A. senegal (L.) Willd. subsp. senegalensis (Houtt.) Roberty var. verek (Guill. & Perr.) Roberty in Candollea 11: 156 (1948)

var. kerensis Schweinf. in Bull. Herb. Boiss. 4, app. 2: 216 (1896); L.T.A.: 828 (1930).
Types: Eritrea, Keren, Schweinfurth 745 (B, syn. †, K, isosyn. !) & Bogu Valley, Schweinfurth 741 (B, syn. †) & near Djuffa, Schweinfurth 998 (B, syn. †).

Spreading bush up to 5 m. high, branching from base. Twigs usually paler and smoother than in var. senegal, appearing as though whitewashed over a purplish background. Leaves usually smaller and inflorescences shorter than in var. senegal.

UGANDA. Mengo District: Beruli, Nakasongola, Nov. 1932, Eggeling 706 in F.H. 1084 !
KENYA. Northern Frontier Province: Dandu, 5 May 1952, Gillett 13052 !; Turkana District: Lokitaung, Feb. 1943, Dale K306 !; Kitui District: Mutha Plains, 24 Jan. 1942, Bally 1637 !
TANGANYIKA/KENYA. Moshi/Teita Districts: Lake Chala, 21 Jan. 1936, Greenway 4438 !
DISTR. U4; K1–4, 7; T ?1 (Michelmore 817), ?2 ; Eritrea, Somaliland
HAB. Probably similar to that of A. senegal; 460–1130 m.

VARIATION. A. senegal is extremely variable, and the present treatment is anything but satisfactory.
    The var. senegal itself shows a wide range of variation in indumentum, armature, flower-size, and general habit. Whether the prickles occur singly or in threes near a node seems of no significance, as both arrangements may occur on one and the same shoot, although some gatherings may show all or nearly all the prickles arranged singly. The type of A. senegal var. leiorhachis Brenan in K.B. 1953: 98 (1953) (Tanganyika, Pare Distr., nr. Same, Greenway 2192 !) differs from var. senegal solely by its glabrous inflorescence-axis, and I do not consider it more than a minor variation. It has also been collected in K4, 6 and T2, 3. Another variant with exceptionally large leaflets and flowers has been collected in Nairobi National Park, Bogdan 1868 (K !); the leaflets are 5–7 × 1·5–2·5 mm., the calyx 3·5 mm. long, and the corolla 4 mm. long; Bally 7134 (K!), new Magadi road, Ngong, is a similar form. Gatherings with unusually long leaflets to 7 mm., with prominent venation beneath, have been made in Kenya at Kwale, Elliott 1373 (K !) and Kwale District, Mrima Hill, Verdcourt 1916 (K !), and in Tanganyika, Korogwe–Handeni road, Faulkner 1474 (K !). Verdcourt 1916 shows unusually long pinnae, up to 4 cm.
    The application of the name var. kerensis is not certain, and I rely mainly on

Schweinfurth's remark (see above reference) " Hier nur in Strauchform ". It seems also probable that var. *kerensis*, as interpreted here, may not be uniform, the bushy habit, which Mr. J. B. Gillett informs me is most distinctive in the field, being perhaps shared by more than one otherwise separate form. The characters other than habit are thus not to be relied on.

At present the status of these forms or variants is quite uncertain. We do not know whether they represent the response to an extreme or unusual habitat, or casual sports in an otherwise normal population, or whether they are distinct local races. Careful observations of living *A. senegal*, supplemented by specimens and habit-sketches or photographs, should give useful evidence towards an ultimate solution of these problems.

18. **A. circummarginata** *Chiov.* in Ann. Bot. Roma 13 : 394 (1915) ; L.T.A. : 834 (1930). Types : Ethiopia, Ogaden, *Paoli* 794, 913 bis, 920, 1010 (FI, syn. !)

Closely related and similar in most of its characters to *A. senegal*, differing mainly as follows :—

Tree 5–9 m. high ; crown flat (T.T.C.L.) or rounded, bark yellow, flaking. Young branchlets subglabrous to puberulous, rather smooth except for lenticels, purplish-brown to purplish-black. Leaves with 3–7 pairs of pinnae, their rhachides 2–6·5 cm. long ; leaflets 8–25 pairs, 3–9 mm. long, 1·2–2 (–3) mm. wide, usually with few to many minute appressed hairs beneath. Axis of flowering spikes glabrous, rarely slightly hairy. Pods (Fig. 14/18, p. 52) usually brown, dehiscent, sparingly to sometimes densely puberulous, oblong to elliptic, usually subacute, sometimes rounded at apex, 7–18·5 cm. long, 1·7–2·8 cm. wide. Seeds 10–11 mm. diam. ; central areole 4–5 × 4–5 mm.

KENYA. Northern Frontier Province: Dandu, 14 May 1952, *Gillett* 13186 ! & Wajir, Jan. 1955, *Hemming* 433 !
TANGANYIKA. Mpwapwa, 2 June 1929, *Hornby* 140 !; Mpwapwa & Gulwe, 26 Apr. 1932, *B. D. Burtt* 3845 !; Dodoma/Iringa Districts: Great Ruaha valley on Iringa–Dodoma road, 19 Aug. 1951, *Trapnell* 2157 !
DISTR. K1; T5, ?6,?7; Ethiopia
HAB. Dry scrub with trees; 730–1180 m.

SYN. [*A. glaucophylla* sensu T.T.C.L.: 330 (1949), *non* [Steud. ex] A. Rich.]
　　 [*A. kinionge* sensu T.T.C.L.: 330 (1949), *non* De Wild.]
　　 [*A. senegal* (L.) Willd. var. *leiorhachis* sensu Brenan in K.B. 1953: 98 (1953), pro parte, quoad spec. *Hornby* 140 & *B. D. Burtt* 3845, *non* Brenan sensu stricto]

NOTE. In keeping this apart from *A. senegal*, I have relied much on the testimony of Mr. J. B. Gillett and on notes (in the Kew Herbarium) by Mr. & Mrs. Hornby, all of whom have seen *A. circummarginata* and *A. senegal* in the field.

The yellow bark is a striking character of the living tree, but in the herbarium the separation of *A. circummarginata* from *A. senegal* may not always be easy. The longer pinnae of the former seem distinctive, as do the rather smooth, purplish twigs, whose internodes are often longer than in *A. senegal*.

It is probable that the Transvaal material cited under *A. senegal* var. *leiorhachis* in K.B. 1953: 98–9 (1953) would be better placed under *A. circummarginata*. This requires further study, however.

19. **A. condyloclada** *Chiov.* in Ann. Bot. Roma 13 : 391 (1915) ; L.T.A. : 854 (1930). Type : Ethiopia, Ogaden, *Riva & Ruspoli* 1079 (FI, holo. !)

Tree to 2–11 m. high ; bark white or yellow, peeling. Young branchlets puberulous, going purplish and then blackish ; internodes rather long, often ± enlarged towards apex. Stipules not spinescent. Prickles just below nodes, singly or in threes, 5–9 mm. long, slightly hooked or straight. Leaves : petiole glandular near base (gland 1·5–3 mm. in diameter) ; rhachis puberulous, eglandular ; pinnae 3–4 pairs, 2–7·5 cm. long ; leaflets 6–9 pairs, 9–20 mm. long, 3·5–9 mm. wide, obliquely oblong to slightly ovate or obovate, puberulous both sides, lateral nerves (except basal) obscure beneath, apex mostly rounded. Flowers sessile ; axis of spikes glabrous. Calyx about 2·5–2·75 mm. long, probably glabrous, when dry whitish with a purplish

stripe up each lobe. Corolla glabrous and 5-lobed, about 3·5 mm. long.
Pods (? not fully mature) (Fig. 14/19, p. 52) grey- or purplish-brown,
densely puberulous, oblong, straight, venose, rounded and apiculate at
apex, 6–10·5 cm. long, 1·8–2·1 cm. wide.

KENYA. Northern Frontier Province: Lag Ola, 45 km. W. of Ramu on Banessa road,
23 May 1952, *Gillett* 13279 !
DISTR. **K**1; Somaliland, Ethiopia
HAB. *Acacia-Commiphora* open scrub on steep limestone slopes; about 450 m.

NOTE. Extremely distinct from its nearest relatives, *A. senegal* and *A. circummarginata*,
in its white peeling bark, eglandular leaf-rhachides and large leaflets.

20. **A. thomasii** *Harms* in E.J. 51 : 366 (1914) ; L.T.A. : 826 (1930) ;
Bogdan in Nature in E. Afr., ser. 2, No. 1 : 12 (1949). Type : Kenya, Kitui
District, Ikutha, *F. Thomas* III 127 (B, holo.†)

Straggling shrub or small tree up to 5 m. high, with elongate whippy
upper twigs. Young branchlets densely pubescent. Stipules not spinescent.
Prickles just below each node, usually blackish, up to 7·5 mm. long, in threes
or occasionally solitary, the central one hooked downwards, the laterals
curved upwards or sometimes nearly straight. Leaves : petiole glandular,
2–11 mm. long including the pubescent eglandular rhachis ; pinnae 1–2(–3)
pairs ; leaflets 7–15 pairs, 3–9 mm. long, 1·5–3 mm. wide, obliquely oblong
or elliptic-oblong, glabrous to somewhat pubescent, rounded to acute or
mucronulate at apex, lateral nerves (other than minute basal ones) not
visible beneath. Flowers cream or lemon, sessile or nearly so, in spikes
4–9 cm. long on peduncles about 0·5–1·2 cm. long. Calyx 3·5–4·5 mm. long,
glabrous. Corolla 6·5–7 mm. long, 5-lobed, glabrous outside. Stamen-
filaments 13–15 mm. long, free ; anthers 0·2 mm. across, with a very caducous apical gland. Ovary glabrous ; stipe very short. Pods (Fig. 14/20,
p. 52) yellow-brown or brown when ripe, dehiscent, puberulous, oblong,
straight or nearly so, venose, coriaceous, subacute to acuminate at apex,
5–10 cm. long, 1–2·3 cm. wide. Seeds ± subcircular-lenticular, 10–12 mm.
diam. ; central areole large, 6–9 × 5–6 mm., markedly impressed, ±
pointed at top.

KENYA. Northern Frontier Province/Meru District: Isiolo–Garba Tula road, 18 July
1952, *Bally* 8220 !; Kitui District: Mutha Plains, Aug. 1938, *Joana in C.M.* 7470 !;
Masai District: Laitokitok, Seret R., 3 Mar. 1948, *Vesey-FitzGerald* 29 !; Teita District:
Voi–Mwatate road, 12 Sept. 1953, *Drummond & Hemsley* 4281 !
DISTR. **K**?1, 4, 6 7; not known elsewhere
HAB. Dry scrub with trees; said to be on lava and limestone; altitude range uncertain,
recorded at 620 m.

NOTE. The holotype having been destroyed, the application of the name to the above
species is based on the description of *A. thomasii*. Harms described it as a tree up to
15 m. high with a calyx 3–3·5 mm. long; the bark was said to be smooth. It seems more
likely than not that further material will cause these apparent discrepancies to
disappear.
    The Tanganyika specimen recorded as *A. thomasii* in T.T.C.L.: 330 (1949) is in fact
A. senegal.

21. **A. mearnsii** *De Wild.*, Pl. Bequaert. 3 : 62 (1925). Types : Kenya,
near Thika, *Mearns* 1092 (BR, lecto. !, BM, isolecto. !)

Tree 2–15 m. high, unarmed ; crown conical or rounded ; all parts (except
flowers) ± densely pubescent or puberulous. Leaves : petiole 1·5–2·5 cm.
long, with a gland above ; rhachis usually 4–12 cm. long, with numerous
raised glands all along its upper side ; pinnae (8–)12–21 pairs ; leaflets
usually in 16–70 pairs, linear-oblong, 1·5–4 mm. long, 0·5–0·75 mm. wide.
Flowers pale yellow, fragrant, in heads 5–8 mm. in diameter on peduncles
2–6 mm. long, panicled. Pods (Fig. 15/21, p. 66) ± grey-puberulous,
jointed, almost moniliform, dehiscing (in Australia forms with less monili-

form, almost glabrous pods occur), usually about 3–10 cm. long and 0·5–
0·8 cm. wide, with 3–12 joints. Seeds black, smooth, elliptic, compressed,
5 mm. long, 3·5 mm. wide ; caruncle conspicuous ; areole 3·5 mm. long,
2 mm. wide.

UGANDA. Ankole District: S. of Ruborogoto, Oct. 1932, *Eggeling* 682 *in F.H.* 1056 !
DISTR. U2; T7; becoming naturalized in K4 & T3 (*fide* Greenway); native of Australia,
    introduced into the Old World
HAB. Roadsides etc., usually planted, sometimes in wild situations; altitude range
    uncertain: *Eggeling* 1056 at 1615 m. Frequently naturalized in upland grassland
    around Njombe at about 1800 m.

SYN. [*A. mollissima* sensu T.T.C.L.: 333 (1949); et auct. al. mult., *non* Willd.]
    A. decurrens Willd. var. *mollis* Lindl. in Bot. Reg., t. 371 (1819); Gilb. & Bout. in
        F.C.B. 3: 168 (1952); Dale, Introd. Trees Uganda: 2 (1953). Type: a cultivated
        plant (location of holo. unknown, perhaps not preserved, not at CGE)
    [*A. decurrens* sensu L.T.A.: 853 (1930) saltem pro parte, *non* Willd. sensu stricto]

NOTE. This is the well-known Australian " Black Wattle ", which is economically
    important on account of its tan bark, and is also used for building and firewood. This
    alone among the various introduced wattles seems to have made itself sufficiently at
    home in East Africa to merit inclusion among the wild species.
        It is commonly planted in East Africa, as in K4 and T2 and 3, whence I have seen
    specimens.

22. **A. brevispica** *Harms* in N.B.G.B. 8 : 370 (1923) ; L.T.A. : 853
(1930) ; T.T.C.L. : 332 (1949) ; Consp. Fl. Angol. 2 : 287 (1956) ; Bol. Soc.
Brot., sér. 2, 31 : 108 (1957). Type : Tanganyika, Lushoto District, Kitivo,
Holst 606 (B, holo.†)

    Shrub or small tree 1–7 m. high, often semi-scandent and forming coppice.
Young branchlets densely pubescent or puberulous and with many minute
reddish glands. Prickles scattered, recurved or spreading, arising from
longitudinal bands along the stem which are usually paler than the intervening lenticellate bands. Leaves : petiole 0·4–1·3(–1·5) cm. long ; pinnae
6–18 pairs, mostly 1–4 cm. long, straight or slightly curved ; leaflets
numerous, linear or linear-oblong, midrib nearer one margin at base. Flowers
white or yellowish-white, in heads 10–15 mm. in diameter, racemosely
arranged, or aggregated into a rather irregular terminal panicle up to about
15(–30) cm. long. Stipules at base of peduncles small, 0·75–1(–2) mm.
wide, inconspicuous, soon caducous, not subcordate at base. Calyx eglandular outside, puberulous or glabrous. Pods (Fig. 15/22, p. 66) subcoriaceous, oblong, 6–15 cm. long, (1·2–) 1·5–3·3 cm. wide, glabrous or
puberulous, with many minute reddish glands. Seeds brown, smooth,
elliptic, compressed, 6–13 mm. long, 6–7 mm. wide ; areole 3–9 mm. long,
2–3 mm. wide.

UGANDA. Karamoja District: R. Kanyao, 27 May 1939, *A. S. Thomas* 2825 !; Kigezi
    District: Kamwezi, Feb. 1948, *Purseglove* 2586 !; Teso District: Kyere, Feb. 1933,
    Chandler 1117 !
KENYA. West Suk District: Moribus, May 1932, *Napier* 2037 !; Meru District: Isiolo, 12
    Mar. 1945, *Mrs. J. Adamson* 67 *in Bally* 4368 !; Machakos District: near Bondoni, 12
    Mar. 1953, *Trump* 50 !; Kisumu-Londiani District: near Kisumu, 18 Nov. 1951,
    Trapnell 2183 !
TANGANYIKA. Shinyanga, Samui Hills, Mar. 1936, *B. D. Burtt* 5673 !; Pare District:
    Kisangara [Kisangiro], May 1928, *Haarer* 1360 !; Morogoro District: about 26 km. E.
    of Morogoro, 25 Nov. 1955, *Milne-Redhead & Taylor* 7377 !
DISTR. U1–4 ; K1–7 ; T1–6 ; Belgian Congo, the Sudan, Ethiopia, Somaliland Pro-
    tectorate, Angola ; apparently also in Portuguese East Africa, Natal and Cape
    Province (Pondoland)
HAB. Bushland, thickets, scrub; 170–1830 m.

SYN. [*A. pennata* sensu L.T.A.: 853 (1930); T.S.K.: 68 (1936); Bogdan in Nature in
    E. Afr., ser. 2, No. 1: 11 (1949); T.T.C.L.: 332 (1949); I.T.U., ed. 2: 212 et verisim. t.
    9 (1952); *omnes pro parte*; Gilb. & Bout. in F.C.B. 3: 154 (1952); *non* (L.) Willd.]

NOTE. Much of what has been called *A. pennata* in East Africa is placed here. *A. brevi-
    spica* is distinguished from *A. pentagona* (p. 100) by its short petioles, differently

coloured twigs, usually shorter and straighter pinnae, its habit—it does not seem to become a true liane like *A. pentagona*—and in not being a forest plant. For the differences between *A. brevispica* and *A. adenocalyx*, see under the latter species, below

A. brevispica is rather uniform except for the indumentum on the inflorescence and the width of the pods. The indumentum may be either pubescent and comparatively long, of spreading hairs exceeding in length the reddish or dark glands with which the inflorescence-axes are sprinkled; or puberulous and comparatively short, of curved hairs which are about as long as or shorter than the glands. There is no obvious geographical separation of the two sorts, although puberulous inflorescences seem to occur more frequently in Kenya and Tanganyika than in Uganda, and intermediates occur. Of the specimens cited above, *A. S. Thomas* 2825 and *Haarer* 1360 are well-marked puberulous examples. It is hard to say whether the variation in the width of the pod is hereditary or just casual.

The following two specimens from Tanganyika, Lindi District: Lake Lutamba, 14 Nov. 1934, *Schlieben* 5624 (BM!) & Tendaguru, 27 Dec. 1930, *Migeod* 1093 (BM!) are perhaps hybrids between *A. brevispica* and *A. schweinfurthii*, having longer petioles and rather smaller heads than in the former, and narrower leaflets than in the latter. *Savile* 3 (EA!), from Tanganyika, Singida, Munya valley, 22 July 1938, is very probably the same.

23. **A. adenocalyx** *Brenan & Exell* in Bol. Soc. Brot., sér. 2, 31 : 115 (1957). Type : Tanganyika, Tanga District, Kange Estate, *Faulkner* 855 (K, holo.!)

Compact shrub or small tree 1–4·5 m. high, sometimes low and spreading, or even scandent. Young branchlets puberulous and with very many minute brown glands. Prickles scattered, deflexed, arising from longitudinal bands along the wholly blackish-brown stems. Leaves : petiole 0·5–1·2 cm. long ; pinnae 10–19 pairs, 0·6–3·5 cm. long ; leaflets very numerous and neat, linear-oblong, 0·3–0·75 mm. wide ; midrib subcentral at base. Flowers white, in heads about 8–10 mm. in diameter, often irregularly paniculate. Stipules at base of peduncles small, 0·3–0·5 mm. wide, inconspicuous, soon caducous, not subcordate at base. Calyx-lobes with many minute brown glands outside (use × 20 lens). Pods (Fig. 15/23, p. 66) subcoriaceous or stiffly papery, oblong, dehiscent, 6·5–14 cm. long, 1·6–3·6 cm. wide, puberulous or glabrous and with very many minute brown glands. Seeds black, smooth, elliptic, compressed, 8–9 mm. long, 5·5–6 mm. wide ; areole 5–6 mm. long, 2·5–3 mm. wide.

KENYA. Kilifi District: without locality, 7 Sept. 1945, *Jeffery* K311! & Sokoke, 29 July 1913, *Battiscombe* 781! & Kibarani, 18 Mar. 1946, *Jeffery* K495!
TANGANYIKA. Handeni District: Kangata, Dec. 1949, *Semsei in F.H.* 2930!; Bagamoyo District: Mandera, Feb. 1899, *Sacleux* 698!; Lindi District: Sudi, 12 Dec. 1942, *Gillman* 1129!
DISTR. **K**7; **T**3, 6, 8; Portuguese East Africa
HAB. Imperfectly known: said to grow in bush and secondary thickets and to be " very common in old native cultivations on sandy soils " (*Semsei* 2930, see above); 45–450 m.

NOTE. The short petioles enable this and *A. brevispica* to be separated from *A. kamerunensis*, *A. pentagona* and *A. latistipulata*. *A. adenocalyx* differs from all of them in having small brown glands on the outside of the calyx-lobes, and the midrib of the leaflets subcentral at base and not to one side. In addition the wholly blackish-brown twigs will readily distinguish *A. adenocalyx* from *A. brevispica*. *A. adenocalyx* is related also to *A. taylorii*, the differences being given under the latter (p. 100).
All the localities of *A. adenocalyx* are either on or quite near the coast.

24. **A. latistipulata** *Harms* in E.J. 51 : 367 (1914) ; L.T.A. : 853 (1930) ; T.T.C.L. : 332 (1949). Types : Tanganyika, Kwa-Mkopo on the Ruvuma R., *Busse* 103 (B, syn.†, EA, iso.!) ; Tanganyika, Uzaramo District, *Stuhlmann* 7025 (B, syn. †) & 7048 (B, syn. †)

Arborescent or scandent shrub up to 5 m. high. Young branchlets densely pubescent or puberulous, eglandular. Prickles scattered, recurved ; leaves often large. Leaves : petiole about 1·8–3·2 cm. long ; pinnae 10–26 pairs, about 3–8 cm. long ; leaflets very numerous, linear to linear-oblong, 0·8–2

(–3) mm. wide, midrib nearer one margin at base.   Heads of flowers about 8 mm. in diameter, in an ample terminal panicle.   Stipules at base of peduncles comparatively large and conspicuous, 5–9 mm. long, 3–4·5 mm. wide, ovate, acute at apex, subcordate at base, pubescent or puberulous.   Calyx puberulous, eglandular.   Pods (Fig. 15/24, p. 66) subcoriaceous, oblong, dehiscent, 5–19 cm. long, 2·5–4·2 cm. wide, glabrous except for some glands, umbonate over seeds.   Seeds dark brown, smooth, elliptic, compressed, 9–11 mm. long, 6–7 mm. wide ; areole 5–6 mm. long, 2·5–3 mm. wide.

TANGANYIKA.   Kilwa/Lindi District: Mbemkuru R. (Liwale–Nachingwea road), 22 Feb. 1956, *Nicholson* 6 !, 7 !, 8 !; Lindi District: Rondo Plateau, 27 May 1953, *Parry* 214 !; Newala District: Kitangari, 25 May 1943, *Gillman* 1478 !
DISTR. T6, 8; Portuguese East Africa
HAB. Evergreen bushland and coastal secondary forest with *Chlorophora* standards; 380–820 m.

NOTE. A little-known but apparently distinct species, about which more information is desired.
    It is probable that *A. makondensis* Engl., *nomen nudum*, in V.E. 1 (1): 401 (1910) (see T.T.C.L.: 338 (1949)), is synonymous with *A. latistipulata*.

25. **A. kamerunensis** *Gandoger* in Bull. Soc. Bot. Fr. 60 : 459 (1913). Type : British Cameroons, between Victoria and Bota, *Winkler* 447 (LY, holo. !)

Scandent shrub.   Young branchlets subglabrous or sparsely and minutely puberulous, glandular or not, soon yellow-brown and then grey.   Prickles deflexed, scattered, arising from narrow brown longitudinal bands usually darker than the intervening ones.   Leaves : petiole normally 1·5–3·5 cm. long, with a flat or rather convex gland 1–4·5 mm. long and 0·5–1·5 mm. wide ; pinnae (10–)15–27(–36) pairs, (0·8–)1·5–4·5 cm. long ; glands on rhachis between the top 3–10 pairs ; leaflets linear, very numerous, 0·3–0·8 mm. wide, glabrous or very inconspicuously ciliolate, midrib nearer one margin at base.   Flowers white, in small heads 4–8 mm. in diameter in ample panicles.   Stipules at base of peduncles small, 0·3–0·75 mm. wide, inconspicuous, soon caducous, not subcordate at base.   Calyx eglandular outside. Pods (Fig. 15/25, p. 66) subcoriaceous, oblong, brown or pale brown, flat, dehiscent, 8–14 cm. long, 1·7–2·8 cm. wide, margins not strongly thickened. Seeds dark brown, smooth, elliptic, compressed, 7–11 mm. long, 5–8 mm. wide ; areole small, 4–5 mm. long and 1·5–2·5 mm. wide.

UGANDA.   Mengo District: Mukono, May 1915, *Dummer* 2446 ! & Kajansi Forest, Kampala–Entebbe road, Apr. 1935, *Chandler* 1239 !
DISTR.   U4; Sierra Leone to Ubangi-Shari, French Cameroons, Belgian Congo and Uganda
HAB. Lowland rain-forest; 1190–1220 m.

SYN. [*A. pennata* sensu auct. pro parte, *non* (L.) Willd.]

NOTE. *A. kamerunensis* differs from *A. pentagona* (p. 100) in its narrow leaflets, thinner paler dehiscent pods, and seeds with a small central areole; from *A. schweinfurthii* (p. 99) in its narrow almost glabrous leaflets, normally smaller prickles, differently coloured young twigs, more numerous glands on the leaf-rhachis, the larger less gibbous gland on the petiole, and the more flattened seeds with a small central areole. Also it is a forest not a deciduous woodland or bushland species.

26. **A. monticola** *Brenan & Exell* in Bol. Soc. Brot., sér. 2, 31 : 125 (1957). Type : Uganda, Murole Hill, *Purseglove* 2693 (K, holo. !, EA, iso. !)

Scandent shrub to 30 m. high.   Young branchlets ± densely pubescent with fulvous hairs and many red-purple glands mixed, dark brown, later going blackish.   Prickles deflexed, scattered, arising from longitudinal bands usually darker than the intervening ones.   Leaves : petiole 1·5–3·5 cm. long ; pinnae 7–19 pairs, 3–4·5(–5·5) cm. long ; leaflets linear-oblong, 0·5–1·25 mm. wide, glabrous or margins sparsely and inconspicuously ciliolate, midrib nearer one margin at base. Flowers cream or white, in heads 10–15 mm. in diameter usually in pyramidal panicles. Stipules at base of peduncles small, 1–1·5 mm. wide, inconspicuous, soon caducous, not subcordate at base. Calyx puberulous and eglandular outside. Corolla puberulous outside. Pods (Fig. 15/26, p. 66) subcoriaceous, oblong, dark brown, dehiscent, 8–18 cm. long and 3–4·5 cm. wide, with margins 1–1·5 mm. wide and not very thickened. Seeds brown or black, smooth, elliptic, compressed, 9–12 mm. long, 6–7 mm. wide ; areole small, 5–7 mm. long and 2·5–3·5 mm. wide.

UGANDA. Kigezi District: Murole Hill, Apr. 1948, *Purseglove* 2693! & Kashambya Swamp, Gombolola side, 16 Oct. 1952, *Mrs. Norman* 53!
KENYA. N. Kavirondo District: Kakamega, Mar. 1944, *Carroll* 15!
TANGANYIKA. Lushoto District: Monga, 24 Nov. 1906, *Zimmermann* 6982! & 24 Nov. 1916, *Peter* K309!; Mpwapwa District: Kiboriani Mt., 17 Sept. 1938, *Hornby* 864!; Morogoro District: Uluguru Mts., 12 Dec. 1932, *Schlieben* 3087!
DISTR. U2; K5; T3, 5, 6; Belgian Congo and Nyasaland
HAB. Upland rain-forest; 1000–2130 m.

NOTE. This until recently would have been included under " *A. pennata* ". *A. monticola* is most closely related to *A. pentagona* (p. 100) and *A. kamerunensis* (p. 98), differing from both in the dense indumentum of hairs and also glands clothing the young branchlets. The leaflets of *A. monticola* are usually narrower than in *A. pentagona*, and the ovary is always pubescent while in *A. pentagona* it is frequently glabrous. Very important differences are found in the pod, which is dehiscent in *A. monticola* with much less thickened margins than those of the indehiscent pods of *A. pentagona*; in addition the areole on the seed is small, not large. *A. pentagona* is typical of lowland rain- and swamp-forest, occurring only doubtfully in upland rain-forest, while *A. monticola* is typical of the latter.
The pods of *A. kamerunensis* are dehiscent, like those of *A. monticola*, but are narrower (1·7–2·8 cm. as against 3–4·5 cm.) and are paler in colour

27. A. schweinfurthii *Brenan & Exell* in Bol. Soc. Brot., sér. 2, 31 : 128 (1957). Type : the Sudan, Gubbiki, *Schweinfurth* 2206 (BM, holo.!, K, iso.!)

Scandent shrub to 12 m., or sprawling, or a small spreading tree. Young branchlets puberulous and glandular, olive-green or pale brown, later olive-brown. Prickles deflexed, scattered, arising from brownish longitudinal bands darker than the intervening yellowish to grey ones. Leaves : petiole 2·6–5·5 cm. long, with a gibbous gland 1–1·8 mm. long and 0·5 mm. wide ; pinnae 9–17 pairs, 3·5–7 cm. long ; glands on rhachis between the top 1–3 pairs ; leaflets numerous, linear or linear-oblong, 0·8–2 mm. wide, margins ciliolate with whitish appressed hairs, midrib nearer one margin at base. Flowers white or palest yellow, in heads 8–12 mm. in diameter in ± pyramidal panicles. Stipules at base of peduncles small, 0·3–1·2 mm. wide, inconspicuous, soon caducous, not subcordate at base. Calyx eglandular outside. Pods (Fig. 15/27, p. 66) coriaceous or subcoriaceous, oblong, ± transversely plicate and umbonate over the seeds, 9·5–17 cm. long, 1–2·9 cm. wide, margins not strongly thickened. Seeds blackish or dark brown, smooth, elliptic, 9–11 mm. long, 6·5–8 mm. wide ; areole largish, 6–8 mm. long and 3–5 mm. wide.

var. **schweinfurthii**; Brenan & Exell in Bol. Soc. Brot., sér. 2, 31: 130 (1957)

Leaflets glabrous beneath except for the ciliolate margins.

TANGANYIKA. Moshi District: Weru Weru R. to the Kikafu Bridge, 14 Feb. 1914, *Peter* K12!; Morogoro, 31 Dec. 1934; *E. M. Bruce* 395!
DISTR. T2, 3, 6; the Sudan southwards to Bechuanaland, the Transvaal and Natal
HAB. Riverine forest ; about 610 m.

var. **sericea** Brenan & Exell in Bol. Soc. Brot., sér. 2, 31 : 131 (1957). Type : Tanganyika, *Hornby* 56 (K, holo.!, EA, iso.!)

Leaflets ± appressed-silky-pubescent beneath.

Tanganyika. Mpwapwa, 12 Nov. 1928, *Hornby* 56 !
Distr. **T**5, ? 6; Portuguese East Africa
Hab.   Deciduous woodland and bushland

Note.   This species, like various of its relatives, has hitherto been wrongly included under
A. pennata. By reason of its long petioles, *A. schweinfurthii* is closest to *A. pentagona*
(below) and *A. kamerunensis* (p. 98) among these relatives, the distinctions being given
in the notes after the two last-named species.
 Tanner 3368 ! from Tanganyika, Pangani, Nkaramo, Mkwaja, 8 Jan. 1957, is intermediate between vars. *schweinfurthii* and *sericea*, the leaflets varying from glabrous
to slightly hairy beneath.

28. **A. pentagona** (*Schumach. & Thonn.*) *Hook. f.* in Niger Fl.: 331 (1849).
Type : Ghana, Jadofa, *Thonning* (C, holo., K, photo. !)

 An often tall liane.   Young branches sparsely puberulous to glabrous
and eglandular, very rarely with inconspicuous sessile glands, red-brown to
deep purplish.   Prickles deflexed, scattered, arising from longitudinal bands
usually darker than the intervening ones.   Leaves : petiole (1·5-)2–6 cm.
long ; pinnae 8–15 pairs, 2·5–9 cm. long ; leaflets linear or linear-oblong,
0·7–1·8(–2) mm. wide, glabrous or nearly so, midrib nearer one margin at
base.   Flowers white, in heads 8–10(–12) mm. in diameter usually in ample
panicles.   Stipules at base /of peduncles small, 0·75–1·5 mm. wide, inconspicuous, soon caducous, not subcordate at base.   Calyx eglandular outside.
Corolla glabrous or rarely sparingly puberulous outside.   Pods (Fig. 15/28,
p. 66) thick, hard, oblong, dark brown, 7·5–16 cm. long, 1·8–3·5 cm. wide,
indehiscent and with markedly thickened margins 2–4 mm. wide.   Seeds
black, smooth, ellipsoid, thick but somewhat compressed, 10–13 mm. long,
6–8 mm. wide; areole large, 7–10 mm. long and 4·5–6 mm. wide.

Uganda.   Busoga District: near Jinja, May 1931, *Harris* 25 *in* F.H. 85 !; Mengo District:
 Kiagwe, Namanve Forest, Apr. 1932, *Eggeling* 405 *in* F.H. 687 ! & Kitabe near
 Entebbe, Feb. 1935, *Chandler* 1168 !
Kenya.   Teita District: Taveta, 22 Jan. 1936, *Greenway* 4475 !
Tanganyika.   Bukoba District: Kaigi, Sept.–Oct. 1935, *Gillman* 620 !; Kilimanjaro,
 H. H. Johnston !; Lushoto District: Sigi–Pandeni, 1 May 1915, *Peter* K828 ! &
 Longuza, 7 Apr. 1922, *Soleman* !
Distr. **U**3, 4; **K**7; **T**1–3, ?5, 6; from French Guinea and Sierra Leone to the Belgian
 Congo and Uganda, southwards to Angola, Northern and Southern Rhodesia and
 Portuguese East Africa
Hab.   Lowland rain-forest and swamp-forest, possibly occurring also in upland rain-
 forest; 400–1450 m.

Syn.   *Mimosa pentagona* Schumach. & Thonn., Beskr. Guin. Pl.: 324 (1827)
 Acacia pentaptera Welw. in Ann. Conselho Ultram. 1858 : 584 (1859) ; Consp.
  Fl. Angol. 2 : 287 (1956)
 [*A. pennata* sensu auct., e.g. T.T.C.L.: 333 (1949), pro parte, *non* (L.) Willd.]
 A. silvicola Gilb. & Bout. in B.J.B.B. 22: 179 (1952), pro parte, sensu stricto &
  in F.C.B. 3: 155 (1952). Type: Belgian Congo, between Lilanga and Yangole,
  Louis 6978 (BR, holo., K, iso. !)

Note.   *A. pentagona* is typically a liane of evergreen forest. From *A. kamerunensis* and *A.
taylorii* it is distinguished by the normally dark twigs, larger leaflets, and by the thick,
normally indehiscent pods. The seeds also have a large areole.   From *A. schweinfurthii*
it is likewise distinct in the darker twigs, the pods, and in the ciliolation of the leaflets
being very inconspicuous or absent.   From the other East African species of the *A.
pennata* complex it is readily separable by the long petioles.

29. **A. taylorii** *Brenan & Exell* in Bol. Soc. Brot., sér. 2, 31 : 139 (1957).
Type : Tanganyika, Lindi District, *Milne-Redhead & Taylor* 7588 (K,
holo. !)

 Scrambling shrub up to 4·5 m. high.   Young branchlets densely pubescent
with small arcuate-ascending hairs ; glands almost absent.   Prickles
scattered, deflexed, arising from longitudinal bands along the wholly grey
or grey-brown stems.   Leaves : petiole 1·7–3·9 cm. long ; pinnae mostly
9–11 pairs ; leaflets numerous, 17–34 pairs, narrowly oblong, 0·6–1 mm.

wide ; midrib subcentral at base. Flowers in heads about 7–8 mm. in diameter, irregularly paniculate. Stipules at base of peduncles small, up to 1·5 mm. wide, inconspicuous, soon caducous, not subcordate at base. Calyx and petals greenish-cream. Calyx-lobes glabrous and eglandular or almost so outside. Stamen-filaments creamy-white. Pods (Fig. 15/29, p. 66) stiffly papery, oblong, dehiscent, 11–15 cm. long, 1·5–2·7 cm. wide, glabrous except for scattered minute sessile brown glands. Seeds purple-brown, smooth, elliptic, flattened, 8–9 mm. long, 6–6·5 mm. wide ; areole small, 4–5 mm. long, 1·5–2·5 mm. wide.

TANGANYIKA. Lindi District: about 6·5 km. N. of Lindi, 8 Dec. 1955, *Milne-Redhead & Taylor* 7588 !
DISTR. T8; known only from this gathering.
HAB. Edge of dry water-course leading down to shore in coastal deciduous bushland at about sea-level

NOTE. The midrib subcentral at base of leaflet is found only in *A. adenocalyx* among the closely related species. That, however, differs in having densely glandular young branchlets, dark brown twigs, petioles only 0·5–1·2 cm. long, and glands on the outside of the calyx.

30. **A. macrothyrsa** *Harms* in E.J. 28 : 396 (1900) ; L.T.A. : 851 (1930) ; T.S.K. : 68 (1936) ; Bogdan in Nature in E. Afr., ser. 2, No. 1 : 12 (1949) ; T.T.C.L. : 336 (1949) ; Gilb. & Bout. in F.C.B. 3 : 157 (1952) ; I.T.U., ed. 2 : 210, fig. 48a (1952) ; Coates Palgrave, Trees Centr. Afr. : 246–9 (1956) ; F.W.T.A., ed. 2, 1 : 501 (1958). Type : Tanganyika, Iringa, *Goetze* 653 (B, holo.†, K, iso. !)

Small or medium tree 2–12(–15 *fide* F.C.B.) m. high ; bark rough, fissured, grey (or brown *fide* F.C.B.). Stipules spinescent, stout, brown, glossy, compressed, up to 1·6(–2·8) cm. long. Leaves large, 10–20 cm. wide ; rhachis (with petiole) 10–37 cm. long ; pinnae mostly 9–16(–27) pairs ; leaflets 12–40 pairs (in our area, elsewhere up to 70 pairs), (4–)6–11(–20) mm. long, 1–3·5(–6) mm. wide, rather stiff and glossy above, glabrous (or ciliolate on margins, but apparently not in our area). Flowers orange or yellow, strongly and sweetly scented, in heads 8–13 mm. in diameter, in a panicle up to about 45 cm. long and 30 cm. wide, whose branches (and usually also main axis) are leafless. Pods (Fig. 15/30, p. 66) coriaceous, glossy, glabrous, oblong, straight, blackish, blackish-purple or brown, 8–20 cm. long and 1·5–2·5 cm. wide. Seeds deep brown, smooth, elliptic to subcircular, compressed, 8–10 mm. long, 7–8 mm. wide ; areole 4–5 mm. long, 3–3·5 mm. wide.

UGANDA. Lango District: Lira, July 1937, *Eggeling* 3353 !; Teso District: Serere, July 1932, *Chandler* 797 !; Mbale District: Siroko valley, 10 Nov. 1933, *Tothill* 2248 !
KENYA. Turkana District: Kacheliba escarpment, May 1932, *Napier* 2008 !; Uasin Gishu District: Kipkarren, Aug. 1934, *Dale* 3287 !; S. Kavirondo District: Gori R., Suna, Sept. 1933, *Napier* 5296 !
TANGANYIKA. Mwanza District: Geita, Jan. 1949, *Watkins* 183 in *F.H.* 2621 !; Kahama District: Ushirombo, 23 Feb. 1937, *B. D. Burtt* 5458 !; Morogoro, 3 Sept. 1930, *Greenway* 2503 ! & Apr. 1951, *Eggeling* 6070 !
DISTR. U1–4; K2, 3, 5; T1, 2, 4–8; Ghana, Nigeria, Belgian Congo, the Sudan, Portuguese East Africa, Nyasaland. Northern and Southern Rhodesia
HAB. Deciduous woodland, wooded grassland, sometimes on rocky hillsides; 600–1830 m.

SYN. *A. buchananii* Harms in E.J. 30: 76 (1901); L.T.A.: 852 (1930); T.S.K., ed. 2: 68 (1936). Type: Nyasaland, *Buchanan* 256 (B, holo. †)
    A. prorsispinula Stapf in J.L.S. 37: 513 (1906); Battiscombe, Cat. Trees Kenya Col.: 54 (1926). Types: Uganda, Acholi District, *Dawe* 856 (K, syn.!); Kenya, Nandi country, Sibu, *Evan James* (K, syn. !)

VARIATION. The young branchlets, inflorescence-axes, leaf-rhachides etc. are usually glabrous in East Africa. Elsewhere they may be puberulous, as in all West African material (including the type of *A. dalzielii* Craib) and in some specimens from the

Belgian Congo, Portuguese East Africa and Northern Rhodesia. Similar forms apparently occur rarely with us: *Busse* 868 (EA !), without precise locality, is an example.

The leaflets vary much in size and number, but the variation does not seem to follow any clear pattern of geography.

The large leaves coupled with the robust panicles of yellow or orange flower-heads make *A. macrothyrsa* easy to recognize.

31. **A. zanzibarica** (*S. Moore*) *Taub.* in P.O.A. C : 195 (1895) ; L.T.A. : 843 (1930) ; T.S.K. : 69 (1936) ; Bogdan in Nature in E. Afr., ser. 2, No. 1 : 12 (1949) ; T.T.C.L. : 338 (1949) ; U.O.P.Z. : 102 (1949). Type : Kenya, Mombasa, *Hildebrandt* 1939 (K, holo. !)

Tree 3–9 m. high ; bark yellow-green to yellow or whitish, turning cinnamon-coloured and powdery with age. Young branchlets glabrous or nearly so, mostly brown or grey-brown, older ones with minutely flaking or powdery, yellowish to brown bark. Stipules spinescent, mostly straight, grey, 1·2–7·5 cm. long, some fused at base into ± deeply bilobed blackish " ant-galls," each rounded or fusiform lobe to about 2–2·5 cm. in diameter ; other prickles absent. Leaves glabrous or nearly so ; petiole 4–7(–15) mm. long, often glandular at middle or top ; rhachis 0–1·5(–6·5) cm. long ; pinnae 1–4 (–6) pairs, mostly 1–2 cm. long ; leaflets 3–10 pairs, oblong to ± obovate, mucronate or acute at apex, 2–13(–20) mm. long, (0·5–)2–6(–10) mm. wide ; venation usually raised, especially beneath. Flowers bright yellow, in heads, sweetly scented ; involucel at or shortly above the base of the glabrous or subglabrous peduncle. Calyx 0·8–1·5 mm. long, puberulous or glabrous. Corolla 2·25–4 mm. long, glabrous. Pods (Fig. 15/31, p. 66) linear, falcate to almost straight, flattened, subcoriaceous, glabrous, closely venose, 5·5–12 cm. long, 4–7 mm. wide, blackish–brown. Seeds probably ± quadrate-oblong.

var. **zanzibarica** ; Brenan in K.B. 1957 : 75 (1957).

Leaflets 2–6 (–9) mm. wide, with venation raised especially beneath.

KENYA. Mombasa, 24 Nov. 1951, *Bogdan* 3303 !; District uncertain: Marereni, *R. M. Graham* L119 *in F.H.* 1620 *in C.M.* 13991 !
TANGANYIKA. Lushoto District: Mnazi–Kivingo, SW. Umba plain, 10 Jan. 1930, *Greenway* 2049 !; Tanga District: Kibuguni, 25 Nov. 1936, *Greenway* 4780 !; Dar es Salaam, 9 km. N. on Bagamoyo road, 25 Apr. 1933, *B.D. Burtt* 4462 !
DISTR. **K**1, 7; **T**3, 6; not known elsewhere
HAB. Woodland and wooded grassland; occurring gregariously in clay soil flats of shallow drainage areas in coastal districts (T.T.C.L.: 338); marginal to saline-water swamp-forest, possibly an indicator of saline soils (Greenway); 0–460 (–914) m.

SYN. *Pithecolobium ? zanzibaricum* S. Moore in J.B. 15: 292 (1877)
     Acacia leucacantha Vatke in Oesterr. Bot. Zeitschr. 30: 276 (1880). Type: Kenya, Kwale District. Maji ya Chumvi, *Hildebrandt* 2332 (? B, holo. †)

NOTE. Except for *A. seyal* var. *fistula*, *A. zanzibarica* is the only " gall-acacia " in our area with bright yellow flowers. The rather large leaflets (of var. *zanzibarica*) combined with few pinnae will separate it from all the other " gall-acacias " except *A. burttii*, which has considerably larger flowers and pods of an utterly different shape, and is not a coastal species. The sides of the ovaries of *A. zanzibarica* are occupied by areas of minute, densely packed, sessile glands, which are not found in the other " gall-acacias " of East Africa, except in *A. seyal* var. *fistula*

var. **microphylla** *Brenan* in K.B. 1957: 75 (1957). Type: Kenya, Northern Frontier Province, Turbi, *Gillett* 13803 (K, holo. !, EA, iso. !)

Leaflets 0·5–1·5 mm. wide, with normally only the midrib somewhat raised beneath.

KENYA. Northern Frontier Province: Wajir, 24 Dec. 1943, *Bally* 3732 ! & Turbi, 11 Sept. 1952, *Gillett* 13803 ! District uncertain: Nzui and Lorian Swamp, 23 Dec. 1942, *Bally* 1967 !
DISTR. **K**1 and very possibly **K**4; also in Somalia and Ethiopia
HAB. Edge of semi-desert, and in scrub; 550–850 m.

NOTE. The var. *microphylla* is rather similar to *A. seyal* var. *fistula*; its most important distinguishing features are the shorter calyx, the corolla usually only 2·5 mm. long, the involucels 1–2 mm. long and basal or up to 3 mm. above the base of the peduncle, the heads of flowers only 8–10 mm. in diameter and characteristically although not always arranged along short lateral branchlets, and the normally more triangular-pointed apices of the bracteoles. The flower-heads of *A. seyal* var. *fistula* are sometimes racemosely arranged along lateral branchlets, but not so neatly and regularly so as in var. *microphylla*. The above characters apply to var. *microphylla*, which is found inland and at higher altitudes than usual in var. *zanzibarica*, and may be worth higher rank than that of variety, but they do not necessarily hold with var. *zanzibarica*.

A specimen, *Hemming* 87 !, from Kenya, Meru District, Isiolo, in mixed bush with *Acacia tortilis* dominant, may well be var. *microphylla*, but is only in bud and very young leaf.

32. **A. seyal** *Del.*, Fl. Egypt. : 142, t. 52, fig. 2 (1813) ; L.T.A. : 844 (1930) ; T.S.K. : 68 (1936) ; B.D. Burtt in Journ. Ecol. 30 : 95 (1942) ; T.T.C.L. ; 337 (1949) ; Bogdan in Nature in E. Afr., ser. 2, No. 1 : 12 (1949) ; I.T.U., ed. 2 : 213, fig. 48 h (1952). Type : Egypt, *Delile* (MPU, ? holo.)

Tree 3–9 (–12) m. high ; bark on trunk powdery, white to greenish-yellow or orange-red. Young branchlets almost glabrous and with numerous reddish sessile glands (rarely, and not in E. Africa, rather densely puberulous) ; epidermis of twigs becoming reddish and conspicuously flaking off to expose a greyish or ± reddish powdery bark. Stipules spinescent, up to 8 cm. long ; " ant-galls " present or not ; other prickles absent. Leaves often with a rather large gland on the petiole and between the top 1–2 pairs of pinnae ; pinnae (2–)3–7(–8) pairs ; leaflets (7–)11–20 pairs, 3–8(–9) mm. long, 0·75–1·5(–3) mm. wide, in our area sparingly ciliolate to glabrous ; lateral nerves invisible beneath. Flowers bright yellow, in axillary pedunculate heads 10–13 mm. in diameter ; involucel in lower half of peduncle, 2–4 mm. long ; apex of bracteoles rounded to elliptic, sometimes pointed. Calyx 2–2·5 mm. long, inconspicuously puberulous above. Corolla 3·5–4 mm. long, glabrous outside. Pods (Fig. 15/32, p. 66) dehiscent, linear, ± falcate, ± constricted between the seeds, finely longitudinally veined, glabrous except for some sessile glands, (5–)7–20(–22) cm. long, 0·5–0·9 cm. wide. Seeds olive to olive-brown, faintly and minutely wrinkled, elliptic, compressed, 7–9 mm. long, 4·5–5 mm. wide; areole 5–6 mm. long, 2·5–3·5 mm. wide.

var. **seyal**

" Ant-galls " absent.

UGANDA. Lango District, Aug. 1932, *Kennedy* 10 *in F.H.* 900 !
KENYA. Northern Frontier Province: Moyale, 14 July (fl.), Oct. 1952 (fr.), *Gillett* 13587 !; Machakos, 4 Feb. 1953, *Trump* 38 !; S. Kavirondo District: Kanam, foot of Homa Mt., 22 Nov. 1934, *Allen Turner in C.M.* 6754 !; Masai District: between the Rift Valley and Narok, 2 Oct. 1954, *Verdcourt* 1155 !
TANGANYIKA. Moshi District: Engare Nairobi, 4 July 1943, *Greenway* 6711 !; Singida District: Iwumbu R., 14 Aug. 1927, *B. D. Burtt* 753 !; about 1·5 km. N. of Iringa, 15 July 1956, *Milne-Redhead & Taylor* 11208 !
DISTR. **U**1; **K**1, 3–6; **T** ?1, 2, 3, 5, 7, 8; widespread in northern tropical Africa, extending to Egypt

var. **fistula** (*Schweinf.*) *Oliv.*, F.T.A. 2: 351 (1871); L.T.A.: 844 (1930); T.T.C.L.: 338 (1949). Types: the Sudan, Gedaref region, and on Mount Gule in the province of Sennar, *Schweinfurth* (B, syn. †)

Some pairs of spines fused at base into " ant-galls " 0·8–3 cm. in diameter which are greyish or whitish, often marked with sienna-red and with a longitudinal furrow down the centre, making the " galls " ± bilobed.

UGANDA. Karamoja District: R. Lotisan, 31 Oct. 1939, *A. S. Thomas* 3130 !; Teso District: Serere, Lobori, July 1926, *Maitland* 1344 !; Mengo District: Beruli, Naka-songola, Nov. 1932, *Eggeling* 705 *in F.H.* 1083 !
KENYA. Northern Frontier Province: Eil Lass near Wajir, Jan. 1949, *Dale* K735 !; Baringo District: Lake Baringo, near mouth of Tiggeri R., 12 Mar. 1950, *Bally* 7742 !;

Meru/N. Nyeri Districts: Ngare Ndare, lower northern slopes of Mt. Kenya, July 1933, *Gardner in F.H.* 3158 *in C.M.* 6928!

Tanganyika. Mwanza District: Nyarwigo, Buhumbi, 3 Feb. 1953, *Tanner* 1195!; Mbulu District: Basotu Lake, 5 Sept. 1932, *B. D. Burtt* 4274!; Morogoro District, Nov.–Dec. 1933, *Gillman* 11!; Mbeya District: Buhoro Flats, Dabajira, Utengule, 26 Aug. 1956, *Disney* 56/8!

Distr. U1, 3, 4; K1, 3–6; T1, 2, ?3, 4–7; the Sudan, Somaliland Protectorate, Portuguese East Africa, Nyasaland and Northern Rhodesia

Hab. (of species as whole). Wooded grassland, especially on seasonally flooded black cotton soils along water-courses; 600–1830 m.

Syn. *A. fistula* Schweinf. in Linnaea 35: 344 (1867–8)

Note. For the differences between *A. seyal, A. zanzibarica* var. *microphylla* and *A. hockii* De Wild., see under the two latter.

" Ant-galls " may occur on trees with white, greenish or reddish bark, but are rare with the last colour, and, in some areas at least, absent. Thus on *Gillett* 14161, Kenya, Northern Frontier Province, Moyale, it is noted that var. *fistula* with greenish-white smooth bark is the chief acacia on black cotton soil in the plains, while the var. *seyal* with reddish bark and no " ant-galls " occurs commonly in the hills. Clear segregation like this is not, however, general. In West Africa var. *fistula* is not known, but var. *seyal* with greenish-white- and red-barked variants occurs, and in East Africa a similar range of bark colour may be found within var. *seyal*. The matter may be further complicated by the fact that *Cooke* 94, from the Sudan, notes that individual trees with red or white bark may change from one colour to the other, and that *Lynes* P.r.9, from Tanganyika, Iringa, states that the bark is red when old and yellow or green on the saplings. The bark-colour can vary due to grass-fires—charred round the base of the bole, orange-red above, and greenish-yellow on the upper branches (Greenway).

There is a tendency in East Africa for the flowers to have the corolla-lobes reflexed in var. *seyal* and suberect in var. *fistula*, but this is not true of other parts of Africa.

33. **A. hockii** *De Wild.* in F.R. 11 : 502 (1913) ; L.T.A. : 849 (1930); F.W.T.A., ed. 2, 1 : 500 (1958). Type : Belgian Congo, Katanga, Luafu Valley, *Hock* (BR, holo. !)

Shrub or tree (1–) 2–6 (–12) m. high ; bark not powdery, red-brown to greenish or rarely pale yellow, peeling off in papery layers when not burned. Young branchlets ± densely puberulous, rarely glabrous, with ± numerous reddish sessile glands, usually elongate and slender with reddish or brownish bark which does not peel to expose a powdery layer as in *A. seyal*. Stipules spinescent, mostly short, straight, suberect or spreading, to 2 (rarely to 4) cm. long, subulate or flattened on upper side ; " ant-galls " and other prickles absent. Leaves often with a gland on the petiole and between the top 1(–3) pairs of pinnae ; pinnae (1–)2–11 pairs ; leaflets 9–29 pairs, 2–6·5 mm. long, 0·5–1(–1·25) mm. wide, usually (at least in East Africa) ± densely ciliolate, lateral nerves invisible beneath. Flowers bright yellow, in axillary pedunculate heads 5–12 mm. in diameter ; involucel ⅓–⅔-way up peduncle, 1·5–3 mm. long. Apex of bracteoles rounded to rhombic, sometimes pointed. Calyx (1–)1·5–2 mm. long, glabrous except above. Corolla 2·5–3·5 mm. long, glabrous outside. Pods (Fig. 15/33, p. 66) as in *A. seyal*, except for being often ± puberulous, (4–)5–14 cm. long, 0·3–0·6(–0·8) cm. wide. Seeds olive-brown, smooth, elliptic, compressed, 5–7 mm. long, 3–4 mm. wide ; areole 3·5–4·5 mm. long, 2–2·5 mm. wide.

Uganda. Ankole District: Mbarara, 26 Apr. 1941, *A. S. Thomas* 3835!; Teso District: Lake Kioga, Lale, 13 Oct. 1952, *Verdcourt* 836!; Masaka District: Kiagwe, Bukasa, June 1932, *Eggeling* 447 *in F.H.* 781!

Kenya. Uasin Gishu District: Eldoret, 15 Oct. 1951, ·G. R. Williams 9295!; Elgon, Oct.–Nov. 1930, *Lugard* 75!; Masai District: Bakitabuk, 29 June 1948, *Vesey-Fitz-Gerald* 175!

Tanganyika. Shinyanga District: Mwantine Hills, 18 June 1931, *B. D. Burtt* 3309!; Tanga District: Ngomeni, 25 Aug. 1944, *Greenway* 7031!; Morogoro District: without locality, 30 Nov. 1932, *Wallace* 517!

Distr. U1–4 ; K1, 3–6 ; T1–7 ; from French Guinea, the Ivory Coast, Ghana and the Sudan in the north to Angola, Northern Rhodesia and Portuguese East Africa in the south

HAB. Deciduous woodland, wooded grassland and deciduous and semi-evergreen bushland; said to be common on hills in *Brachystegia* woodland in Tanganyika (T.T.C.L.:
338); 0–2300 m.

SYN. [*A. stenocarpa* sensu auct. mult. e.g. F.T.A. 2: 351 (1871), pro max. parte; L.T.A.:
845 (1930); T.S.K.: 68 (1936); F.P.N.A. 1: 389 (1948), pro parte; Bogdan in
Nature in E. Afr., ser. 2, No.1: 13 (1949), *non* [Hochst. ex] A. Rich.]
A. holstii Taub. in P.O.A. C: 194, t. 21, fig. C (1895) pro parte, excl. legumina;
*nom. rejic.** I.T.U.: 113, fig. 36k (1940). Types: Tanganyika, Lushoto District,
Mashewa, *Holst* 8744 (B. holo. †, K, P, iso.!). The Kew isotype is entirely *A.
hockii*, but lacks pods.
A. chariensis A. Chev. in Bull. Soc. Bot. Fr. 74: 958 (1927); L.T.A.: 845 (1930).
Types: Shari, Koddo, *Chevalier* 6432 (P. syn.!, K, isosyn.!) & Dar-Banda, *Chevalier* 6661 (P, syn.! K, photo.!)
A. seyal Del. var. *multijuga* [Schweinf. ex] Bak. f., L.T.A.: 844 (1930); T.T.C.L.:
338 (1949); I.T.U., ed. 2: 213, fig. 48k, t.11 (1952); Consp. Fl. Angol. 2: 284
(1956). Type: not cited; the Sudan, *Schweinfurth* 1091. 2061, 2627 (BM,
annotated by *Schweinfurth* and presumably syntypes!, 1091 & 2627 at K,
isosyn.!)
[*A. seyal* sensu Gilb. & Bout. in F.C.B. 3: 160 (1952), saltem pro max. parte
non Del.]

NOTE. *A. hockii* occupies a wide range both of habitat and altitude, and is also widespread in tropical Africa. The plant is correspondingly variable, and it may later be
possible to divide it into races.
In the past it has been confused with *A. seyal*, but, although the two are closely
related, it seems preferable to maintain them as distinct species. *A. hockii* differs from
A. seyal primarily by having non-powdery bark. The twigs are usually (not always)
more elongate and slender, with reddish or brownish bark which does not peel to expose
the inner layer so characteristic of *A. seyal*. The young branchlets are usually clothed
with a more or less dense puberulence which is not found in *A. seyal*; rarely, however,
the branchlets are glabrous except for sessile glands (*Holst* 2158 from Tanganyika,
Tanga District, Gombelo; *Snowden* 1058 & *Tweedie* 708, from Uganda, Elgon). The
spines are never " ant-galled " and usually short, but may occasionally be up to 4 cm.
long and whitish even on the flowering twigs (*Bally* 5288 from Kenya, Masai District;
Welch 40 from Tanganyika, Shinyanga District; and *Doggett* 122 from Tanganyika,
Nzega District)
In T.T.C.L.: 337–8 it is stated that the corolla is divided as far as or beyond the
calyx in *A. hockii* (as *A. seyal* var. *multijuga*), but not as far as the calyx in *A. seyal*.
This is incorrect: the two species are not distinguishable in this way, although both
the calyx and corolla are rather shorter in *A. hockii* than in *A. seyal*; for measure-
ments see the descriptions.
The internode-length varies considerably in different gatherings, often resulting in
great differences in general appearance, which may however be mainly or entirely due
to the various habitats. There is evidence also that the bark may vary in colour,
perhaps in the way that it does in *A. seyal*; but careful observation of this in the field is
still wanted. *Gillett* 13705 (Kenya, Northern Frontier Province, Moyale) is said to
have pale yellow papery-peeling bark (a most unusual colour for *A. hockii*), but to
lose it and develop grey-brown bark when the tree gets old; this specimen is also abnormal in having a dense, almost pubescent, indumentum extending even over the
surface of the leaflets.
A strange tendency of *A. hockii*, shared also by *A. seyal* and *A. nilotica*, is for a few
flowers to arise in the involucel on the peduncle, sometimes giving the appearance of
a smaller secondary capitulum below the main one.

34. **A. ancistroclada** *Brenan* in K.B. 1958: 412 (1959). Type : Kenya,
Masai District, Amboseli, *Knight & Thomas* H 344/58 (K, holo.!)

Shrub or small tree up to 7·5 m. high ; trunk usually branching a short
distance above ground ; bark peeling off in large papery pieces, reddish- or
greenish-yellow. Young branchlets glabrous except for minute sessile
reddish glands which soon disappear, purplish at first, soon grey- to red-
brown ; bark of twigs not flaking off. Stipules spinescent, mostly short,
2·5–8 mm. long and downwardly hooked, but some (especially on older
twigs) elongate, 2–5·8 cm. long and nearly straight ; " ant-galls " and other
prickles absent. Leaves : pinnae 1–2 pairs, a small gland between the top
(or only) pair ; rhachis 0–8 mm. long ; leaflets 3–9 pairs, mostly 2–5 mm.
long, 0·75–1·5 mm. wide, glabrous or margins very sparsely and incon-

* See note under 42, *A. etbaica* on p. 116.

spicuously ciliolate ; lateral nerves invisible beneath. Flowers bright yellow, in axillary pedunculate heads ; involucel about $\frac{1}{3}$–$\frac{2}{3}$-way up the peduncle. Calyx (except for puberulous lobes) and corolla glabrous outside. Pods dehiscent, linear, falcate, slightly constricted between the seeds, finely longitudinally or somewhat obliquely veined, glabrous, 6–15 cm. long, 0·5–0·6 cm. wide. Seeds olive or olive-brown, smooth, elliptic or elliptic-oblong, 6–7 mm. long, 3–3·5 mm. wide ; areole 4·5–5 mm. long, 1·5–2 mm. wide.

Kenya. Masai District : Ngom watercourse near Selengai, 4 Nov. 1948, *Vesey-FitzGerald* 214 ! & Amboseli, Apr. 1958, *D. B. Thomas* 628 ! ; Masai/Teita Districts : Tsavo National Park, near Nzima Springs, 3 May 1952, *Trapnell* 2213 !
Tanganyika. Masai District : Naberera, 9 Oct. 1934, *Hornby* M1106/No. 6 ! & between Mgera and Kibaya, *B. D. Burtt* 4934 ! & between Longido and Namanga, 13 Nov. 1958, *Trapnell* 2431 ! ; ? Handeni District : E. of Loskitu Mt., 18 Sept. 1933, *B. D. Burtt* 4916 !
Distr. **K**6, ?7 ; **T**2, ?3 ; not known elsewhere
Hab. Scattered-tree grassland, ? dry scrub with trees ; about 820–1310 m.

Note. The bright yellow flowers and pods are similar to those of *A. seyal*, while the spines are reminiscent of those of *A. tortilis*. The papery-peeling bark is different from both.

35. **A. kirkii** *Oliv.* in F.T.A. 2 : 350 (1871) ; L.T.A. : 848 (1930) ; T.T.C.L. : 333 (1949) ; Consp. Fl. Angol. 2 : 285 (1956). Type : Northern Rhodesia, Southern Province, Batoka country, *Kirk* (K, holo. !)

Tree 2·5–15 m. high, flat-crowned ; bark green, peeling or scaling. Young branchlets pubescent to sometimes subglabrous, with numerous reddish sessile glands ; twigs grey, brown or plum-coloured, not showing yellow bark. Stipules spinescent, straight or almost so, varying in length, up to 8 cm. long ; " ant-galls " and other prickles absent. Leaves : rhachis 3–8 cm. long, normally rather densely pubescent above ; pinnae 6–14 pairs (some leaves always with 8–9 pairs or more) ; leaflets numerous, small, narrowly oblong or oblong-linear, 2–5 mm. long, 0·5–1(–1·25) mm. wide. Flowers with red corolla and white stamen-filaments, in axillary heads whose involucels are conspicuous, 2–3 mm. long and near base of or $\frac{1}{5}$–$\frac{1}{2}$-way up the peduncle ; peduncles rather densely pubescent and with sessile glands throughout, rarely sparingly pubescent. Pods indehiscent, narrowly oblong, straight (or only bent in a plane at right-angles to the flattened plane of the pod), 3·5–9 cm. long, 0·8–2·1 cm. wide, often ± moniliform with the segments mostly as wide or wider than long. Seeds blackish-olive, smooth, subcircular to elliptic, compressed, 5–7 mm. long, 4·5–5·5 mm. wide ; areole 3·5–5 mm. long, 2·5–3 mm. wide.

Key to intraspecific variants

Joints of pod with a medium or small wart-like projection up to 4–5 mm. high in the centre of each
  of their flat sides (subsp. **kirkii**) :
    Pods 1–1·2 cm. wide .   .   .   .   .   . var. **sublaevis**
    Pods 1·2–2·5 cm. wide .   .   .   .   .   . var. **intermedia**
Joints of pod without any central projections .   . subsp. **mildbraedii**

subsp. **kirkii**; Brenan in K. B. 1957: 363 (1958)

Pods with prominent or obscure veins; each joint of the pod bearing in the middle of each of its flat sides a small medium or large wart-like projection up to 2–5 mm. high; some joints sometimes almost lacking the projections; stipe of pod 0·5–2·5 cm. long.

Note. The var. *kirkii*, so far found only in Northern Rhodesia, differs from the two following varieties in having the pods 1–1·3 cm. wide, with the projections from the joints prominent, (2–)3–5 mm. high.

var. **intermedia** *Brenan* in K.B. 1957: 363 (1958).   Type: Kenya, Athi Plains, *van Someren in C.M.* 2700 (K, holo. !)

Pods (Fig. 15/35A, p. 66) 1·2–2·5 cm. wide.

UGANDA.   Acholi District : Chua, Rom, Dec. 1935, *Eggeling in F.H.* 2424 !
KENYA.   Kiambu District: Kabete, 19 May 1947, *Bogdan* 536 !; Nairobi to Ngong Hills, 9 Dec. 1947, *Bogdan* 1425 !; Nairobi, Athi Plains, Aug. 1933, *van Someren in C.M.* 2700 !
TANGANYIKA.   Mbulu District: Tlawi Hills, 30 Aug. 1932, *B. D. Burtt* 4279 !; Nzega District: about 8 km. S. of Igunga, 22 July 1949, *Doggett* 125 !; Dodoma District: Logi " mbuga ", Chipogoro [? Chipogolo], 104 km. S. of Dodoma, 18 Apr. 1956, *Disney* 56/4 B !
DISTR.   **U**1; **K**4, 6; **T**2, 4, 5; Bechuanaland, Northern Rhodesia and Angola
HAB.   Riverine or ground-water forest; 1520–1980 m.

SYN.   [*A. mildbraedii* sensu Bogdan in Nature in E. Afr., ser. 2, No. 1: 14 (1949), *non* Harms sensu stricto]
   [*A. kirkii* sensu T.T.C.L.: 333 (1949), *non* Oliv. sensu stricto]

NOTE.   Specimens without pods, Tanganyika, Shinyanga District, Uduhe, between Kisapu and Mango, *B. D. Burtt* 5518 !, and Tanganyika, Dodoma District, Fufu, *Wigg* 983 ! may be referable to this variety.

var. **sublaevis** *Brenan* in K.B. 1957:363 (1958).   Type: Uganda, Aswa R., *Eggeling* 775 in *F.H.* 1161 (K, holo. !, EA, iso. !)

Pods 1–1·2 cm. wide, with small projections; stipe of pod about 0·5–1·3 cm. long.

UGANDA. ·Acholi District: Aswa R., Gulu–Kitgum road, *Eggeling* 775 in *F.H.* 1161 !
DISTR.   **U**1; probably also in the Belgian Congo
HAB.   Uncertain

NOTE.   A specimen at Kew from Kenya, Machakos, riverside on calcareous red-brown soil in *Acacia-Commiphora* grassland, 11 Mar. 1953, *Trump* 39 !, has a single fragment of an ancient pod suggesting those of var. *sublaevis*.   More material is required from here.
   The plant described as *A. kirkii* by Gilbert & Boutique in F.C.B. 3: 163 (1952) is probably var. *sublaevis*.

subsp. **mildbraedii** (*Harms*) *Brenan* in K.B. 1957: 364 (1958).   Types: Ruanda-Urundi, between Issenji [Kisenyi] and Mpororo, *Mildbraed* 343; Belgian Congo, Kwenda, *Mildbraed* 1887; Tanganyika, Bukoba District, *Holtz* 1712 (all B, syn. †)

Pods (Fig. 15/35B, p. 66) prominently veined, (0·8–)1–1·5 cm. wide; joints of the pod entirely lacking any wart-like central projections; stipe of pod about 0·5–1 cm. long.

UGANDA. Kigezi District: Kebisoni, Sept. 1949, *Dale* U697 ! & Rubabu, Jan. 1951, *Purseglove* 3524 ! ; Masaka/Mengo Districts: Katonga R. on Masaka–Kampala road, 2 Dec. 1951, *Trapnell* 2184 !
TANGANYIKA.   Bukoba District: between Itara and Kakindu, by the Kagera R., July 1906, *Holtz* 1712
DISTR.   **U**2, 4 ; **T**1 ; Belgian Congo and Ruanda-Urundi
HAB.   Riverine and swamp-forest; 1140–1370 m.

SYN.   *A. mildbraedii* Harms in Z.A.E.: 234 (1910); L.T.A.: 850 (1930); I.T.U.: 114, fig. 36e (1940) & ed. 2 : 211, fig. 48e, photo 37 (1952) ; T.T.C.L. : 333 (1949) ; Gilb. & Bout. in F.C.B. 3 : 163, t. 12 (1952)

35 × 32. **A. kirkii** *Oliv.* × **seyal** *Del.*

Flat-crowned tree resembling *A. kirkii*, with green papery peeling bark. Young branchlets thinly pubescent and glandular.   Leaf-rhachis slightly pubescent above ; pinnae 4–9 pairs.   Flowers (from dried specimen) apparently with slightly reddish corolla and yellow stamens.   Peduncles sparsely pubescent (except for glands).   Pods apparently indehiscent, oblong-linear, ± falcate, irregularly constricted here and there, or with margins almost straight, gradually attenuate at base and apex, veined as in *A. kirkii*, scabrid with very numerous reddish sessile glands, otherwise glabrous, 0·7–0·9 cm. wide ; segments of pod with here and there small inconspicuous projections in centre ; stipe of pod 3–5 mm. long.

Nearer to *A. kirkii*, differing in the narrower, falcate, very glandular, less moniliform pods attenuate at ends, in the shorter stipe to the pod, and probably in the yellow anthers. Differs from *A. seyal* in the bark and the scabrid very glandular pods with shorter segments and with closer more transverse veins.

KENYA. Machakos District: near Kitandi, 5 Mar. 1952, *Trapnell* 2234!
DISTR. **K**4; not known elsewhere
HAB. On grey clay by streamside in hill country; 1520 m.

NOTE. I am indebted to the collector for suggesting the parentage of this remarkable and convincing hybrid. According to Mr. Trapnell, both putative parents occur in the area. Which of the varieties and subspecies of *A. kirkii* was involved in the cross is uncertain.
    Rammell 560 *in. C.M.* 9334 (EA!), Kenya, Masai District, Mara R., Feb. 1932, may possibly be a product of a similar cross. It is described as being a tree 5 m. high with golden flowers. In most ways it is near *A. seyal*, but the calyces are only about 1–1·5 mm. long, and the foliage is suggestive of *A. kirkii*. The material is not good enough for certainty.

36. **A. xanthophloea** *Benth.* in Trans. Linn. Soc. 30 : 511 (1875) ; L.T.A. : 851 (1930) ; T.S.K. : 70 (1936) ; T.T.C.L. : 334 (1949) ; Bogdan in Nature in E. Afr., ser. 2, No. 1 : 12 (1949). Types : Nyasaland, E. end of Lake Shirwa [Chilwa], *Meller* (K, syn.!) & Portuguese East Africa, Sena, *Kirk* (K, syn.!)

Tree 10–25 m. high ; bark on trunk lemon-coloured to greenish-yellow. Young branchlets brown to plum-coloured, almost glabrous and with some sessile reddish glands ; twigs showing conspicuous pale yellow powdery bark. Stipules spinescent, straight or almost so, varying in length, up to 7 (–8·5) cm. long ; " ant-galls " and other prickles absent. Leaves : rhachis (2·5–)3–7 cm. long, glabrous to sparingly pubescent ; pinnae 3–6(–8) pairs (on juvenile shoots sometimes to 10 pairs) per leaf ; leaflets rather numerous, 2·5–6·5 mm. long, 0·75–1·75 mm. wide ; lateral nerves invisible beneath. Flowers varying from white or purplish to yellow or golden (see note below). Peduncles sparingly (rarely rather densely) pubescent to subglabrous, and glandular below and sometimes also above the involucel, usually (at least) on abbreviated lateral shoots whose axes do not elongate and are represented by clustered scales, the peduncles thus appearing to be in lateral fascicles on the older often yellow-barked twigs whose leaves have fallen ; involucel conspicuous, 3–3·5 mm. long, near base of to about half-way up peduncle. Calyx 1–1·5 mm. long. Pods (Fig. 16/36, p. 67) indehiscent, linear-oblong, straight or slightly curved, ± moniliform with segments mostly longer than wide, often breaking up, pale brown or brown, reticulate-venose, eglandular or sparingly glandular, (3–) 4–13·5 cm. long, 0·7–1·4 cm. wide. Seeds olive to blackish-olive, smooth or nearly so, subcircular to elliptic, compressed, 4·5–5·5 mm. long, 3·5–4 mm. wide ; areole 3–3·5 mm. long, 2 mm. wide.

KENYA. Northern Frontier Province: Malal [? Maralal] to Kisima, 4 Nov. 1932, *D. C. Edwards* 1865!; Naivasha District: Lake Naivasha, Oct. 1930, *Napier* 458 *in C.M.* 1413! & Sept. 1933, *van Someren* 2722 *in C.M.* 5143!; Kiambu District: foot of Limuru Escarpment, 14 Jan. 1951, *Bogdan* 2880!
TANGANYIKA. Masai District: Ngorongoro Crater bottom, 19 Sept. 1943, *Lindeman* 846!; Kondoa District: Bubu R., 9 Nov. 1927, *B. D. Burtt* 698!; Mbeya (or ? Chunya) District: Songwe Valley to Lake Rukwa, 17 Sept. 1936, *B. D. Burtt* 6040!
DISTR. **K**1, 3, 4, 6, 7; **T**2–5, 7; Portuguese East Africa, Nyasaland, Southern Rhodesia, Transvaal and Zululand
HAB. Ground-water and riverine forest; 600–1980 m.

SYN. *A. songwensis* Harms in E.J. 30: 317 (1901). Type: Tanganyika, Mbeya/Chunya District, Songwe Valley, *Goetze* 1054 (B, holo. †, BM, drawing !)

NOTE. *A. xanthophloea*, the famous Fever Tree with its pallid yellow bark, is apparently unique among the East African acacias in having flowers either white to pinkish or purplish, or else yellow to golden. White to purplish flowers are noted for Kenya and Tanganyika (see, e.g., Bogdan, A List of Kenya Acacias with Keys for Identification: 1, 4 (1956) and T.T.C.L.: 334 (1949)), but nowhere, so far, south of these territories; while to the south of Tanganyika yellow to golden flowers are noted by all collectors who mention flower-colour for *A. xanthophloea*. The only evidence for the yellow-flowered variant in our area is that the flowers of the type of the synonymous *A. songwensis*, collected perhaps in the same locality as *Burtt* 6040 (see above), are described as yellow; also that *Semsei* 2097, from Tanganyika, Pare District, near Same, is described as having bright yellow flowers. The geographical dividing line between the colour-variants is thus presumably in Tanganyika, but beyond that it is quite uncertain, presenting an interesting problem for observers in the field. I have so far failed to find any differences except colour between the two although closer observation may yet reveal them.

White-flowered *A. xanthophloea* is closely akin to *A. kirkii*, especially subsp. *mildbraedii*, while the yellow-flowered form resembles *A. seyal* var. *seyal*. The differences are given in the keys. An apparent hybrid between *A. seyal* var. *fistula* and *A. xanthophloea* has been collected in Nyasaland (*Greenway* 6349) and crosses between the same two species should be looked for in East Africa.

37. A. nilotica (*L.*) [*Willd. ex*] *Del.*, Fl. Aegypt. Ill. : 79 (1813) ; A. F. Hill in Bot. Mus. Leafl. Harvard Univ. 8 : 97 (1940). Type : Egypt, *Herb. Linnaeus* 1228. 28 (LINN, syn.!)

An exceedingly variable species. Tree (1·2–) 2·5–14 m. high ; bark on trunk rough, fissured, blackish, grey or brown, neither powdery nor peeling. Young branchlets from almost glabrous to subtomentose ; glands inconspicuous or absent ; bark of twigs not flaking off, grey to brown. Stipules spinescent, up to 8 cm. long, straight or almost so, often ± deflexed ; " ant-galls " and other prickles absent. Leaves often with 1(–2) petiolar glands and others between all or only the topmost of the 2–11 pairs of pinnae ; leaflets 7–25 pairs, 1·5–7 mm. long, 0·5–1·5 mm. wide, glabrous to pubescent ; lateral nerves invisible beneath. Flowers bright yellow, in axillary pedunculate heads 6–15 mm. in diameter ; involucel from near base to about half-way up peduncle. Calyx 1–2 mm. long, subglabrous to pubescent. Corolla 2·5–3·5 mm. long, glabrous to ± pubescent outside. Pods especially variable, indehiscent, straight or curved, glabrous to grey-velvety, ± turgid, (4–) 8–17(–22) cm. long, 1·3–2·2 cm. wide. Seeds deep blackish-brown, smooth, subcircular, compressed, 7–9 mm. long, 6–7 mm. wide ; areole 6–7 mm. long, 4·5–5 mm. wide.

KEY TO INTRASPECIFIC VARIANTS

Pods necklace-like, narrowly and regularly constricted
    between seeds, white-tomentellous   .   .   .   subsp. **indica**
Pods not necklace-like, their margins straight or
    crenate, or if narrowly constricted, then only irre-
    gularly so here and there :
Branchlets ± densely pubescent to subtomentose; pods
    (at first at any rate) pubescent to subtomentose :
Pods persistently subtomentose all over, often
    rather wide, (1·3–2·2 cm.), their margins
    straight or slightly crenate   .   .   .   .   subsp. **subalata**
Pods glabrescent and later ± shining on the raised
    part over each seed, mostly rather narrow, 1–
    1·7 cm. wide, their margins slightly crenate   subsp. **kraussiana**
Branchlets glabrous or nearly so, sometimes
    puberulous ; pods glabrous or subglabrous
    even when young, rather narrow, 1–1·3 cm.
    wide   .   .   .   .   .   .   subsp. **leiocarpa**

subsp. **indica** (*Benth.*) *Brenan* in K.B. 1957: 84 (1957). Types: India, " East India ". *Roxburgh* & Oungein, without collector's name (? *Jacquemont*) in *Herb. Bentham* (both K, syn.!)

Young branchlets thinly pubescent, sometimes almost glabrous. Pods (Fig. 16/37A, p. 67) necklace-like, narrowly and regularly constricted between the ± round seed-containing segments, whitish- or grey-tomentellous.

TANGANYIKA. Mwanza, solitary trees on lake-shore, July 1932, *Rounce* 193 !
DISTR. T1; native of India only
HAB. Said to occur planted in an avenue and as solitary trees on the shore of Lake Victoria; 1130 m. Also occurs at Malampaka in Maswa District (*Willan* 100 !) and in cultivation in Zanzibar

SYN.   *A. arabica* (Lam.) Willd. var. *indica* Benth. in Hook., Lond. Journ. Bot. 1: 500 (1842)
        A. nilotica (L.) Del. var. *indica* (Benth.) A.F. Hill in Bot. Mus. Leafl. Harvard Univ. 8: 99 (1940)
        [*A. arabica* sensu T.T.C.L.: 333 (1949), *non* (Lam.) Willd. sensu stricto]

NOTE. In P.O.A. C: 194 (1895) *A. arabica* is said, with doubt, to occur in Zanzibar. The pod illustrating this species on t. 20 fig. F resembles that of *A. nilotica* var. *indica*. The trees described as *A. nilotica* in U.O.P.Z.: 102 (1949) and said to occur in Zanzibar and Pemba may be the same, but the true *A. nilotica* var. *nilotica*, resembling var. *indica* in its branchlets and pods, except that the latter are glabrous not tomentellous, is cultivated in Zanzibar at Mgambani, *Greenway* H 11/29 !

subsp. **subalata** (*Vatke*) *Brenan* in K.B. 1957: 85 (1957). Type: Kenya, Teita District, Ndi, *Hildebrandt* 2589 (B ?, holo. †)

Young branchlets densely pubescent to subtomentose. Pods (Fig. 16/37B p. 67) not necklace-like, oblong, densely and persistently subtomentose all over, often rather wide (1·3-2·2 cm.), their margins straight or slightly crenate.

UGANDA. Acholi District: Agoro, Chua, 12 Nov. 1945, *A. S. Thomas* 4359 !; Karamoja District: Moroto, 27 Oct. 1939, *A. S. Thomas* 3074 !; Mbale District: Greek River Camp, Bugishu, Jan. 1936, *Eggeling* 2502 !
KENYA. Rift Valley District: Kamasia, 15 July 1945, *Bally* 4541 !; Machakos District: below Mua Hills, 22 Jan. 1952, *Trapnell* 2201 !; Masai District: Bakitabuk near Narok, 29 Nov. 1948, *Vesey-FitzGerald* 172 !
TANGANYIKA. Shinyanga District: Ngunga, 6 July 1949, *Doggett* 115 !; Mbulu District: Mbulumbul, 16 July 1943, *Greenway* 6808 !; Nzega, 14 June 1954, *F. G. Smith* 1180 !; Mpwapwa, 28 Aug. 1930, *Greenway* 2479 !
DISTR. U1, 3; K1-4, 6; T1-5, 7*; doubtfully also in the Sudan
HAB. Wooded grassland, deciduous bushland, dry scrub with trees, and possibly in semi-desert grassland; 15-1830 m.

SYN.   *A. subalata* Vatke in Oesterr. Bot. Zeitschr. 30: 276 (1880); P.O.A. C: 194, t. 21, fig. D (1895); Harms in N.B.G.B. 4: 203, fig. 4 (1906); L.T.A.: 850 (1930); I.T.U.: 117, fig. 36/1 (1940) & ed. 2, 215, fig. 48/1 (1952); T.T.C.L.: 334 (1949), pro parte; Bogdan in Nature in E. Afr., ser. 2, No. 1: 13 (1949)
        A. taitensis Vatke in Oesterr. Bot. Zeitschr. 30: 278 (1880). Type: Kenya, Teita District, Ndi, *Hildebrandt* 2591 (? B, holo. †). A doubtful synonym.
        [*A. arabica* sensu L.T.A.: 849 (1930); T.S.K.: 69 (1936), saltem pro parte, *non* (Lam.) Willd.]
        [*A. benthami* sensu T.S.K.: 69 (1936), saltem pro parte, *non* Rochebr. sensu stricto]
        "*A. sp. nr. abyssinica* " T.S.K.: 70 (1936), probably; the specimen, *Graham* V322 in *F.H.* 1822 is certainly *A. nilotica*, but lacks pods.

NOTE. A common and not very variable plant. The two most noteworthy extremes are the following: *Bally* 4633 ! from Kenya, Baringo District, Lake Hannington, Kamasia, with neatly falcate pods and shorter indumentum than usual; and *Gillett* 12570 ! & 13047 ! both from Kenya, Northern Frontier Province, Dandu, with unusually small flower-heads 6-8 mm. in diameter and a rather crenate-margined pod only 1·2—1·3 cm. wide. *B. D. Burtt* 1793 ! from Tanganyika, Dodoma District, Kilimatinde, has similar pods up to 1·4 cm. wide. Gillett's specimens carry the interesting comment that, unlike most acacias, this species keeps sporadically in flower for a long time.

subsp. **kraussiana** (*Benth.*) *Brenan* in K.B. 1957: 84 (1957). Type: Natal, *Krauss* 69 (K, holo. !)

Young branchlets ± densely pubescent. Pods not necklace-like, oblong, ± pubescent all over at first, with the raised parts over the seeds becoming glabrescent shining and black when dry, 1·0-1·7 cm. wide, their margins ± shallowly crenate.

*The districts given are those from which specimens with adequately mature pods have been seen. Specimens without pods, quite possibly subsp. *subalata*, have also been collected in K7 and T6 and 8.

TANGANYIKA. Ufipa District : Zimba, 2 Oct. 1949, *Silungwe in Bullock* ! ; District uncertain, Rukwa Rift, 2 Nov. 1933, *Michelmore* 715 ! ; Songea District : about 2·5 km. S. of Gumbiro, 29 June 1956, *Milne-Redhead & Taylor* 10922 !

DISTR. T4, 6, ?7, 8; Portuguese East Africa, Nyasaland, Northern and Southern Rhodesia, Angola, the Transvaal and Natal; a specimen from Ethiopia (*Mooney* 5576 !) is apparently referable to subsp. *kraussiana*

HAB. Uncertain, but apparently similar to that of var. *subalata*

SYN. *A. arabica* (Lam.) Willd. var. *kraussiana* Benth. in Hook., Lond. Journ. Bot. 1: 500 (1842). *non A. kraussiana* [Meisn. ex] Benth.
     A. benthami Rochebr., Toxicol. Afr. 2: 192 ()1898, *non* Meisn. Type as *A. nilotica* subsp. *kraussiana* (Benth.) Brenan
     A. nilotica L. var. *kraussiana* (Benth.) A. F. Hill in Bot. Mus. Leafl. Harvard Univ. 8: 98 (1940)
     [*A. subalata* sensu T.T.C.L.: 334 (1949), pro parte, *non* Vatke sensu stricto]

subsp. **leiocarpa** *Brenan* in K.B. 1957: 84 (1957). Type: Kenya, Lamu District, *Dale* 3832 *in C.M.* 13988 (K, holo. !, EA, iso. !)

Young branchlets glabrous or nearly so except for inconspicuous glands, or with sparse puberulence (sometimes the puberulence may be denser). Pods not necklace-like, oblong, glabrous or almost so even when young (rarely and perhaps not in our area slightly puberulous), 1–1·3 cm. wide, margins straight to slightly crenate.

KENYA. Kilifi District: Malindi, 18 Aug. 1949, *Bogdan* 2612 !; Lamu District: Patta [Patte] Is., Oct. 1937, *Dale* 3832 ! & Kiunga, 23 Oct. 1947, *Mrs. J. Adamson* 437 *in Bally* 6136 !

TANGANYIKA. Morogoro District: Kingolwira, July 1935, *B. D. Burtt* 5169 !; Kilwa District: Kilwa Kivinje, 4 Dec. 1955, *Milne-Redhead & Taylor* 7545 !

DISTR. K7; T6; Somaliland Protectorate, Somalia, and doubtfully Swaziland

HAB. Uncertain, in Tanganyika in *Acacia* grassland near the coast, also in woodland

NOTE. The specimens showing somewhat greater puberulence have so far been found mainly in Somaliland Protectorate and Somalia, but one specimen with puberulous young branchlets has been collected on Patta Is., Lamu District, *Greenway & Rawlins* 8878 !

SYN. (of *A. nilotica*).
     Mimosa nilotica L., Sp. Pl.: 521 (1753)
     Mimosa scorpioïdes L., Sp. Pl.: 521 (1753). Type uncertain.
     Mimosa arabica Lam., Encycl. 1: 19 (1783). Types: Arabia and Africa, *Sonnerat* (P–LA, syn.)
     Acacia arabica (Lam.) Willd., Sp. Pl. 4: 1085 (1806)
     A. scorpioïdes (L.) W. F. Wight in Contr. U.S. Nat. Herb. 9: 173, *in adnot.* (1905); A. Chev. in Bull. Soc. Bot. Fr. 74: 954 (1927) & in Rev. Bot. Appliq. 8: 199 (1928)

DISTR. (of species as whole). Widespread in tropical and subtropical Africa and Asia, as far eastwards as India

38. **A. farnesiana** (*L.*) *Willd.*, Sp. Pl. 4 : 1083 (1806) ; L.T.A. : 835 (1930) ; T.T.C.L. : 334 (1949) ; U.O.P.Z. : 102 (1949) ; Gilb. & Bout. in F.C.B. 3 : 164 (1952) ; Consp. Fl. Angol. 2 : 278 (1956) ; F.W.T.A., ed. 2, 1 : 499 (1958). Type : uncertain, based primarily on Linnaeus, Hort. Ups. : 146 (1748).

Shrub 1·5–4 m. high. Young branchlets glabrous or nearly so, purplish to grey ; epidermis not obviously peeling off ; glands (as on peduncles) few and inconspicuous. Stipules spinescent, usually short, up to 1·8(–3) cm. long, never inflated ; other prickles absent. Leaves with a small gland on petiole and sometimes one on the rhachis near the top pair of pinnae ; pinnae 2–7 pairs ; leaflets 10–21 pairs, 2–7 mm. long, 0·75–1·75 mm. wide, very rarely larger, with both midrib and lateral nerves visible and somewhat raised beneath. Flowers bright golden-yellow, sweetly scented, in axillary pedunculate heads ; involucel at apex of peduncle. Calyx and corolla glabrous outside except for extreme tips of lobes. Pods (Fig. 16/38, p. 67) indehiscent, straight or curved, subterete and turgid, dark brown to blackish, glabrous, finely longitudinally striate, 4–7·5 cm. long, 0·9–1·5(–2) cm. in diameter. Seeds chestnut-brown, smooth, elliptic, thick, only slightly compressed, 7–8 mm. long, 5·5 mm. wide ; areole 6·5–7 mm. long, 4 mm. wide.

UGANDA. Mbale District: Tororo, Mar. 1935, *Cree* 21 !

DISTR. **U**3; probably native of tropical America, doubtfully so in Africa (not in our area) and Australia; widely introduced in the tropics and often becoming wild. Cultivated elsewhere in our area, e.g. in **U**4 and **T**3

HAB. Uncertain, probably only planted or as an escape from cultivation

SYN. *Mimosa farnesiana* L., Sp. Pl.: 521 (1753)

NOTE. Grown for ornament and for its fragrant flowers which are used to make perfume. The pods of *A. farnesiana* are most distinctive and make the species easy to recognize. If they are absent, then it may be helpful to recall that no African acacia but this has the following combination of features:— absence of " ant-galls ", leaflets with the lateral nerves raised and somewhat prominent beneath, and bright yellow flowers in non-paniculate heads.

A further outstanding but less easily seen difference is that the anthers of *A. farnesiana* lack, even in bud, the small often caducous apical gland which is present in all the other capitate-flowered acacias native of our area.

39. **A. abyssinica** [*Hochst. ex*] *Benth.* in Hook., Lond. Journ. Bot. 5 : 97 (1846) ; Oliv., F.T.A. 2 : 347 (1871) ; L.T.A. : 839 (1930) ; Pichi-Serm., Miss. Stud. Lago Tana, Ric. Bot. 1 : 52 (1951) ; Brenan in K.B. 1957 : 81 (1957). Type : Ethiopia, near Mendel, *Schimper*, Sect. 3, 1813 (K, holo. !)

Flat-crowned tree 6–15(–20) m. high ; bark rough and fissured, brown to nearly black ; epidermis not peeling on the twigs ; bark on young trees papery. Indumentum of branchlets variable, pubescent to shortly villous, grey or somewhat yellowish. Stipules spinescent, other prickles absent ; spines variable, absent, short or up to 3·5 cm. long, straight, ashen when elongate, never inflated. Leaves : petiole 2–5 mm. long ; pinnae of well-developed leaves of mature shoots 15–36 pairs (reduced leaves with fewer pairs usually also present) ; leaflets up to 4 mm. long and 0·75 mm. wide. Flowers in heads ; stamens white ; calyx and corolla red (? always). Corolla glabrous or inconspicuously puberulous on lobes outside. Pods (Fig. 16/39, p. 67) subcoriaceous, straight or slightly curved, grey or brown, longitudinally veined, ± glandular and sometimes puberulous, narrowed at base and sometimes at top, 5–12 cm. long, 1·2–2·1(–2·8) cm. wide. Seeds oblique in the pod, olive-brown, smooth, elliptic, compressed, 7–9 mm. long, 4–5 mm. wide ; areole 6–7 mm. long, 2·5–3·5 mm. wide.

subsp. **calophylla** *Brenan* in K.B. 1957: 82 (1957). Type: Kenya, S. Kavirondo District, Mugunga, *Greenway* 7860 (K, holo.!, EA, iso. !)

Pinnae very closely set, 0·4–1·5 cm. long; leaflets extremely small, up to 2·5 mm. long, 0·25–0·4(–0·5) mm. wide.

UGANDA. Karamoja District: Napak, Feb. 1938, *Sangster* 413 !; Kigezi District: Ruzumbura, Ruhinda, May 1946, *Purseglove* 2049 !; Elgon, Kaburon, *Eggeling* 2475 ! & 2488 !

KENYA. Laikipia District: Rumuruti–Thomson's Falls road, *Wimbush* 3 *in F.H.* 2592 !; Naivasha, 29 Sept. 1953, *Greenway & Pearsall* 8792 !; Kiambu District: Muguga, 25 Sept. 1951, *Trapnell* 2122 !

TANGANYIKA. Masai District: S. side of Ngorongoro at Kamsiya Nyoka, 24 Sept. 1932, *B. D. Burtt* 4285 ! & Ngorongoro, 12 Dec. 1956, *Greenway* 9161 !; Arusha District: Olmotonyi, Meru Forest Reserve, Jan. 1955, *Nuru Salimu* 1 ! & *Athanes Beseko* 1 !; Songea District: Matengo Hills, Miyau, 13 Jan. 1956, *Milne-Redhead & Taylor* 8321 !

DISTR. **U**1–3; **K**3–5; **T**2, 8; Belgian Congo, the Sudan, Portuguese East Africa, Nyasaland and Southern Rhodesia

HAB. Woodland and wooded grassland; 1500–2300 m.

SYN. [*A. abyssinica* sensu T.S.K.: 70 (1936); Bogdan in Nature in E. Afr., ser. 2, No. 1: 13 (1949), *non* [Hochst. ex] Benth. sensu stricto]
　　[*A. xiphocarpa* sensu I.T.U.: 117 (1940) & ed. 2: 215, fig. 48m, photo 39 (1952) *non* [Hochst. ex] Benth. sensu stricto]
　　[*A. rehmanniana* sensu T.T.C.L.: 336 (1949), pro parte, quoad *Burtt* 4285, *non* Schinz]

NOTE. Three gatherings from Meru Forest Reserve, Tanganyika, are described as yellow-flowered. This may mean that the flower colour varies (as in *A. xanthophloea*) or, more probably, that a creamy-white was noted as yellow. More observation of this matter is required.

Two specimens from central and southern Tanganyika are close to subsp. *calophylla* but have longer pinnae than usual (about 1·5–2·2 cm.), due perhaps merely to being unusually robust or juvenile, or to the presence of another species such as *A. rehmanniana* Schinz, which has not so far been certainly found in our area, although *Greenway* 5798 was placed under this name in T.T.C.L.: 336 (1949), probably wrongly; they are not *A. abyssinica* subsp. *abyssinica*. These specimens are:—

Mpwapwa District: Kiboriani Mts., 3 Oct. 1938, *Greenway* 5798 !; Iringa District: Sao highlands, 58 km. N. of Ikwera, 17 June 1938, *Pole Evans & Erens* 788 !

More material from both localities is wanted.

40. **A. pilispina** *Pichi-Serm.*, Miss. Stud. Lago Tana, Ric. Bot. 1 : 205, t. 43 (1951). Type : Ethiopia, Atghebà Ghiorghis, *Pichi-Sermolli* 2696 (FI, holo.)

Shrub or tree 1–15 m. high (in our area usually a tree 5 m. or more high) ; crown flat or spreading ; bark on trunk grey to brown and rugose or ± smooth. Young branchlets densely clothed with long grey to slightly yellowish spreading hairs mostly 0·5–2 mm. long, red-brown beneath the hair ; epidermis falling away to expose a yellow or sometimes greenish powdery bark on the twigs. Stipules spinescent, mostly short, straight or nearly so, up to 7 mm. long, hairy except towards tips, sometimes (? on young shoots) longer, grey, to 4·5 cm. ; " ant-galls " and other prickles absent. Leaves : rhachis mostly 2·5–6 cm. long, hairy ; pinnae mostly 8–16(–26) pairs, 1–2·2 cm. long ; leaflets 14–28 pairs, 1·5–3·75(–4) mm. long, 0·5–1 mm. wide, ciliate. Flowers cream or tinged red outside, in heads on axillary peduncles (0·5–)1·5–2·5 cm. long and ± hairy but eglandular, whose involucel is basal or in the lower fifth. Corolla glabrous outside. Pods (Fig. 16/40, p. 67) straight or slightly curved, flattened, papery or subcoriaceous, grey to grey-brown or purple-brown, dehiscent, narrowed at base and sometimes at top, 7–12·5 cm. long, 1·4–2·9 cm. wide, finely and ± longitudinally veined, glabrous or nearly so.

TANGANYIKA. Bukoba District: Bugera, Jan. 1949, *Watkins* 289 !; Buha District: Mpembe R., 14 Aug. 1950, *Bullock* 3166 !; Ufipa District: Ufipa Plateau, Kizombwe, 13 Dec. 1934, *Michelmore* 1055 !

DISTR. T1, 4; Belgian Congo (Katanga), Ethiopia, Northern Rhodesia and Nyasaland

HAB. Wooded grassland and deciduous woodland; characteristically on stream margins and edges of river flood-plains; recorded also on termite-mounds in *Brachystegia-Julbernardia paniculata* woodland very near our border in Northern Rhodesia; 1220–1830 m.

NOTE. Pichi-Sermolli described *A. pilispina* as a shrub 1–3 m. high, with up to 10 pairs of pinnae and 8–16 pairs of leaflets per pinna. These differences seem very possibly due to the higher altitude at which the species was growing in Ethiopia, but this theory needs checking.

A. pilispina, which has a distribution along the western side of Tanganyika, should be carefully looked for in Uganda, where it probably occurs.

The pods are somewhat variable, but in Tanganyika are usually rather broad, slightly curved or oblique, and grey when old.

41. **A. elatior** *Brenan* in K.B. 1957 : 94 (1957). Type : Kenya, Tana R., Garissa, *Greenway* 8857 (K, holo. !)

Large tree up to 18–25 m. high ; crown rounded, with pendulous branchlets ; bark brown to almost black, deeply longitudinally fissured. Young branchlets glabrous to pubescent, grey-brown. Stipules spinescent, straight or nearly so, some short, to about 7 mm. long, others long, whitish, to about 9 cm. long, sometimes modified to inflated fusiform " ant-galls." Leaves : rhachis 1–6 cm. long ; pinnae 5–13 pairs, some leaves normally with 8 or more pairs ; leaflets (7–)13–25 pairs per pinna, 1·25–4 mm. long, 0·5–1·25 mm. wide, glabrous or ciliate. Flowers greenish-white to white or very pale yellow, in heads on peduncles 2–5 cm. long, whose involucel is about $\frac{1}{3}$–$\frac{1}{2}$-way above base. Pods straight, narrowly oblong, dehiscent, 3–12 cm. long,

1·2–1·8 cm. wide, shortly attenuate at base, acuminate or rounded at apex, brown or purplish-brown, finely and mostly obliquely veined, glabrous or ± pubescent near base. Seeds oblique in the pod, olive-brown, smooth, subcircular, compressed, 6–7 mm. diam. ; areole 3–5 mm. long, 3·5–4 mm. wide.

subsp. **elatior**; Brenan in K.B. 1957: 95 (1957)

Branchlets, leaves and peduncles glabrous or almost so. Spines sometimes inflated, fusiform, to about 1·5 cm. wide. Pods quite glabrous.

KENYA. Northern Frontier Province: Uaso Nyiro, 15 May 1945, *Mrs. J. Adamson* 75 *in Bally* 4437 !; Tana River District: Tana R., Garissa, 4 Feb. 1956, *Greenway* 8857 ! & 26 Sept. 1957, *Greenway* 9235 !; Kitui District: Nyali, Thua R., 1953, *L. C. Edwards* E258 !
DISTR. **K**1, 4, 7; not known elsewhere

subsp. **turkanae** *Brenan* in K.B. 1957: 95 (1957). Type: Kenya, Lodwar, *Hemming* 250 (K, holo.!, EA, iso.!)

Branchlets, spines, rhachides of the leaves and peduncles ± densely pubescent with spreading hairs. Inflated spines not yet seen. Pods (Fig. 16/41, p. 67) pubescent near base and on the stipe.

UGANDA. Karamoja District: Moroto R., at base of escarpment, Feb. 1936, *Eggeling* 2970 !
KENYA. NW. Turkana: R. Gapuss, Apr. 1944, *Dale* K376 ! & Oropoi, Apr. 1944, *Dale* K377 ! & Lodwar, 31 Mar. 1954, *Hemming* 250 !; Baringo District: Lake Baringo, 16 Jan. 1957, *Bogdan* 4364 !
DISTR. **U**1; **K**2, 3; also in the Sudan

SYN. [*A. etbaica* sensu I.T.U., ed. 2: 209 (1952), pro parte, quoad *Eggeling* 2970 saltem, non Schweinf.]

HAB. (of species as whole). Locally common on sandy river-banks, often associated with *Acacia tortilis;* about 180–1070 m.

NOTE. This is closely related to *A. etbaica,* but differs from all the races of that species in being a large riparian tree, and by the occasional (at least) production of " ant-galls ". *A. elatior* is more easily distinguished from the individual subspecies of *A. etbaica:* from subsp. *uncinata* by the pubescent not puberulous indumentum (when present), straight spines and broader pods; from subsp. *australis* again by the indumentum and the straight spines of varying length; and from subsp. *platycarpa* by the spines all—short and long—being straight or nearly so, and also by the more numerous pinnae (in subsp. *platycarpa* there are up to about 5–6 pairs per leaf).
  More information is required about the flowers and the production and occurrence of " ant-galls " in *A. elatior;* in addition the characters of subsp. *turkanae* should be rigorously tested in the field.

42. **A. etbaica** *Schweinf.* in Linnaea 35 : 330 (1867–8) ; L.T.A. : 840 (1930) ; Bogdan in Nature in E. Afr., ser. 2, No. 1 : 14 (1949) ; I.T.U., ed. 2 : 209, fig. 48c (1952) ; Brenan in K.B. 1957 : 90 (1957). Type : the Sudan, Soturba Mts., *Schweinfurth* 1994, 1995 (K, isosyn.!) *

Normally a tree 2·4–12 m. high with a well-defined trunk and a flat or rounded crown ; bark rough, brown to almost black. Young branchlets glabrous to puberulous ; older ones glabrous or glabrescent, grey to brown. Stipules spinescent, all short and hooked or straight, up to about 7 mm. long, or with long straight spines up to 6 cm. long intermixed ; " ant-galls" and other prickles absent. Leaves : rhachis variable, 0·4–5 cm. long ; pinnae 1–9 pairs, some leaves usually with 4 or more pairs ; leaflets 4–35 pairs per pinna, 0·5–4 mm. long, 0·25–1·25 mm. wide, glabrous or ± puberulous. Flowers white or cream, in small axillary heads on peduncles 0·7–2·5 cm. long whose involucel is $\frac{1}{3}$–$\frac{2}{3}$-way up, or sometimes near base. Pods

* Schweinfurth (*l.c.*) did not cite actual specimens but said that *A. etbaica* occurred in the mountains of Elba and Soturba in the Sudan. The true syntypes were no doubt destroyed at Berlin during the war, but it seems probable that the two Kew specimens cited are isosyntypes.

straight, linear-oblong to oblong, dehiscent, 2–12 cm. long, (0·6–)0·7–2·2 cm. wide, attenuate at base, acuminate to rounded at apex, grey to brown or deep purplish, marked with fine mostly oblique or longitudinal veins and glabrous or ± puberulous especially near base. Seeds transverse or oblique in the pod, brown, smooth, subcircular, compressed, 8 mm. diam. ; areole 4–4·5 mm. long, 4·5–5 mm. wide.

KEY TO INTRASPECIFIC VARIANTS

Indumentum puberulous (examine young branchlets, leaf-rhachides and peduncles) :

Pods (6–)7–11(–12) mm. wide ; spines often long and short mixed ; in Uganda and Kenya   .   .   subsp. **uncinata**

Pods 11–15 mm. wide ; spines all short ; in Tanganyika   subsp. **australis**

Indumentum pubescent, of short distinct spreading hairs (examine rhachis of leaves and pinnae) ; pods (10–) 11–15(–22) mm. wide ; spines mostly long and short mixed   .   .   .   .   .   .   subsp. **platycarpa**

subsp. **uncinata** *Brenan* in K.B. 1957: 91 (1957). Type: Somaliland Protectorate, Erigavo, *McKinnon* 8/220 (K, holo.!)

Spines usually ± hooked or curved, sometimes almost straight, mostly short and up to about 7 mm. long, occasionally with long straight ones intermixed; twigs sometimes almost unarmed. Branchlets grey-puberulous. Rhachides of leaves and pinnae puberulous mostly all round. Pods (6–)7–11(–12) mm. wide.

UGANDA. Karamoja District: Moroto, Feb. 1936, *Eggeling* 2976 ! & 5 Mar. 1936, *Michelmore* 1255 ! & Pirre, 10 Nov. 1939, *A. S. Thomas* 3258 !
KENYA. Mombasa, Mariakani, 7 Oct. 1949, *Hornby* 3108 !
DISTR. U1 ; K7 ; Eritrea, Ethiopia and Somaliland Protectorate
HAB. Wooded grassland, " forming dense thickets on overgrazed lands " (Hornby 3108); 270–1524 m.

NOTE. *Greenway* 9485 from Kenya, Tana River District, Garsen, 16 km. S. on the Mallindi road, resembles subsp. *uncinata* but is said to be a virgately branched shrub to 4–5 m. high with several whippy stems arising from ground-level and then forming a much branched flattish crown. Pods are lacking and its identity thus doubtful, but the combination of characters of *A. etbaica* subsp. *uncinata* with a habit recalling *A. reficiens* subsp. *misera* perhaps implies that these two closely related species may ultimately have to be merged into one.

subsp. **australis** *Brenan* in K.B. 1957: 92 (1957). Type: Tanganyika, Ngomeni, *Greenway* 7034 (K, holo.!, EA, iso.!)

Spines all ± curved, short, up to 4 mm. long. Indumentum puberulous. Branchlets and rhachides of leaves and pinnae puberulous (typically) or almost glabrous. Pods 11–15 mm. wide.

TANGANYIKA. Pare District: Kisiwani, 26 June 1942, *Greenway* 6498 !; Tanga District: Ngomeni, 25 Aug. 1944, *Greenway* 7034 !
DISTR. T3; not known elsewhere
HAB. Scattered-tree grassland and perhaps deciduous bushland; 120–610 m.

subsp. **platycarpa** *Brenan* in K.B. 1957: 93 (1957). Type: Kenya, Moyale, *Gillett* 13641 (K, holo.!, EA, iso.!)

Spines mostly of two sorts, some short ± hooked or curved and up to about 4 mm. long, others long straight and up to about 6 cm. long. Indumentum on the leaves pubescent with short but distinct ± spreading hairs. Pods (Fig. 16/42, p. 67) (10–)11–15 (–22) mm. wide.

KENYA. Northern Frontier Province: Moyale, 26 July 1952, *Gillett* 13641 !; Meru District: Isiolo, Mar. 1950, *Dale* K790 !; Masai District: 19 km. N. of Namanga, 30 Jan. 1952, *Trapnell* 2205 !; Kwale District: between Samburu and Mackinnon Road, 2 Sept. 1953, *Drummond & Hemsley* 4114 !
TANGANYIKA. Masai District: Kitumbeine, 6 Jan. 1936, *Greenway* 4265 !; Lushoto District: Mkomazi, 30 Nov. 1935, *B. D. Burtt* 5326 ! & Mashewa, 6 Mar. 1952, *Faulkner* 924 !
DISTR. K1, 2, 4, 6, 7; T2, 3; not known elsewhere

HAB. Deciduous bushland, dry scrub with trees and semi-desert scrub and grassland; 460–1520 m.

SYN. *A. holstii* Taub. in P.O.A. C: 194, t. 21, fig. C (1895), pro parte, quoad legumina tantum; *nom. rejic.*, vide notam infra. Type: see note below.

DISTR. (of species as whole) the Sudan and Eritrea southwards to Tanganyika

NOTE.—Subsp. *platycarpa* is the most distinct of the three subspecies occurring in our area, separable from subsp. *uncinata* by the wider pods and different indumentum, and from subsp. *australis* by the indumentum and usually also by the mixed armature. Although a rather wide range of width is given for the pods of subsp. *platycarpa*, pods of a single gathering are so variable that they will cover much of it: thus *Gillett* 13641, the holotype of the subspecies, has pods ranging from 10–19 mm. wide. *Burtt* 5326 from Tanganyika, Mkomazi, is alone in having the characters of subsp. *platycarpa*, notably the indumentum, but with the pods, which are seemingly quite mature, only 10–11 mm. wide. It is possible that further collecting at Mkomazi may show that the pods of subsp. *platycarpa* have a greater range of width in this locality, but *Burtt* 5326 has at least 10 pods either in their entirety or as valves, and appears to be representative. The occurrence of narrow pods in this gathering and of pods in subsp. *australis* as wide as those in subsp. *platycarpa* must at present be taken as evidence for making *platycarpa* a subspecies and not a separate species.

This is an appropriate place to consider the identity of *A. holstii* Taub., which was based on *Holst* 8744 from Tanganyika, Lushoto District, Mashewa. There is an isotype of this at Kew, which is certainly referable to *A. hockii* De Wild., but which lacks pods. Now pods were described and illustrated by Taubert for *A. holstii*, but were evidently quite different from those of *A. hockii*. In 1952 Mrs. Faulkner collected *A. etbaica* subsp. *platycarpa* at Mashewa, the pods of which agree well with those of *A. holstii*, which I have not matched elsewhere among the East African species. It seems clear that *A. holstii* was a mixture of the vegetative parts and flowers of *A. hockii* with the pods of *A. etbaica* subsp. *platycarpa*, and that the name *A. holstii* Taub. must be rejected.

43. A. reficiens *Wawra* in Sitz. Math. Akad. Wien, 38 : 555 (1860) ; L.T.A. : 841 (1930) ; Consp. Fl. Angol. 2 : 283 (1956), pro parte ; Brenan in K.B. 1957 : 89 (1957). Type : Angola, between Benguela and Catumbela, *Wawra* (W, holo., K, fragm. !)

Bush 1–5(–6) m. high, obconical, branching from base. Young branchlets shortly puberulous to pulverulent, as are the leaf-rhachides and peduncles ; older branchlets glabrescent, grey to brown, often rather slender and almost straight. Stipules spinescent, all short, hooked, 2–6 mm. long, coloured like the twigs ; "ant-galls" and other prickles absent. Leaves : rhachis short, 0·3–1·2 cm. long ; pinnae 1–3 pairs (at least in our area) ; leaflets 5–11 pairs per pinna, 2–4·5 mm. long, 0·5–1·25 mm. wide, glabrous or slightly puberulous. Flowers whitish, in small axillary heads ; involucre on peduncle basal or in lower third. Pods straight, linear-oblong, dehiscent, 3–6·5 cm. long, attenuate at base, acuminate to obtuse at apex, brown or deep purplish, marked with fine mostly longitudinal veins and ± pulverulent to glabrous or nearly so. Seeds longitudinal in the pod, pale olive-brown, smooth, mostly elliptic, compressed, 5–7·5 mm. long, 4–5 mm. wide ; areole 4–4·5 mm. long, 2·5–3 mm. wide.

subsp. **misera** (*Vatke*) *Brenan* in K.B. 1957: 90 (1957). Type: Somaliland Protectorate, Meid, *Hildebrandt* 1394 (B, holo. †, K, iso. !)

Peduncles 0·4–1(–1·3) cm. long. Pods (Fig. 16/43, p. 67) 0·6–0·95 cm. wide.

UGANDA. Karamoja District: Kanamugit, Feb. 1936, *Eggeling* 2921 ! & 28 Oct. 1939, *A. S. Thomas* 3082 !
KENYA. Northern Frontier Province: Buna, Jan. 1949, *Dale* K734 ! & Dandu, 16 May & 29 June 1952, *Gillett* 13209 !; Turkana District: Moroto to Lodwar, 9 Oct. 1952, *Verdcourt* 805 !; Teita District: Voi, Waller's Camp, 2 Feb. 1952, *Trapnell* 2209 !
DISTR. U1; K1, 2, ? 4, 7; also in the Sudan, Somalia and Somaliland Protectorate
HAB. Semi-desert scrub and dry scrub with trees; 80–1220 m.

SYN. *A. misera* Vatke in Oesterr. Bot. Zeitschr. 30: 275 (1880); L.T.A.: 843 (1930); I.T.U., ed. 2: 211, fig. 48n (1952)
" *A. sp. near etbaica* Schweinf."; Bogdan in Nature in E. Afr., ser. 2, No. 1: 14 (1949)

NOTE.—Typical subsp. *reficiens* is so far known only from Angola.

44. **A. tortilis** (*Forsk.*) *Hayne*, Arzneyk. 10, t. 31 (1827) ; I.T.U., ed. 2 : 215, fig. 48j, photo 38 (1952) ; Consp. Fl. Angol. 2 : 284 (1956) ; Brenan in K.B. 1957 : 86 (1957). Type : Arabia, " Mons Soudân prope Hás," *Forskål* (C, holo. !)

Tree 4–21 m. high, occasionally (probably not in our area) a bush 1 m. high ; crown flat or spreading ; * bark grey to black, fissured. Young branchlets glabrous to densely pubescent, going brown to purplish-black. Stipules spinescent, some short ± hooked and up to about 5 mm. long, mixed with other long straight whitish ones to about 8(–10) cm. long ; " ant-galls " and other prickles absent. Leaves : rhachis short, 2 cm. long or less ; pinnae 2–10 pairs, 2–17 mm. long ; leaflets 6–19 pairs per pinna, usually very small, 0·5–2·5(–6) mm. long, ciliate to glabrous. Flowers cream or whitish, in axillary heads 5–10 mm. in diameter on peduncles 0·4–2·4 cm. long. Pods contorted or spirally twisted, longitudinally veined, tomentellous to glabrous. Seeds olive- to red-brown, smooth, elliptic, compressed, 7 mm. long, 4·5–6 mm. wide ; areole 5–6 mm. long, 3–4 mm. wide.

subsp. **raddiana** (*Savi*) *Brenan* in K.B. 1957: 87 (1957). Type: Egypt, *Raddi* (K, iso. !)

Young branchlets and leaves glabrous, subglabrous or pubescent. Pods (6–)7–9 mm. wide, glabrous or appressed-pubescent, eglandular.

var. **raddiana**

Young branchlets and leaves glabrous or subglabrous. Pods glabrous.

KENYA. Lamu District: Fazi Is., 1929, *R. M. Graham* W309 *in F.H.* 1800 ! & Manda Is., Takwa, 4 Nov. 1957, *Greenway* 9444 !
DISTR. (of var.) **K**7; Algeria, Egypt, Senegal, Nigeria
HAB. Bushland

SYN. *A. raddiana* Savi, Alc. Acazie Egiz. : 1 (1830) ; F.W.T.A., ed. 2, 1 : 500 (1958)

NOTE. The status of this in our area—whether native or introduced—is unknown. More specimens and information are wanted. A specimen from Kenya, Lamu, *R. M. Graham* X298 *in F.H.* 1804 (EA !, K !), lacks pods but from its subglabrousness may well be subsp. *raddiana*.

subsp. **spirocarpa** ([*Hochst. ex*] *A. Rich.*) *Brenan* in K.B. 1957: 88 (1957). Types: Ethiopia, *Schimper* 502, 612, 658 (all K, isosyn. !)

Young branchlets ± densely pubescent. Petioles and leaf-rhachides pubescent. Pods (Fig. 16/44, p. 67) 6–9(–13) mm. wide, tomentellous or pubescent with spreading or curved hairs, among which are numerous dark red glands clearly visible through a hand-lens.

UGANDA. Karamoja District: between Greek R. and Loporokocho, *Eggeling* 2512 ! & Moroto, 5 Mar. 1936, *Michelmore* 1253 !
KENYA. Northern Frontier Province: Moyale, 16 July 1952, *Gillett* 13599 !; Machakos District: Kima, *Napier* 59** !; Teita District: Taveta, 1 Feb. 1952, *Trapnell* 2208 !
TANGANYIKA. Musoma, 1933, *Emson* 316** !; Pare District: Kisiwani–Gonja, 4 Feb. 1930, *Greenway* 2125 !; Ufipa District: Milepa, 29 May 1951, *Bullock* 3920 !
DISTR. **U**1; **K**1, 2, 4, 6, 7; **T**1–5, doubtfully in 7; Eritrea and the Sudan southwards to Southern Rhodesia, Portuguese East Africa and Angola
HAB. Deciduous woodland, wooded grassland, deciduous bushland and semi-desert scrub; 600–1500 m.

SYN. *A. spirocarpa* [Hochst. ex] A. Rich., Tent. Fl. Abyss. 1: 239 (1847); Harms in N.B.G.B. 4: 200, fig. 3 (1906); L.T.A.: 842 (1930); T.S.K.: 70 (1936); Bogdan in Nature in E. Afr., ser. 2, No. 1: 14 (1949); T.T.C.L.: 334 (1949)
   A. spirocarpa [Hochst. ex] A. Rich. var. *major* Schweinf. in Linnaea, 35: 323 (1867–8)
   [*A. tortilis* sensu B. D. Burtt in Journ. Ecol. 30: 93 (1942); I.T.U., ed. 2: 215 (1952), *non* (Forsk.) Hayne sensu stricto]

NOTE.—*Ivens* 634 (EA !) from Kenya, Machakos District, Makueni & *B. D. Burtt* 2191 (BM !, K !) from Tanganyika, Kondoa District, Sambala, have glandular but **very**

Smith 1085 (EA !), Tanganyika, Masai District (*A. tortilis* subsp. *spirocarpa*), is described as having a rounded crown.
   ** These specimens lack pods.

sparingly hairy pods, showing thus a close approach to subsp. *heteracantha* (Burch.) Brenan, to which indeed they may be referable. More material is required.

A specimen from Kenya, Turkana District, Lodwar, *Lake Rudolf Expedition* 152 *in C.M.* 13990 (EA!) has the glands on the pods very sparse, approaching thus to the typical subsp. *tortilis*.

Occasionally the pods of subsp. *spirocarpa* are ± densely clothed with whitish spreading hairs about 1–3 mm. long. Plants showing this feature are referable to var. *crinita* Chiov. in Res. Sci. Miss. Stefanini-Paoli 1: 71 (1916). Type: Somalia, between Doriànle and Oneiátta, *Paoli* 907 (FI, holo.!). *Paulo* 325! from Tanganyika, Arusha District, Olbalbal, 20 Apr. 1958; *Trapnell* 2156! from Tanganyika, Mpwapwa/Iringa Districts, Great Ruaha R. on Iringa–Dodoma road, 760 m., 19 Aug. 1951; *Milne-Redhead & Taylor* 11224! from the same place, in Iringa District; *B. D. Burtt* 2192! from Tanganyika, Kondoa District, Sambala, 16 Oct. 1929; and *Willan* 26! from Tanganyika, Dodoma District, 80 km. E. of Dodoma, 26 July 1952, are examples; *Pole Evans & Erens* 1098! from Kenya, Machakos District, Yatta Plains, 2 July 1938, is also referable to the variety, although less marked. The variety is known otherwise only from Somalia.

The typical subspecies *tortilis* occurs in Egypt, the Sudan, Arabia, Aden and perhaps Palestine. It has the following synonyms:—

SYN. (of subsp. *tortilis*). *Mimosa tortilis* Forsk., Fl. Aegypt.-Arab. : 176 (1775)
   Acacia spirocarpa [Hochst. ex] A. Rich. var. *minor* Schweinf. in Linnaea, 35 : 323 (1867–8)

DISTR. (of species as whole). Algeria to Egypt (? Palestine) and Arabia southwards to South Africa

45. A. clavigera *E. Mey.*, Comm. Pl. Afr. Austr. : 168 (1836). Type : South Africa, near Port Natal [Durban], *Drège* (K, iso.!)

Tree 5–25 m. high ; crown flat or spreading ; bark on trunk grey to dark brown, fissured or sometimes smooth. Young branchlets usually (always in our area) glabrous, eglandular, becoming grey to grey-brown, sometimes grey-purplish ; bark of branchlets lenticellate, otherwise rather smooth, neither flaking off nor fissuring to expose red under-bark. Stipules spinescent, straight or very slightly curved, mostly short, up to 7 mm., sometimes longer, to 6(–9) cm. ; " ant-galls " and other prickles absent. Leaves : rhachis (2·5–)3–7 cm. long ; pinnae (2–)3–8(–10) pairs ; leaflets 9–27 pairs (with us usually 13 or more pairs), (2–)3·5–6·5 mm. long, 1–3·5 mm. wide, glabrous or ciliolate on margins, oblong. Flowers white, very sweetly scented, profuse, in heads on axillary, shortly pubescent or puberulous, eglandular or very inconspicuously glandular peduncles, whose involucel is shortly above base or in lower third of peduncle. Corolla glabrous outside. Pods falcate, dehiscent, glabrous, linear, 10·5–19 cm. long, 0·7–1·7 cm. wide ; valves rather thin, grey- to deep red-brown, ± longitudinally veined, otherwise smooth, attenuate to base. Seeds dark blackish-olive, smooth, quadrate, compressed, 8–11 mm. long, 5–6·5 mm. wide ; areole 6·5–7·5 mm. long, 3·5–4·5 mm. wide.

subsp. clavigera; Brenan in K.B. 1957: 367 (1958)

Leaf-rhachis typically ± densely pubescent; leaflets (1·75–)2–3·5 mm. wide. Pods 1·3–1·7 cm. wide.

NOTE.—Typical subsp. *clavigera* occurs in Portuguese East Africa, Southern Rhodesia and South Africa, but is connected with the following subspecies by intermediates (found additionally in Northern Rhodesia). Typical *clavigera* has not been found with us, but two or three of the intermediates have. *Busse* 2337 (EA!) from Tanganyika, Lindi, with a pubescent leaf-rhachis, the leaflets varying from 1·5–2·2 mm. in width, and pods 0·9–1·0 cm. wide; and *Gerstner* 7207 (EA!) from Tanganyika, Masasi District, Ndanda, with a pubescent leaf-rhachis, leaflets 2–3·25 mm. wide, and pods 0·9–1·1 cm. wide. *Schlieben* 5432 (BM!) from Tanganyika, Lindi District, Lake Lutamba, is similar to *Busse* 2237, but shows no pods.

subsp. usambarensis (*Taub.*) Brenan in K.B. 1957: 369 (1958). Types: Tanganyika, Lushoto District, Simbili, *Holst* 2362 (B, syn. †, K, isosyn.!) & Mashewa, *Holst* 8820 (B syn. †, K, isosyn.!) & Bwiti, *Holst* 2386 (B, syn. †, K, isosyn.!)

Leaf-rhachis typically glabrous; leaflets 1–2 mm. wide. Pods (Fig. 16/45, p.67) 0·7–1·2 cm. wide.

KENYA. Kitui District: Tiva R. crossing near Kitui, 1 Apr. 1955, *Wilson* 45 !; Kilifi District: Mida, Sept. 1929, *R. M. Graham* A673 *in F.H.* 2063 ! & Kilifi, *Jeffery* K292 !
TANGANYIKA. Lushoto District: Mashewa, *Gillman* 835 !; Mpwapwa, *Mrs. Hornby* 52 !; Morogoro, 20 Jan. 1951, *Wigg in F.H.* 954 !; Nachingwea, 6 Mar. 1953, *Anderson* 856 !
DISTR. K4, 6, 7; T2, 3, 5, 6, 8; Portuguese East Africa
HAB. Riverine and ground-water forest, wooded grassland; 0–1370 m.

SYN. *A. usambarensis* Taub. in P.O.A. C: 195, t. 20 H (1895); Harms in N.B.G.B. 4: 206, fig. 5 (1906); L.T.A.: 846 (1930); T.S.K.: 68 (1936); B. D. Burtt in Journ. Ecol. 30: 91 (1942); T.T.C.L.: 338 (1949); Bogdan in Nature in E. Afr., ser. 2, No. 1: 14 (1949)
" *A.* sp. nr. *xanthophloea* " T.S.K.: 70 (1936)
A. sacleuxii A. Chev. in Rev. Bot. Appl. 27: 509 (1947). Type: Tanganyika, Tanga, *Sacleux* 2455 (P, holo. !)

NOTE.—Subsp. *usambarensis* shows little variation in our area. A single fruiting specimen from an apparently unusual habitat, *B. D. Burtt* 5180 (K !) from Morogoro, summit of Mbokwa Hill, July 1935, has the pods with rather marked constrictions between the seeds; I have not seen similar pods in any other gathering of this species. The seeds seem well developed. More material is wanted.

46. **A. gerrardii** *Benth.* in Trans. Linn. Soc. 30 : 508 (1875) ; L.T.A. : 846 (1930) ; T.S.K. : 69 (1936). Type : Natal, *Gerrard* 1702 (K, holo. !)

Shrub or more usually a tree 3–15 m. high ; crown flat, umbrella-shaped or irregular (I.T.U., ed. 2 : 209) ; bark on trunk grey, blackish-brown or black, rough, fissured. Young branchlets ± densely grey-pubescent, rarely glabrous or nearly so ; epidermis usually splitting or falling away to expose a rusty-red inner layer. Stipules spinescent, usually straight or nearly so, sometimes recurved, rarely hooked, mostly short, to about 1 cm. long, rarely to about 6 cm. long and then usually grey : " ant-galls " and other prickles absent. Leaves : rhachis (1·5–)2–7 cm. long, ± densely pubescent ; pinnae (3–)5–10(–12) pairs ; leaflets (8–)12–23(–28) pairs, 3–7·5 mm. long, 1–2 mm. wide, ± ciliate on margins at least near base, otherwise glabrous or nearly so, sometimes hairy on surface. Flowers white or cream, scented, in heads on axillary densely grey-pubescent eglandular or inconspicuously glandular, occasionally strongly glandular peduncles ; involucel at or shortly above base or sometimes to one-third way up peduncle. Corolla glabrous or only slightly and inconspicuously pubescent outside. Pods falcate, dehiscent, linear or linear-oblong, (4·5–)7–16(–22) cm. long, mostly 0·6–1·1 sometimes to 1·7 cm. wide ; valves rather thin, ± grey-puberulous to -tomentellous, rarely subglabrous or glabrous. Seeds olive-brown, smooth, ± irregularly quadrate, compressed, 9–12 mm. long, 7 mm. wide ; areole 6·5–7 mm. long, 3·5–4·5 mm. wide.

KEY TO INTRASPECIFIC VARIANTS
Young branchlets ± densely pubescent :
    Pods 0·6–1·1(–1·2) cm. wide ; spines generally straight
      or nearly so .   .   .   . var. **gerrardii**
    Pods 1·2–1·7 cm. wide ; spines often hooked or recurved   var. **latisiliqua**
Young branchlets glabrous or almost so ; pods 1·2–1·4 cm.
    wide ; spines often hooked or recurved .   .   . var. **calvescens**

var. **gerrardii**; Brenan in K.B. 1957: 369 (1958)

Young branchlets ± densely pubescent. Spines straight, occasionally recurved. Pods (Fig. 16/46, p. 67) 0·6–1·1(–1·2) cm. wide.

UGANDA. Karamoja District: Moroto, 5 Mar. 1936, *Michelmore* 1254 !; Ankole District: Ruizi R., 15 May 1950, *Jarrett* 44 !; Masaka District: Kabula, 14 Mar. 1936, *Michelmore* 1313 !
KENYA. Laikipia District: 3 km. S. of Rumuruti, *Wimbush* 2 *in F.H.* 2604 !: Nairobi District: Karura, *Rammell* 1511 !; Masai District: Mara R., Feb. 1932, *Rammell* 2745 !

TANGANYIKA. Mbulu District: Mbulumbul, 14 July 1943, *Greenway* 6773!; Lushoto District: near Mashewa, 7 July 1953, *Drummond & Hemsley* 3212!; Iringa, 20 May 1935, *Emson* 496!
DISTR. U1-4; K1-6; T1-7; Natal to the Sudan, westwards to Nigeria
HAB. Woodland and wooded grassland, widespread and in some areas dominant and common; (450-)900-2130 m.

SYN.   *A. hebecladoïdes* Harms in E.J. 36: 208 (1905); L.T.A.: 846 (1930); T.S.K.: 71 (1936); T.T.C.L.: 337 (1949); Bogdan in Nature in East Africa, ser. 2, No. 1: 14 (1949); Gilb. & Bout. in F.C.B. 3 : 162 (1952) ; I.T.U., ed. 2 : 209, fig. 48d (1952) ; F.W.T.A., ed. 2, 1 : 500 (1958). Type : Tanganyika, " Masai Steppe " in the Kilimanjaro region, *Merker* (B, holo.†)

NOTE. The pods of var. *gerrardii* are usually puberulous, but may sometimes be sub-glabrous, e.g. in *Davidson* 522, from Kenya, Nyanza Province (EA!).
See note under species No. 67 (p. 136).

    var. **latisiliqua** *Brenan* in K.B. 1957: 369 (1958). Type: Kenya, Machakos, *Trapnell* 2215 (K, holo.!)

    Young branchlets ± densely pubescent. Spines often hooked or recurved, sometimes straight. Pods 1·2-1·7 cm. wide.

KENYA. Nairobi, Mar. 1934, *Napier* 6026!; Machakos, Wilson's farm, 7 Feb. 1952, *Trapnell* 2215!; Masai District: Ngarika near Kajiado, 3 Nov. 1948, *Vesey-Fitz-Gerald* 213!
TANGANYIKA. Kondoa District: near Thlawa, 17 Dec. 1927, *B. D. Burtt* 860!; Mbeya, 25 Feb. 1934, *Michelmore* 969! & 972!
DISTR.: K4, 6; T5, 7; not known elsewhere
HAB. Wooded grassland; 1370-1750 m.

NOTE. The Kenya specimens characteristically show the hooked or curved spines. The Tanganyika ones have them straight or but slightly curved, and may prove taxonomi-cally worth separation. The pods of var. *latisiliqua* are usually glabrescent at maturity. Those of *Burtt* 860, however, have rather dense grey puberulence all over, similar to that usually shown by var. *gerrardii*.

    var. **calvescens** *Brenan* in K.B. 1957: 370 (1958). Type: Tanganyika, Mbulu, *Eggeling* 6689 (K, holo.!, EA, iso.!)

    Young branchlets glabrous or almost so. Spines mostly hooked or recurved. Pods 1·2-1·5 cm. wide, glabrous or almost so.

KENYA. Teita Hills, 1934, *Bally in C.M.* 12013!
TANGANYIKA. Mbulu, Aug. 1953, *Eggeling* 6689!; Dodoma District: Kazikazi, 18 Aug. 1931, *B. D. Burtt* 1784! & Usule, 15 Dec. 1933, *Michelmore* 824!
DISTR. K7; T1, 2, 5; not known elsewhere
HAB. Riverine forest and probably wooded grassland; 1160-1770 m.

NOTE. The material cited under var. *calvescens* may not be homogeneous. Further good collections are wanted.

    47. **A. lasiopetala** *Oliv.*, F.T.A. 2 : 346 (1871) ; L.T.A. : 847 (1930) ; T.T.C.L. : 337 (1949). Type : Nyasaland, Pemba [Impemba] Mt., *Kirk* (K, holo.!)

    Small tree 2-6 m. high ; bark rusty-red. Branchlets persistently grey-tomentellous (indumentum sometimes yellowish when young) ; then epidermis flaking away to expose red bark. Stipules spinescent, to 2·3 cm. long, never inflated ; other prickles absent. Leaves with gleaming, silky, pale golden indumentum when young, grey-pubescent when older ; petiole 5-8 mm. long (to 1·7 cm. in juvenile leaves) ; pinnae of well-developed leaves of mature shoots 15-40 pairs (reduced leaves with fewer pairs usually also present), mostly 2-3·5 cm. long ; leaflets very numerous, 2·5-5 mm. long, 0·7-1 mm. wide. Flowers white, in heads ; involucel at base of the nearly or quite eglandular peduncle. Corolla densely pubescent on lobes outside, about 1½ times as long as calyx. Pods (Fig. 16/47, p. 67) dehiscent, mostly arcuate, grey-tomentellous, ± turgid over the seeds, with constric-tions between them about 1-2 cm. apart, about 9-15 cm. long, 0·8-1·1 cm. wide. Seeds olive-brown, smooth or nearly so, irregularly quadrate or elliptic, compressed, usually 8–9 mm. long, 5–6 mm. wide ; areole 5·5–6 mm. long, 2·5–3 mm. wide.

TANGANYIKA. Ufipa District: Namwele, 20 Oct. 1949, *Silungwe* 23 !; Mbeya District: Mbozi plateau, near descent to Mbeya, 16 Oct. 1936, *B. D. Burtt* 6051 ! & Ivuna–Mbozi, 25 Oct. 1950, *Bullock* 3453 !
DISTR. T4, 7; Belgian Congo, Portuguese East Africa and Nyasaland
HAB. Deciduous woodland (*Brachystegia*) on hills and wooded grassland ; 1500–1830 m.

48. **A. drepanolobium** [*Harms ex*] Sjöstedt, Schwed. Zool. Exped. Kilimandjaro 8 : 116–117, t. 6, fig. 7–8, t. 7, fig. 2–3 (1908) ; Harms in E.J. 51 : 361 (1914) ; L.T.A. : 846 (1930) ; T.S.K. : 71 (1936) ; B. D. Burtt in Journ. Ecol. 30 : 89, 95 (1942) ; Bogdan in Nature in E. Afr., ser. 2, No. 1 : 14 (1949) ; T.T.C.L. : 336 (1949) ; I.T.U., ed. 2 : 207, fig. 48b (1952). Type * : Tanganyika, Kilimanjaro, between Kwagogo and Moshi, *Engler* 1688 (B, holo.†, K, drawings !)

Bush or small tree 1–5(–7·5) m. high, with short radiating branches from main stem, sometimes spreading at top. Young branchlets shortly pubescent to puberulous, rarely glabrous, grey then going brown ; no powdery inner bark on twigs. Old bark black or grey, usually rough, sometimes smoothish. Stipules spinescent, mostly 1·5–4·5(–7·5) cm. long (some shorter ones often also present), straight, grey or whitish, some fused at base into ± round "ant-galls" 1–3·5 cm. in diameter, grape-purple going blackish. Leaves : petiole 2–5 (very rarely to 10) mm. long, glandular at the lowest of the 3–13 pairs of pinnae ; rhachis 0·8–4·5(–9) cm. long, glandular between the top 1–6 pairs of pinnae ; leaflets 11–22 pairs, glabrous or minutely ciliolate, subacute or acute sometimes obtuse at apex, 1·5–5·5 mm. long, 0·7–1·25 (–1·75) mm. wide. Flowers white or sometimes cream, in heads ; involucel at or rarely a short way above base of the glabrous puberulous or shortly pubescent peduncle. Calyx 0·75–1·5(–2·5) mm. long, glabrous or ciliolate. Corolla glabrous outside, sometimes puberulous on lobes, 3–4 mm. long. Pods (Fig. 16/48, p. 67) falcate or annular, thinly coriaceous, finely longitudinally venose, glabrous or ± puberulous, mostly attenuate or even acuminate at ends, 4–7 cm. long, 0·5–1·0 cm. wide. Seeds mottled whitish-grey and dark brown, smooth, irregularly quadrate or elliptic, compressed, 10–12 mm. long, 4·5–5·5 mm. wide ; areole 5–6 mm. long, 3–3·5 mm. wide.

UGANDA. Karamoja District: near Kangole, July 1930, *Liebenberg* 187 ! & near Moroto, 6 Mar. 1936, *Michelmore* 1266 !; Mbale District: Agu Swamp, Pallisa, *Eggeling* 746 in F.H. 1147 !
KENYA. Naivasha District: Longonot Crater, 25 Aug. 1940, *Greenway* 5996!; Machakos District: Mua Hills, 12 Mar. 1953, *Trump* 49 !; Masai District: Kapiti Plains; Nairobi–Kajiado road, 21 Feb. 1953, *Drummond & Hemsley* 1247 !
TANGANYIKA. Shinyanga District: near Kizumbi, 4 Jan. 1933, *B. D. Burtt* 4513 !; Moshi District: Engare Nairobi, 4 July 1943, *Greenway* 6702 !; Dodoma District: Mwitikira, 14 Aug. 1928, *Greenway* 775 !
DISTR. U1, 3 ; K1–6 ; T1–5 ; Belgian Congo, the Sudan, Ethiopia and Somalia
HAB. Shrub or dwarf-tree grassland; " gregarious, usually on alkaline hard-pan grey soils with *Lannea humilis* and *Commiphora schimperi*, or in fringing ' mbuga ', or on dark clay cracking lime-accumulating soils " (T.T.C.L.: 336); 600–2680 m.

SYN. *A formicarum* Harms in E.J. 51: 362, 363 (fig. 2) (1914); L.T.A.: 846 (1930); T.T.C.L.: 337 (1949). Type: Tanganyika, Moshi District, Engare Nairobi, *Endlich* 721 (B, syn. †, K, drawing !); Tanganyika, between Kilimanjaro and Mt. Meru, *Merker* (B, syn. †)

NOTE. *A. drepanolobium* is both rather widely distributed and variable. It is not clear, however, how much of the variation is due to heredity and how much to such causes as exposure and burning. The indumentum, particularly of the branchlets, peduncles and flowers, may be absent or comparatively dense; the pods may be glabrous to puberulous, and they vary in width. *A. formicarum* appears to be simply a form of

* Sjöstedt's publication cited no specimens, but mentioned that the species occurred near Kahe in Moshi District and near the W. Usambaras. In the circumstances I have given as the holotype the only specimen cited by Harms in 1914.

A. drepanolobium with glabrous or subglabrous pods and peduncles, and with involucels above the base. Correlation between these characters appears, however, to be absent.

The habit likewise varies: *Greenway* 5996 is a gnarled, condensed form with stout stems, while *Eggeling* 746 is an opposite extreme. The range of altitude of *A. drepanolobium* is rather wide, and the species may comprise different ecotypes.

Bally from Kenya, foot of Ngong Escarpment, 1947, bears an "ant-gall" approximately 6 cm. across (not counting the spines). Has a bigger one ever been found?

49. **A. pseudofistula** *Harms* in E.J. 51 : 363 (1914) ; L.T.A. : 847 (1930) ; T.T.C.L. : 337 (1949). Types : Tanganyika, Tabora District, Goweko, *Holtz* 2801 (B, syn.†, K, drawing !) ; Dodoma District, Kilimatinde, *Holtz* 1358 (B, syn.†)

Shrub or small "columnar" tree 1·8–6(–9) m. high, with horizontal branches all the way up the 1–3 main stems ; bark red-brown. Young branchlets grey-puberulous or pubescent ; then epidermis flaking away to expose rusty-red, powdery inner bark (but see note below). Stipules spinescent, long, straight, whitish, about 2–5(–9) cm. long, some fused at base into round, blackish "ant-galls" up to 2·5(–3) cm. in diameter. Leaves with grey, inconspicuous puberulence on the 2–4 mm. long petiole and the 2–10 cm. long rhachis ; pinnae of well-developed leaves of mature shoots 15–22 pairs (reduced leaves with fewer pairs usually also present), 1–3 cm. long ; leaflets very numerous, 1·5–4·5 mm. long, 0·75–1 mm. wide, acute or subacute at apex, ciliolate near base or glabrous. Flowers creamy-white, in heads ; involucel at base of the tomentellous to densely puberulous, nearly or quite eglandular, 0·75–1 mm. thick peduncle. Calyx 1·75–2 mm. long. Corolla glabrous, or slightly puberulous only near apex outside, 4–5 mm. long, 2–3 times as long as calyx. Pods (Fig. 16/49, p. 67) falcate, grey-puberulous, acute at both ends, 2–8 cm. long, 0·4–1·0 cm. wide. Seeds grey, smooth, elliptic with an irregular outline, compressed, often curved, 9–11 mm. long, 4–6 mm. wide ; areole 6–7 mm. long, 3 mm. wide.

TANGANYIKA. Singida District: Ushola, Sept. 1935, *B. D. Burtt* 5255 !; Dodoma District: Kazikazi, 5 Sept. 1931, *B. D. Burtt* 3331 !; Mbeya District: Ruiwa, 11 Aug. 1951, *Trapnell* 2154 !
DISTR. **T**4, 5, 7; not known elsewhere
HAB. Shrub or dwarf-tree grassland, locally gregarious and abundant in black soils of valleys; 900–1500 m.

SYN. [*A. formicarum* sensu B. D. Burtt in Journ. Ecol. 30: 96, 143 etc., t. 16, photo. 29 (1942), *non* Harms]

NOTE. B. D. Burtt remarks (Journ. Ecol. 30: 97) that "in the rainy season the long feathery leaf of fresh green colour will at once distinguish this gall-acacia from *Acacia drepanolobium* and *A. malacocephala*, whose leaves are olive-green ".
According to *B. D. Burtt* 3390 ! from Tanganyika, Dodoma District, swamp of Bubu R., SE. of Kilimatinde scarp, 5 Sept. 1931, *A. pseudofistula* is said to have in this locality creamy-buff young bark, but does not seem to be otherwise different.
B. D. Burtt 6052 !, from Tanganyika, Mbeya District, Great North Road at foot of scarp leading to Mbozi, 26 Aug. 1936, is without flowers or pods, but suggests *A. pseudofistula* in most ways (including the bark). It differs, however, in having the internodes very short, the spines closely set along the stems, and, especially, in the leaves each having only 3–8 pairs of pinnae. Without fuller material, the status of this plant is uncertain.

50. **A. bullockii** *Brenan* in K.B. 1957 : 77 (1957). Type : Tanganyika, Buha District, *Bullock* 3144 (K, holo. !)

Small tree 1–3 m. high, with rather stout simple or only slightly branched stems ; epidermis probably ultimately flaking away to expose inner bark whose colour is uncertain. Stipules spinescent, mostly 1·5–3·2 cm. long, straight or nearly so, whitish at first, soon blackish, some fused at base into ± rounded "ant-galls" 1·2–3 cm. in diameter. Leaves glabrous, or the rhachides and sometimes the leaflets subglabrous ; petiole 4–6 mm. long; rhachis (1·4–)10–27 cm. long ; pinnae of well-developed leaves of mature shoots 15–28 (or ? more) pairs (reduced leaves with fewer pairs usually also present), 3–8·5 cm. long ; leaflets 21–49 pairs, 3–7 mm. long, 0·75–1·4 mm. wide, acute at apex.   Flowers cream, in heads ;   involucel at base of the rather stout, 1·25–3 mm. thick peduncle. Calyx 2·5–2·7 mm. long. Corolla 4·5–5·5 mm. long. Pods falcate, glabrous to puberulous, mostly 6–10 cm. long, 0·3–0·5 cm. wide.

var. **bullockii**; Brenan in K.B. 1957: 78 (1957)

Young stems, leaflets, peduncles and flowers glabrous. Bracteoles sparsely ciliolate at apex only.

TANGANYIKA. Buha District: Kaberi " mbuga ", 12 Aug. 1950, *Bullock* 3144 ! & Mpemvi R., *Bullock* 3144 A !
DISTR. **T4**; not known elsewhere
HAB. Same as *A. erythrophloea* (p. 124); 1160–1220 m.

var. **induta** *Brenan* in K.B. 1957: 78 (1957). Type: Tanganyika, Kigoma District, *C. H. N. Jackson* 117 ! (K, holo. !, BM, iso. !)

Young stems densely puberulous. Leaflets ciliolate on the anticous margin at base. Peduncles sparsely puberulous to densely pubescent. Calyx and corolla ciliolate on lobes or puberulous above outside. Bracteoles densely ciliolate above.

TANGANYIKA. Kigoma District: Tandala in Uvinza, Aug. 1935, *C. H. N. Jackson* 117 ! & Oct. 1935, *C. H. N. Jackson* 117 A ! & 117 B !
DISTR. **T4** ; not known elsewhere
HAB. Same as last; ? 1130 m.

SYN. *Acacia sp. nov.* (in Kigoma) — B. D. Burtt in Journ. Ecol. 30: 143 (1942)

NOTE. The unusual habit of this species is well shown by a fine series of photographs of it taken by Mr. A. A. Bullock. It may be separable by this from *A. pseudofistula*, but insufficient is known of the habit of its other close relative, *A. erythrophloea*. *A. bullockii* also differs in its usually exceptionally large leaves (not clearly shown however, by var. *induta*, see below), longer leaflets, large flowers on stouter peduncles and longer pods (normally 7–10 as against 3–6 cm.).
   The status of var. *induta* is doubtful. Jackson 117 shows a greater development of indumentum than in var. *bullockii*. The leaves are also in general smaller, with usually fewer than 15 pairs of pinnae, but are borne, however, on abnormal-looking shoots, evidently the results of very severe burning, and these differences may thus be more apparent than real. More material and more field observations are wanted both of var. *induta* and var. *bullockii*.

51. **A. erythrophloea** *Brenan* in K.B. 1957 : 76 (1957). Type : Tanganyika, Tabora District, Kakoma, *Glover* 186 (K, holo. !)

Small tree up to 3·6 m. high. Young branchlets very shortly grey-puberulous, then epidermis flaking away to expose the powdery, intensely brick-red inner bark. Stipules spinescent, nearly straight but rather short, black, 0·3–1·5 cm. long (on young coppice up to 2 cm. and whitish), some fused at base into round or ovoid " ant-galls " 1–2 cm. in diameter. Leaves with grey, inconspicuous puberulence on the 3–5 mm. long petiole and the 2–13 cm. long rhachis ; pinnae of well-developed leaves of mature shoots 15–31 pairs (reduced leaves with fewer pairs usually also present), 1–2·9 cm. long ; leaflets very numerous, 2–3 mm. long, about 0·75–0·9 mm. wide, acute or subacute at apex, ciliolate towards base or glabrous. Flowers white, in heads ; involucel at base of the ± puberulous, nearly or quite eglandular, 0·5–0·75 mm. thick peduncle. Calyx 1–1·5 mm. long. Corolla ciliolate on lobes, 3 mm. long. Pods (Fig. 16/51, p. 67) falcate, grey-puberulous, acute at both ends, 3·5–6 cm. long, 0·4–0·8 cm. wide (? not fully mature).

TANGANYIKA. Tabora District: Kakoma, Feb. & Aug. 1934, *C. H. N. Jackson* 43 ! & 44 ! & 10 Aug. 1938, *Glover* 186 ! & 24 June 1949, *Hoyle & Greenway* 1037 !
DISTR. **T4**; not known elsewhere

HAB. Shrub or dwarf-tree grassland, very locally gregarious in black soils of valleys; about 1130 m.

SYN. *Acacia sp. nov.*: " new species of gall-acacia "—B. D. Burtt in Journ. Ecol. 30: 98, 143 (1942)

NOTE. According to field-notes with *Jackson* 44, this is said to differ from *A. pseudofistula* in its white not cream flowers, in being smaller, and in having more delicate and slender fruits. I have not been able to check these differences. The following remarks relating to *A. erythrophloea* are extracted from Journ. Ecol. 30: 98 (1942):—" The heavy clay ' mbugas ' [of Tabora District] are usually waterlogged for several months in the rainy season, and support a new species of gall-acacia . . . flowering in the dry season. This gall-acacia has very long leaves of a rich dark green colour, contrasting with the deep purply brown of the younger branches. The galls support paired stumpy thorns which are never silvery (as they are on the other gall-acacias) except in quite young coppice. In the north and east of the District [Tabora] *A. formicarum* [i.e. *A. pseudofistula*] dominates in the ' mbugas ', but 40 miles S. of Tabora the new species replaces it as soon as the hilly country gives place to the great flat plateau."

52. **A. malacocephala** *Harms* in E.J. 51 : 364 (1914) ; L.T.A. : 848 (1930) ; B. D. Burtt in Journ. Ecol. 30 : 96 (1942), pro parte, excl. loc. Mbulu, Basotu, Basodesh ; T.T.C.L. : 337 (1949). Type : Tanganyika, Shinyanga District, between Samuye and Kizumbi, *Holtz* 1548 (B, holo.†, K, fragment and drawing !)

Small tree 2·5–4·5(–6) m. high ; stems brown or sometimes black. Young branchlets grey-puberulous with hairs about 0·1 mm. long or less ; older branchlets brown or blackish but without any powdery inner bark. Stipules spinescent, 1·5–5·5 cm. long, a few shorter, straight, pale grey or whitish, some fused at base into ± rounded blackish ant-galls 1·5–3 cm. in diameter. Leaves : petiole 5–10 mm. long, glandular at apex ; rhachis (1·5–)3–6·5 cm. long, puberulous like the petiole, glandular between the top 1–3 pairs of pinnae ; pinnae 3–10 pairs, mostly 1·5–3·5 cm. long ; leaflets 10–22 pairs, glabrous, subacute at apex, 2·5–6 mm. long, 0·8–1·5 mm. wide. Flowers white, in heads ; involucel at base of the densely puberulous peduncle. Calyx 1–1·5 mm. long, densely tomentellous outside. Corolla 3–3·75 mm. long, clothed like the calyx outside (except at base). Pods (Fig. 16/52, p. 67) curved or falcate, thinly coriaceous, densely grey-puberulous, mostly attenuate or acuminate at ends, 4·5–7 cm. long, 0·6–1·1 cm. wide. Seeds grey, smooth, ± elliptic or quadrate, compressed, sometimes curved, 9–11 mm. long, 5·5 mm. wide ; areole 5–7 mm. long, 2·5–3 mm. wide.

TANGANYIKA. Shinyanga District: Wembere region towards Sakamaliwa, Sept.–Oct. 1935, *B. D. Burtt* 5254 ! & between Mango and Sakamaliwa, 24 Jan. 1936, *B. D. Burtt* 5530 !; Nzega District: Ukama & Sakamaliwa, 27 Aug. 1933, *B. D. Burtt* 4938 !
DISTR. T1, 4; not known elsewhere
HAB. Shrub or dwarf-tree grassland on grey calcareous soils of valleys; about 1060–1100 m.

NOTE. Burtt remarks (*l.c. supra*) that this acacia " covers vast expanses of country fringing the Wembere Steppe and extending up some of the tributary valleys ". To this area it is perhaps confined. There is, however, a specimen at Kew, *Doggett* 108 from Mwanza/Kwimba Districts, 16 km. S. of Nyegezi on the Nyambiti road, which appears to be *A. malacocephala* but is uncertain because it is incomplete; more material from this locality is wanted.

For the differences between *A. malacocephala* and *A. mbuluënsis*, to which it is related, see under the latter. (p. 125). *A. malacocephala* much resembles *A. drepanolobium* when not in flower, and on the evidence of herbarium specimens one might be tempted to suggest that they are extremes of one species. However, B. D. Burtt, who knew them both well as living trees, considered them distinct species, and made (Journ. Ecol. 30: 96 (1942)) the interesting distinction that *A. malacocephala* flowers in the later dry season, the flowers disappearing in the first rains, while *A. drepanolobium* flowers in the rains.

53. **A. mbuluënsis** *Brenan* in K.B. 1957 : 79 (1957). Type : Tanganyika, Mbulu District, *B. D. Burtt* 4936 (K, holo. !, BM, EA, iso. !)

Tree 1·2–10·5 m. high with flattened crown and very dark brown, ribbed bark. Young branchlets densely grey-pubescent with hairs 0·3–0·75 mm. long ; older grey or blackish with brown but not at all powdery inner bark showing here and there. Stipules spinescent, mostly 0·4–1 cm. long, a few longer, up to 4 cm., straight or slightly curved, grey, some fused at base into ± round, black, pubescent " ant-galls " 1–2 cm. in diameter. Leaves : petiole 2–5 mm. long, glandular or not ; rhachis 2·5–5 cm. long, pubescent like the petiole, glandular between the top 1–2 pairs of pinnae ; pinnae 10–20 pairs, more than 15 pairs on best-developed leaves, 0·8–1·7 cm. long ; leaflets 9–12 pairs, pubescent on margins, obtuse or subacute, 2–3 mm. long, about 0·75 mm. wide. Flowers in heads ; involucel at or to 3 mm. above base of the densely pubescent or tomentellous peduncle. Calyx 0·75–1·5 mm. long, densely pubescent like the upper half of the 3–4 mm. long corolla. Pods (Fig. 16/53, p. 67) falcate, subcoriaceous, densely grey-puberulous, acute or attenuate at ends, 5–9 cm. long, 0·6–0·9 cm. wide. Seeds grey or grey mottled with dark brown, smooth, oblong or elliptic, compressed, often ± curved, 9–13 mm. long, 5–6 mm. wide ; areole 5–7 mm. long, 3–4 mm. wide.

TANGANYIKA. Mbulu District: Dongobesh Valley, 5 Sept. 1932, *B. D. Burtt* 4271 ! & Ufana, 6 Oct. 1933, *B. D. Burtt* 4936 ! & 4937 !; Moshi District: between Moshi and Engare Nairobi, 8 Oct. 1932, *B. D. Burtt* 4260 !
DISTR. **T2**; not known elsewhere
HAB. Uncertain; 900–1980 m.

SYN. [*A. malacocephala* sensu B. D. Burtt in Journ. Ecol. 30: 96 (1942), pro parte, quoad loc. Mbulu, Basotu, Basodesh, *non* Harms]

NOTE. *A. mbuluënsis* differs from *A. malacocephala* in: young branchlets densely pubescent with hairs 0·3–0·75 mm. long, as against 0·1 mm. or less in *A. malacocephala*; spines mostly short (see above), instead of mostly 1·5–5·5 cm. long; petiole shorter, 2–5 mm. as against 5–10 mm. and, like the rhachis, pubescent not puberulous; the pinnae are often more numerous and, in some leaves at least, in as many as 15–20 pairs; the leaflets of *A. mbuluënsis* are pubescent on the margins, not glabrous, and mostly smaller, those of *A. malacocephala* being 2·5–6 mm. long and 0·8–1·5 mm. wide. The young branchlets of *A. mbuluënsis* are usually stouter, with conspicuous thickenings under the pairs of spines; the leaves are smaller, with the pinnae and leaflets more closely set.

54. **A. burttii** *Bak. f.* in J.B. 71 : 342 (1933) ; B. D. Burtt in Journ. Ecol. 30 : 98, 143, t. 9, photo. 13 (1942) ; T.T.C.L. 335 (1949). Type : Tanganyika, Kahama District, 9 km. along Shinyanga road, *B. D. Burtt* 4501 (BM, holo. !, EA, FHO, K, iso. !)

Altogether glabrous shrub or small tree 2–3 m. high with pole-like stem and short lateral branches giving plant a columnar appearance ; bark buff or fawn. Young branchlets pale grey, sometimes going brown ; epidermis flaking away later to expose a powdery rusty-brown inner layer. Stipules spinescent, mostly 1·5–3·5(–5·5) cm. long, straight, mostly grey or whitish, some fused at base into round or ovoid " ant-galls " 1–2·5 cm. in diameter, which are purplish going black and characteristically spotted or flecked. Leaves : petiole 3–13 mm. long, frequently glandular at apex ; rhachis 0–3·2 cm. long ; pinnae 1–4 pairs, 2–4·5 cm. long ; leaflets 7–16 pairs, acute or subacute and mucronate at apex, (4–)5–13(–17) mm. long, 1·5–4 (–6) mm. wide. Flowers creamy-white, in heads ; involucel at base of peduncle. Calyx 1·75–3 mm. long. Corolla 4·5–5 mm. long. Pods (Fig. 17/54, p. 68) short, half-moon-shaped or reniform, flattened, thin, glabrous, finely net-veined, grey-brown, 1·5–4 cm. long, 1–1·7 cm. wide, one-seeded. Seeds grey, mottled with dark brown, smooth, subcircular or broadly elliptic, compressed, 9–11 mm. long, 8·5–9 mm. wide ; areole 5–7 mm. long, 4–6 mm. wide.

TANGANYIKA. Mwanza District: 40 km. E. of Geita Gold Mine, 11 Apr. 1937, *B. D. Burtt* 6454 !; Kahama District: 24 km. along Kahama–Ushirombo road, 9 Jan. 1933, *B. D. Burtt* 4502 !; SW. Tabora District, 10 Oct. 1934, *C. H. N. Jackson* 45 !

Distr.   T1, 4 ; not known elsewhere

Hab.   Shrub or dwarf-tree grassland; gregarious on brown or black clay soil in valleys in *Brachystegia* country; 1130–1220 m.

Note.   No other species of " gall-acacia " in our area combines creamy-white flowers with such large leaflets and constantly so few pinnae. As the pod-shape is also unique among East African acacias, *A. burttii* is one of the most distinct and easily recognized of our species.

55. **A. arenaria** *Schinz* in Mém. Herb. Boiss. 1 : 108 (1900) ; L.T.A. : 839 (1930) ; Consp. Fl. Angol. 2 : 282 (1956).   Types : South West Africa, Amboland, Olukonda–Oshiheke, *Schinz* 2072 (Z, syn.!) & Omatope, *Schinz* 2071 (Z, syn.!)

Shrub or small tree 2–9 m. high, with very short bole, branching near ground ; bark on bole dark and rough.   Branchlets with short inconspicuous puberulence or pubescence, purplish, soon going grey or sometimes brownish, zig-zag, their epidermis not peeling or flaking away.   Stipules spinescent, to 6 cm. long, never inflated ; other prickles absent.   Leaves with inconspicuous, dull pubescence ; petiole 4–14 mm. long ; rhachis (5–)10–21 cm. long ; pinnae of well-developed leaves of mature shoots 15–35 pairs (reduced leaves with fewer pairs sometimes also present), 0·7–2·2(–3) cm. long ; leaflets 1·5–4·5 mm. long, 0·7–1 mm. wide, glabrous or ciliolate.   Flowers white or pale pink, in heads ; involucel at or above middle or at apex of the pubescent and glandular peduncle.   Corolla glabrous outside, 2–3 times as long as calyx.   Pods (Fig. 17/55, p. 68) dehiscent, arcuate, glabrous and deep red-brown outside, flat or slightly constricted between the seeds, 8–18(–22) cm. long, 0·5–0·8 cm. wide.   Seeds olive-grey, smooth, quadrate or oblong, compressed, 7–9 mm. long, 3–4·5 mm. wide ; areole 3·5–4·5 mm. long, 1·5–2·25 mm. wide.

Tanganyika.   About 5 km. E. of Nzega, 21 July 1949, *Doggett* 120 !; Singida District: Iramba Plateau above Sekenke, 25 July 1931, *B. D. Burtt* 3389 ! & Rift Valley near Manyigi, Oct. 1935, *B. D. Burtt* 5270 !; Kondoa District: Kissesse, below Irangi scarp, 30 June 1929, *B. D. Burtt* 2005 !

Distr.   T2 (*fide* T.T.C.L.), 4, 5; Southern Rhodesia, Bechuanaland, Angola and South West Africa

Hab.   Deciduous bushland and woodland; locally common in transition between sandy alluvial soil and grey hard-pan (T.T.C.L.: 337); 1220–1520 m.

Syn.   *A. hermannii* Bak. f. in J.B. 67: 198 (1929); T.T.C.L.: 337 (1949). Type: Tanganyika, Singida District, near Manyugi [? Manyigi], *B. D. Burtt* 1379 (BM, holo. !, FHO, K, iso. !)

56. **A. fischeri** *Harms* in E.J. 51 : 365 (1914) ; L.T.A. : 838 (1930) ; B. D. Burtt in Journ. Ecol. 30 : 91 (1942) ; T.T.C.L. : 335 (1949).   Types : Tanganyika, without locality, *Fischer* 157 (B, syn.†) ; Tanganyika, " Manjanga Bach " [probably Manyonga River], *Stuhlmann* 672 (B, syn.†)

Low shrub or small tree 1–6 m. high, flat-crowned.   Old bark very dark and rough.   Branchlets densely grey-pubescent or -puberulous, often with minute reddish glands, ultimately brown to blackish ; bark not peeling.   Stipules spinescent, straight, to 6·7 cm. long, never inflated ; other prickles absent.   Leaves with dull pubescence ; petiole 3–8 mm. long ; rhachis 5–18·5 cm. long ; pinnae of well-developed leaves of mature shoots often 15–41 pairs (reduced leaves with fewer pairs sometimes also present), mostly 1·5–4·5 cm. long ; leaflets 1·5–5 mm. long, 0·75–1·5 mm. wide, minutely ciliate, otherwise glabrous to pubescent.   Flowers cream (T.T.C.L.: 335), in heads ; involucel $\frac{1}{4}$–$\frac{3}{4}$-way up the pubescent and glandular peduncle.   Corolla puberulous outside (glabrous *fide* Harms), 2–4 times as long as calyx.   Pods doubtful.

Tanganyika.   Mwanza District: Ujashi, 10 Sept. 1951, *Tanner* 330!; Shinyanga, *Koritschoner* 1739 !; Nzega District: 32 km. N. of Igunga, 23 July 1949, *Doggett* 129 !; Kondoa District: Salia, 23 December 1927, *B. D. Burtt* 1131 !

DISTR. T1, 2, 4, 5; not known elsewhere
HAB. On hard-pan grey soils, normally growing in patches of many trees crowded together; 1220–1520 m.

NOTE. Harms (see above reference) described some fragmentary pods that may (there was doubt about it) belong; these were " flat, lanceolate, brownish, smooth and . . . on both margins with a prominent longitudinal ridge ". *Greenway* 9055 ! (Tanganyika, Masai District, Subiti) has a single complete valve of a pod. It is coriaceous, though not much thickened, about 14 cm. long and 1·4 cm. wide, flattened, smooth, thickened at margins and densely puberulous and glandular. It is abruptly curved, but this may possibly be abnormal.

The comparatively broad rigid-looking leaf-rhachis is a characteristic feature of *A. fischeri.*

57. **A. sieberiana** *DC.,* Prodr. 2 : 463 (1825) ; L.T.A. : 836 (1930) ; Bogdan in Nature in E. Afr., ser. 2, No. 1 : 14 (1949) ; T.T.C.L. : 335 (1949) ; I.T.U., ed. 2 : 214, fig. 48 i (1952) ; Gilb. & Bout. in F.C.B. 3 : 166 (1952). Type : Senegal, *Sieber* 43 (G, holo., K, iso. !)

Tree 5–18 m. high ; bark usually grey and rough on trunk, sometimes light brown, or yellowish and flaking especially on branches. Young branchlets glabrous to tomentose, eglandular, green to grey or yellowish, later grey ; outer bark then usually flaking away to expose an olive or yellow inner layer. Stipules spinescent, straight, up to 9(–12·5) cm. or more long, whitish ; " ant-galls " and other prickles absent. Leaves : rhachis 2·5– 10 cm. long ; pinnae mostly 6–23(–35) pairs ; leaflets 14–45(–52) pairs, 2–6·5 mm. long, (0·5–)0·6–1·5 mm. wide, glabrous to ciliate, narrowly oblong, rounded to obtuse at apex ; midrib, and sometimes small lateral nerves also, somewhat prominent on both surfaces. Flowers white or very pale yellow, in heads on axillary peduncles 1·5–5 cm. long which are variable in indumentum but eglandular, and whose involucel is normally apical or in upper half of peduncle. Pods (Fig. 17/57, p. 68) straight or sometimes ± falcate, flattened but thick and almost woody in texture when dry, very slow in dehiscing, (8–)9–21 cm. long, (1·5–)1·7–3·5 cm. wide, ± smooth and glossy, without raised veins, glabrous or somewhat hairy. Seeds olive-grey, smooth, elliptic to subcircular, compressed, 9–12 mm. long, 7–8 mm. wide ; areole 7–9·5 mm. long, 5–6 mm. wide.

KEY TO INTRASPECIFIC VARIANTS

Young branchlets glabrous or nearly so ; branches of crown usually ascending . . . . . var. **sieberiana**
Young branchlets ± hairy, usually densely so ; branches of crown usually widely spreading :
Indumentum on branchlets usually neither markedly golden nor villous . . . . . . var. **vermoesenii**
Indumentum on branchlets normally villous and markedly golden, especially when young . . var. **woodii**

var. **sieberiana** ; F.W.T.A., ed. 2, 1 : 499 (1958)

Crown with ascending branches, less spreading than in the following. Young branch-lets glabrous or almost so.

UGANDA. West Nile District: Rhino Camp, 27 Mar. 1936, *Michelmore* 1398 ! & Laropi, *Eggeling* 914 *in F.H.* 1260 !; Teso District: Lake Kioga, Sambwa Peninsula, near Serere, 2 Mar. 1936, *Michelmore* 1206 !; Mengo District: Gomba Madu, Mar. 1932, *Eggeling* 300 *in F.H.* 540 !
KENYA. Central Kavirondo District: Alego, 31 Apr. 1944, *Davidson* 218 *in Bally* 4298 !; Mombasa, Feb. 1930, *R. M. Graham* 2281 !
TANGANYIKA. Mpwapwa District: Matamondo, 26 Nov. 1940, *Hornby* 2100 !; Kilosa District: Kidodi, Oct. 1952, *Semsei* 967 !; Masasi District: Likesse, 23 Mar. 1943, *Gillman* 1253 !

DISTR. U1–4; **K**2, 5, 7; **T**2–6, 8; Senegal, Nigeria, French Cameroons, Belgian Congo, the Sudan, Ethiopia, Portuguese East Africa and Nyasaland
HAB. Deciduous woodland, wooded grassland, and also recorded from riverine forest; from near sea-level to 1220 m.

SYN. *A. verugera* Schweinf. in Linnaea 35: 340, t. 9, 10 (1867–8); Taub. in P.O.A. C: 195 (1895). Type: the Sudan, Kassala, by the R. Gasch, *Schweinfurth* 1963 (B, syn. †, EA, K, isosyn. !)
    A. purpurascens Vatke in Oesterr. Bot. Zeitschr. 30: 277 (1877); Bogdan in Nature in East Africa, ser. 2, No. 1: 14 (1949); T.S.K.: 70 (1936); T.T.C.L.: 335 (1949). Type: Kenya, near Mombasa, *Hildebrandt* 1938 (BM, K, iso. !)

NOTE. In Kenya and Tanganyika especially, the epidermis on the smaller twigs falls away quickly, exposing a yellow often very flaky surface, reminiscent of *A. xantho-phloea*. This feature is shown by the type of *A. purpurascens*, but though the peeling is certainly less obvious in *A. sieberiana* var. *sieberiana* of Uganda and West Africa, there are no other differences, and it does not seem possible to draw any clear distinc-tion between the two. Field observations are required to see if the bark of *A. sieberiana* shows any differences correlated with geography or climate.

    var. **vermoesenii** (*De Wild.*) *Keay & Brenan* in K.B. 1950: 364 (1951). Type: Belgian Congo, Boma, *Vermoesen* 1378 (BR, holo. !)

Crown usually with spreading branches, broad, flat or mushroom-shaped. Young branchlets ± hairy, usually densely so; indumentum usually neither markedly golden nor villous. Pods glabrous or nearly so, even when young.

UGANDA. West Nile District: Arua, Mar. 1934, *Tothill* 2539 !; Ankole District: Ruizi R., Feb. 1951, *Jarrett* 33 !; Kigezi District: Katete, Feb. 1950, *Purseglove* 3315 !; Mengo District: Kampala–Jinja road, Sezibwa Falls, Apr. 1932, *Eggeling* 676 !
KENYA. Nakuru District: Rongai, 29 Mar. 1944, *Vet. Dept. in Bally* 3146 *in C.M.* 11814 ! & 32 km. NW. of Nakuru, near Molo R., 18 Sept. 1948, *Bogdan* 2086 !
TANGANYIKA. Mbulu District: Mbulumbul, 24 June 1944, *Greenway* 6959 !; Ufipa District: Kisa, 5 Nov. 1933, *Michelmore* 734 !; Mpwapwa District: on path to Kiboriani, 25 Jan. 1933, *B. D. Burtt* 4535 !
DISTR. U1–4; **K**3, 5; **T**1, 2, 4–7; the eastern side of Africa from the Sudan and Ethiopia southwards to the Rhodesias and Portuguese East Africa; the closely related var. *villosa* in the French Sudan, Ghana and Nigeria
HAB. Woodland and wooded grassland; 950–1830 m.

SYN. *Inga nefasia* [Hochst. ex] A. Rich., Tent. Fl. Abyss. 1: 237 (1847), sensu stricto. Type: Ethiopia, without locality, *Schimper* 940 (P. syn., K, isosyn. !)
    Acacia nefasia ([Hochst. ex] A. Rich.) Schweinf. in Bull. Herb. Boiss. 4, app. 2: 209 (1896)
    A. verugera Schweinf. f. *latisiliqua* Harms in Z.A.E.: 235 (1910). Types: Belgian Congo and Ruanda, *Mildbraed* 587, 1104, 2108 (all B. syn. †); also cited from Mwanza and Bukoba, Tanganyika (*Holtz* 1551, 1630). A probable but doubtful synonym
    A. vermoeseni De Wild., Pl. Bequaert. 3: 69 (1925); T.T.C.L.: 335 (1949)
    [*A. sieberiana* sensu auct. mult., pro parte, *non* DC. sensu stricto]

See note under the following variety.

    var. **woodii** (*Burtt-Davy*) *Keay & Brenan* in K.B. 1950: 364 (1951); Consp. Fl. Angol. 2: 281 (1956); Coates Palgrave, Trees Centr. Afr.: 254–7 (1956). Type: Natal, between Estcourt and Colenso, *Medley Wood* 3528 (K. holo. !)

Crown as in var. *vermoesenii*. Young branchlets ± densely hairy; indumentum normally villous and markedly golden especially when young. Pods densely pubescent when young and usually slightly so even when old.

TANGANYIKA. Kondoa District: near Kinyassi, 2 Jan. 1928, *B. D. Burtt* 941 ! & near Kolo, 5 Jan. 1928, *B. D. Burtt* 1195 ! & Mbereko, 17 Dec. 1953, *F. G. Smith* 1042 !
DISTR. **T**5, ? 7; Portuguese East'Africa, Nyasaland, the Rhodesias, Angola, the Transvaal and Natal; probably occurs in the Belgian Congo
HAB. Probably similar to that of var. *vermoesenii*; 1520–1830 m.

SYN. *A. woodii* Burtt-Davy in K.B. 1922: 332 (1922); T.T.C.L.: 335 (1949), excl. *Wigg* 16
    [*A. abyssinica* sensu T.T.C.L.: 335 (1949), non [Hochst. ex] Benth.]

NOTE. The variations in East Africa of *A. sieberiana* seem to fall into two groups, var. *sieberiana* in one and vars. *vermoesenii* and *woodii* in the other. Mr. C. G. Trapnell writes (*in litt.*) of " the characteristic mushroom-shaped crown [of var. *vermoesenii*], of great width in proportion to the length of bole, which contrasts sharply in the field with the ascendent branching of *A. sieberiana*. Var. *vermoesenii* in west Uganda occupies higher rainfall areas and higher altitudes than the main species, the dividing line in the region in which we were working answering to about the 36 ins. isohyet." As far as can be ascertained the habit and ecology of var. *woodii* are decidedly those of var. *vermoesenii* and not var. *sieberiana*. At present vars. *vermoesenii* and *woodii* do not appear to share any common distinctive characters other than those mentioned in the descriptions above, except tendencies to produce more pinnae and wider pods than in var. *sieberiana*; thus the pods of var. *sieberiana* are up to about 2·5 cm. wide (rarely, as in *Greenway* 9103, from Tanganyika, Masai District, Seronera NE. to Naabi Hill, 2·2–3·2 cm. wide), while those of the other two varieties are often up to 3 and sometimes to 3·5 cm.; there is however much overlapping. Careful field-work is greatly needed in various parts of the range of *A. sieberiana*, which may show that the two groups mentioned in the first sentence are subspecies or even species; if the latter, then *A. nefasia* is the correct specific name for that group comprising *vermoesenii* and *woodii*.

At present it seems more prudent to maintain the three recognized varieties, especially because they are all connected by intermediates. These are particularly frequent between vars. *vermoesenii* and *woodii*, and it is hard to refer them either to one or the other. Examples are Tanganyika, Mpanda District, Kabungu, *Semsei* 59 *in F.H.* 2491!, Mpwapwa District, Mpwapwa, *Hornby* 735! & Kiboriani Mts., *Hornby* 944!, Mbeya District, Mbozi, *Jessel* 67! Intermediates between var. *sieberiana* and var. *vermoesenii* are shown by *Michelmore* 1308! and *Trapnell* 2188! both from Mbirizi in Masaka District, Uganda.

Wigg 16!, referred to *A. woodii* in T.T.C.L.: 335 (1949), is I think better placed under var. *vermoesenii*.

58. **A. nubica** *Benth.* in Hook., Lond. Journ. Bot. 1 : 498 (1842) ; Brenan in K.B. 1953 : 101 (1953). Type : the Sudan, Kordofan, *Kotschy* 407 (K, holo.!)

Shrub 1–5 m. high, with branches from base ; bark green below, usually pale grey to whitish or whitish-green above. Young branchlets glabrous to pubescent with short spreading hairs to 0·5(–0·75) mm. long, greenish, going whitish to grey-brown ; epidermis not peeling or flaking ; lenticels pale, dot-like. Stipules spinescent, straight or almost so (at least in our area), 0·4–1·7(–2·7) cm. long ; " ant-galls " and other prickles absent. Leaves : rhachis mostly (1·5–)2–4(–6) cm. long, rarely shorter, pubescent to subglabrous ; pinnae (2–)3–7(–11) pairs ; leaflets 5–16 pairs, 2·5–6(–9) mm. long, (0·5–)0·75–2·5 mm. wide, ± ciliate to glabrous. Flowers white, cream or greenish (perianth and anthers pink to red), in heads on axillary, pubescent, eglandular peduncles 0·5–1·5 cm. long ; involucel below or sometimes about middle of peduncle. Corolla-lobes conspicuously pubescent outside.   Pods (Fig. 17/58, p. 68) straight or sometimes slightly curved, coriaceous, dehiscent, 4–13 cm. long, 0·9–2·2 cm. wide, puberulous to densely and shortly pubescent, straw-coloured to pale brown or grey-brown ; valves with a convex longitudinally veined central part and (in our area) with a narrow flat wing-like margin 1–3·5 mm. wide. Seeds usually flinty-grey and shallowly and closely wrinkled (under a lens), globose or sometimes ellipsoid, not or scarcely compressed, 4·5–6·5 mm. long, 3·5–6 mm. wide ; areole 4–5 mm. long, 3–3·5 mm. wide.

UGANDA. Karamoja District: Kanamugit, Feb. 1936, *Eggeling* 2951! & Toror–Moroto road, about km. 56, 8 Oct. 1952, *Verdcourt* 798!
KENYA. Northern Frontier Province: Dandu, 4 June 1952, *Gillett* 13220!; Turkana District: 128 km. N. of Lodwar, 21 May 1953, *Padwa* 180!; Teita District: near Maungu, Jan. 1938, *Dale in F.H.* 3882!
TANGANYIKA. Masai District: Longido district, 9 Aug. 1951, *Greenway* 8581! & Engaruka, 29 May 1955, *Disney* 37!; Nzega District: W. edge of " Wembere Steppe ", at Sakamaliwa, 25 July 1931, *B. D. Burtt* 3388!
DISTR. U1, K1, 2, 6, 7; T2–5; in NE. Africa from Egypt southwards to our area; also in Arabia and Persia
HAB. Deciduous bushland, dry scrub with trees, and probably in semi-desert scrub; 600–1370 m.

SYN.　*A. virchowiana* Vatke in Oesterr. Bot. Zeitschr. 30: 275 (1880), pro parte, quoad
　　legumina tantum. Type: Kenya, Teita District, Voi R. and elsewhere, *Hilde-*
　　brandt 2486 (B ?, holo. †, K, iso.!)
　　A. merkeri Harms in E.J. 36: 208 (1905). Type: Tanganyika ?, " Masai Steppe, "
　　Merker (B, holo. †, BM, drawing !)
　　[*A. orfota* sensu auct. mult., e.g. L.T.A.: 839 (1930); T.T.C.L.: 336 (1949); Bogdan
　　in Nature in East Africa, ser. 2, No. 1: 14 (1949); I.T.U., ed. 2: 211, fig. 48 f
　　(1952); *non* (Forsk.) Schweinf.]

NOTE. The living plant is said to give off a strong bad smell when cut. *A. nubica* shows
　　a good deal of variation in our area, particularly in the indumentum and the width
　　of the pod. The young branchlets vary from glabrous to pubescent, the latter con-
　　dition appearing to occur most frequently in Uganda. The pods may be comparatively
　　narrow or wide, and inconspicuously puberulous to strongly pubescent; Uganda speci-
　　mens usually have rather wide pubescent pods. The variation in our area does not,
　　however, show any clear pattern at present.
　　Variants occur elsewhere with the spines somewhat curved, or with the marginal
　　wing to the pod practically absent, but they are unlikely to be found with us.

58 × 59. A. nubica *Benth.* × paolii *Chiov.*

Young branchlets similar to those of *A. nubica*, pubescent with hairs
mostly 0·5–1 mm. long. Stipular spines as in *A. nubica*. Leaves not seen.
Bracteoles densely pubescent. Calyces pubescent on lobes outside. Flowers
otherwise similar to those of *A. nubica*. Pods straight, coriaceous, dehiscent,
9–12 cm. long, 0·8–1 cm. wide, pubescent with rather long hairs 0·5–1 mm.
long, brown to grey-brown, longitudinally veined, attenuate for about
1·5 cm. at base, beaked for 0·5–1·5 cm. at apex; marginal wing 0 or very
narrow, up to about 1 mm. wide. Seeds ellipsoid, apparently well-formed.

Differs from *A. nubica* in the longer hairs on branchlets and pods, in
the very narrow or absent marginal wings to the pods, which have a more
or less obvious beak at apex. Differs from *A. paolii* in the shorter hairs on
branchlets and pods, in the pubescent bracteoles, the calyces pubescent on
the lobes outside, and in the straight or nearly straight pods.

KENYA. Northern Frontier Province: Dandu, 15 June 1952, *Gillett* 13439 !
DISTR. K1; not known elsewhere
HAB. *Commiphora-Acacia* deciduous bushland; 760 m.

NOTE. The collector suspected in the field that 13439 might be a hybrid with the parent-
　　age given above. This suggestion seems most probably correct. 13439 was growing
　　alongside *A. paolii*, with *A. nubica* in the area.
　　13439 is similar in some ways to No. 60, *A.* sp. B (see p. 131) but differs in having
　　pubescent bracteoles and calyces.

59. A. paolii *Chiov.* in Ann. Bot. Roma 13 : 395 (1915) ; L.T.A. : 848

(1930). Types : Ethiopia, Ogaden, between Bardera and Mansur, *Paoli* 578
(FI, syn. !) & Heima, *Paoli* 611 (FI, syn. !)

Usually a shrub up to 1·5–2·4 m. high, branching from the base, obconical
and ± flat-crowned, sometimes a small tree to 5 m. ; bark dark green,
smooth, apparently without papery-peeling. Young branchlets and their
indumentum as in *A. stuhlmannii* except that the hairs are less golden when
young, usually fewer and more quickly disappearing. Stipular spines as in *A.*
stuhlmannii, up to 5 cm. long. Leaves : rhachis 2·5–7 cm. long, pubescent ;
pinnae 4–9 pairs ; leaflets 8–15 pairs, (2–)3–7 mm. long, 1–1·75 mm. wide,
with ± appressed cilia. Flowers in heads on axillary, pubescent, eglandular
peduncles 0·7–2 cm. long ; involucel basal or in lower quarter of peduncle.
Bracteoles and calyces glabrous or almost so. Corolla-lobes conspicuously
hairy or pubescent outside. Stamen-filaments white ; anthers red. Pods
(Fig. 17/59, p. 68) somewhat falcate, coriaceous, dehiscent, attenuate for
about 2–3 cm. at base and apex, (6–)7–12·5 cm. long, 0·6–1·0 cm. wide,
densely clothed with whitish spreading hairs up to 3–4 mm. long, the hairs
not matted and the finely longitudinally veined surface of the valves easily
visible. Seeds olive-brown, minutely punctate, ellipsoid, not or scarcely

compressed, 6–8 mm. long, 4·5–5 mm. wide ; areole 5–6 mm. long, 3–3·5 mm. wide.

Kenya. Northern Frontier Province: Ajao road near Buna Jan. 1949, *Dale* K750 ! & 5 km. SW. of Takabba, 21 May 1952, *Gillett* 13255 ! & about 5 km. W. of Dandu on the Gadaduma road, 15 June 1952, *Gillett* 13438 !; Meru District: Isiolo, 9 Mar. 1952, *Gillett* 12520 !

Distr. K1, 4; Somalia, the Sudan

Hab. Wooded grassland and deciduous bushland, locally frequent or dominant; often on alluvial or colluvial soils; 140–1090 m.

Note. Outstanding by its long narrow dehiscent pods with a profusion of long white whiskery hairs. The latter evidently grow after fertilization, as the ovary in the flower is quite glabrous.

60. A. sp. B

Very similar to *A. paolii*. Obconical bush to 2 m., but said (*Buxton* 1021) to grow sometimes to tree-size. Young branchlets with short hairs to about 0·5(–0·75) mm. long, soon becoming glabrous, olive-brown to purplish-brown (not olive or pallid when dry as usually in *A. paolii*) ; epidermis marked with pale dot-like lenticels, wrinkled when dry but not cracking or peeling. Stipules spinescent, straight, 0·5–3·5 cm. long ; " ant-galls " and other prickles absent. Leaves : rhachis 0·5–2 cm. long, pubescent ; pinnae 1–5 pairs ; leaflets 7–9 pairs, appressed-ciliolate, 2–4 mm. long, 0·8–1·5 mm. wide. Flowers : only withered remains seen, in heads, on axillary pubescent eglandular peduncles 0·4–1·5 cm. long ; involucel basal or in lower quarter of peduncle ; bracteoles glabrous or almost so. Calyx glabrous or almost so. Corolla-lobes apparently hairy outside. Pods (Fig. 17/60, p. 68) slightly falcate, coriaceous, dehiscent, attenuate for about 0·5–1 cm. at base, more shortly beaked at apex than in *A. paolii*, 5–9 cm. long, 0·7–0·9 cm. wide, densely clothed with whitish rather ascending hairs up to 1–1·5 mm. long, the hairs rather matted and mostly concealing the surface of the valves. Seeds apparently as in *A. paolii* (certainly punctate).

Kenya. W. Turkana: Logiriama, Apr. 1944, *Dale* K378 !; " N. Turkana ", June 1934, *Buxton* 1021 !

Distr. K2 ; not known elsewhere

Hab. " Occurring on badly eroded dry soils " (*Dale*); " abundant " (*Buxton*); about 900 m.

Note. This is unquestionably closely related to *A. paolii*, differing in the absence of long hairs on the young branchlets, the usually shorter pods more densely clothed with shorter hairs and less attentuate at either end, and possibly in the smaller leaves with usually fewer pinnae. Not enough is known to decide if this is a distinct species or a race of *A. paolii* (which is not known from K2), but it seems likely that the differences are specific. More material is needed. According to Dale the vernacular name of this acacia in Turkana is " eiyuloit ".

61. A. stuhlmannii *Taub.* in P.O.A. C: 194, t. 21, E, F (1895) ; Harms in N.B.G.B. 4 : 196, fig. 1 (1906) ; L.T.A. : 836 (1930) ; T.S.K. : 71 (1936) ; T.T.C.L. : 334 (1949) ; Bogdan in Nature in E. Afr., ser. 2, No. 1 : 12 (1949). Types : Tanganyika : Dar es Salaam, *Stuhlmann* 6755 (B, syn.†, EA, probable isosyn., but number lacking !) ; Pangani, *Stuhlmann* 282 (B, syn.†) ; Tanga, *Volkens* 189 (B, syn.†) ; Amboni, *Holst* 2202 (B, syn.†, K, isosyn. !) ; Tanganyika/Kenya, Lake Jipe, *Volkens* 2383 (B, syn.†)

1–6(–7·5) m. high, varying from a low spreading bush to a small ± obconical-crowned tree. Young shoots with spreading golden villous hairs up to 1·5–3 mm. long, hairs later going grey ; branchlets becoming glabrescent, olive- to grey-brown, marked with pale dot-like lenticels, longitudinally wrinkled, but epidermis neither cracking nor peeling ; the old stems in the tree-form, however, may have papery-peeling golden-brown bark over a green layer. Stipules spinescent, straight, 0·7–4·5(–6·5) cm. long ; " ant-galls " and other prickles absent. Leaves : rhachis usually 2–5 cm. long, spreading-hairy ; pinnae 4–8(–12) pairs* ; leaflets 7–25 pairs, ciliate, 2–5·5 mm. long, (0·8–)1–1·5(–2) mm. wide. Flowers white with reddish-buff or mauve anthers, in heads on axillary, densely hairy or tomentose, eglandular peduncles 0·4–3 cm. long, often produced when the plant is without leaves ; involucel basal or in lower half of peduncle** ; bracteoles conspicuously ciliate or pubescent. Calyx ± pubescent outside. Corolla-lobes conspicuously pubescent outside. Pods (Fig. 17/61, p. 68) somewhat curved or sometimes straight, thick, hard and woody, indehiscent, usually much attenuate at base, densely clothed with long spreading hairs, (2–)4–9 (–10·5) cm. long, (1·1–)1·2–2·5(–3) cm. wide. Seeds olive, minutely punctate, ellipsoid to subglobose, 6–9 mm. in diameter ; central areole 6–7 mm. long, 4·5–5 mm. wide. Fig. 19.

KENYA. Northern Frontier Province: 60 km. SW. of Mandera, 30 May 1952, *Gillett* 13396 !; Machakos District: between Stony Athi and Machakos, *Trapnell* 2216 !; Mombasa, 25 Nov. 1951, *Bogdan* 3311 !
TANGANYIKA. Shinyanga District: Uduhe, near Mango, 23 Jan. 1936, *B. D. Burtt* 5503 !; Tanga Bay, N. of Tanga, 25 Nov. 1936, *Greenway* 4779 !; Dodoma District: Saranda, 5 Sept. 1931, *B. D. Burtt* 3400 !
DISTR. K1, 4, 6, 7; T1–3, 5, 6; Somaliland Protectorate, Somalia, Bechuanaland and the Transvaal
HAB. Wooded grassland and deciduous bushland; often on heavy alluvial soils; said also to occur on the coast on margins of mangroves and to be a good indicator of saline soils; 0–1740 m.

NOTE. This species is rather variable in our area. The variants may be provisionally grouped into four categories, showing some indications of morphological as well as geographical distinctness.

(1) Kenya, Northern Frontier Province, about 390 m. Shrub with several rather erect stems from base, to about 4 m. tall. Peduncles to about 1 cm. long. Pods small, 2–4·5 × 1–1·4 cm. Example: *Gillett* 13396. A rather similar specimen has been collected in the Transvaal (*Smuts & Gillett* 4035 !).

(2) Kenya Highlands, about 1070–1740 m., on waterlogged black clays (black cotton soils). Spreading low bush to about 1·5 m. tall and 3–6 m. wide. Peduncles up to about 1·5 cm. long. Pods usually wider than (1). Examples: *van Someren in C.M.* 6759, *Bogdan* 1097, *Trapnell* 2216.

(3) Kenya and Tanganyika coast, 0–120 m. (? higher). Bush or small tree about 2·5–6 m. high. Peduncles mostly 1·5–3 cm. long. Pods as in (2). Examples: *R. M. Graham* 1670, *Bogdan* 3311, *Greenway* 4779, *Drummond & Hemsley* 1029. Typical *A. stuhlmannii* is probably referable here.

(4) Tanganyika, inland, about 850–1070 m. Shrub or bush about 2–3 m. high ?) higher). Peduncles as in (3). Pods as in (2). Examples: *Greenway* 786, *Michelmore* 838.

The status of category (1) is uncertain, owing to lack of material; (2) and (3) are rather distinct and separable from one another even in the herbarium by the peduncles and also by the leaflets of (3) being less ciliate than those of (2). Category (4) is hardly separable from (3) except by habit, in which it seems to bridge the difference between (3) and (2).

Battiscombe 272 *in C.M.* 13984 was collected on the Tana R. in Kenya at 90 m. alt. This specimen has bright red-brown branchlets and red-purple spines, but in other ways resembles *A. stuhlmannii*, of which it may be a variant. Since the specimen lacks pods, its identity is uncertain, and more material is wanted. It is said to be " common . . . near the coast."

Two other difficult specimens have been collected in Kenya 53 km. NE. of Garissa in the Northern Frontier Province on 5 Jan. 1958 (*Hemming* 1302, 1303). They are similar vegetatively to *A. stuhlmannii* (of which they are probably a variant) except for the hairs being mostly 1·5 mm. long or less. The pods, however, are narrow, about 4–8 × 1–1·5 cm., only slightly compressed and somewhat longitudinally veined. They differ from those of the Northern Frontier Province form described above in being less compressed and with shorter indumentum.

*Occasional leaves on apparently juvenile non-flowering shoots may have up to 17 pairs of pinnae and a rhachis up to 8 cm. long.
**One specimen from Somaliland Protectorate, *Peck* 72, apparently *A. stuhlmannii*, has the involucel in the upper half of the peduncle or even apical. I have not seen such an abnormality from our area and have not allowed for it in the key to the species.

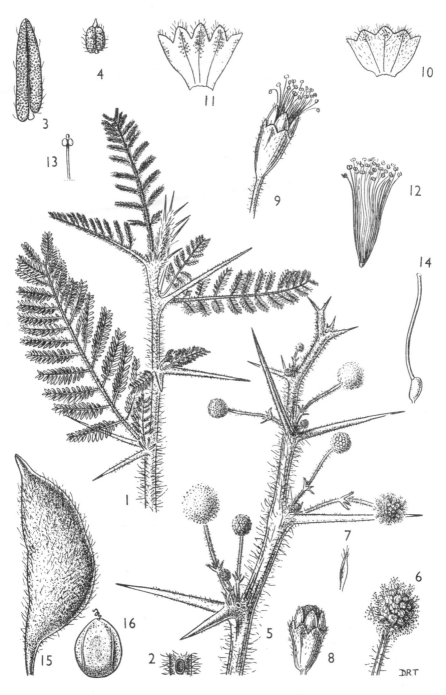

FIG. 19. *ACACIA STUHLMANNII*,—**1**, part of leafy branch, × 1; **2**, gland on petiole, × 6; **3**, leaflet, × 6; **4**, leaflet from lowest part of pinna, × 6; **5**, part of flowering branch, × 1; **6**, flower-head, × 2; **7**, bract subtending flower, × 6; **8**, flower-bud, × 6; **9**, flower, × 6; **10**, calyx, opened out, × 6; **11**, corolla opened out, × 6; **12**, stamens, × 6; **13**, anther with part of filament, × 12; **14**, ovary, × 6; **15**, pod, × 1; **16**, seed, × 2. 1–4, from *B. D. Burtt* 5503; 5–16, from *B. D. Burtt* 3400.

62. **A. edgeworthii** *T. Anders.* in J.L.S. 5, suppl. 1 : 18 (1860). Types :
Aden, *Edgeworth, Hooker & Thomson* (K, syn. !) & *T. Anderson* (K, syn. !)

Shrub 0·3–2 m. high, usually low, flat-topped and up to 2·4–4·5 m. wide,
branching from base or sometimes with a very short main stem exposed,
Young branchlets puberulous to pubescent or sometimes tomentose (hairs.
to 0·5 mm. long), usually glabrescent, grey to grey-brown, sometimes purplish-brown or dark grey ; epidermis not or inconspicuously peeling or
flaking ; lenticels inconspicuous, or pale and dot-like. Stipules spinescent,
straight or nearly so, slightly ascending to slightly deflexed, on mature shoots
(0·4–)1–3·5 cm. long (to 4·5 cm. on robust apparently juvenile shoots) ;
" ant-galls " and other prickles absent. Leaves : rhachis 0·5–4·5 cm. long,
pubescent or tomentose ; pinnae (3–)4–10 pairs ; leaflets 6–15(–20) pairs,
0·75–3·5 mm. long, 0·5–1·5 mm. wide, usually ciliate or pubescent. Flowers
white, in heads on axillary, tomentose or densely pubescent, eglandular
peduncles 0·4–1·5 cm. long ; involucel basal or in lower half of peduncle.
Calyx 1–2 mm. long. Corolla 2·5–3 mm. long ; lobes densely white-pubescent outside. Pods (Fig. 17/62, A & B, p. 68) falcate to straight or nearly
so, thick and woody, ultimately dehiscent, 7–13(–15) cm. long, 1·3–2·5 cm.
wide, densely velvety-pubescent or tomentellous, brown to brownish-
crimson, longitudinally veined, not winged. Seeds large, blackish, minutely
roughened or smooth, ellipsoid to subglobose, 9–13 mm. in diameter ; central
areole 7–10 mm. long, 3–4·5 mm. wide.

KENYA. Northern Frontier Province: Banessa–Ramu and Mandera, 23 May 1952,
    Gillett 13283 ! & N. of El Wak, 26 May 1952, *Gillett* 13333 !
DISTR. K1; Somalia, Somaliland Protectorate, Ethiopia and Socotra
HAB. Dry scrub (*Acacia-Commiphora*); 360–550 m.

SYN. *A. socotrana* Balf. f. in Proc. Roy. Soc. Edinb. 11: 511 (1882); & in Botany of
        Socotra (Trans. Roy. Soc. Edinb. 31): 87, t. 23 (1888); L.T.A.: 836 (1930).
        Types: Socotra, *Balfour, Cockburn & Scott* 191 (K, isosyn. !) & *Schweinfurth*
        260 (K, isosyn. !)
      A. pseudosocotrana Chiov., Fl. Somala 1: 161 (1929). Type: Somalia, Migiurtini,
        Hafun, *Puccioni & Stefanini* 38 (FI, holo. !)
      A. sultani Chiov., Fl. Somala 1: 162 (1929). Type: Somalia, Obbia, Magghiole,
        Puccioni & Stefanini 472 (524) (FI, holo. !)
      A. erythraea Chiov., Fl. Somala 1: 163 (1929). Type: Somalia, between Bulo
        Burti and Garass Hebla Aden, *Puccioni & Stefanini* 138 (157) (FI, holo. !)

NOTE. *Gillett* 13333 has strongly curved pods, while 13283 has them almost straight-
    there seems no other difference. *A. edgeworthii* is rather variable in facies, indumen;
    tum, etc., and it does not seem feasible to separate the Kenya species specifically or
    even varietally from the Socotran and Aden types.

63. **A. turnbulliana** *Brenan* in K.B. 1957 : 370 (1958). Type : Kenya,
23 km. NE. of Wajir, *Gillett* 13364 (K, holo. !, EA, iso. !)

Shrub 0·5–1·5 m. high, very flat-topped, spreading to 3·6 m. Young
branchlets densely hairy (hairs up to 0·5(–0·75) mm. long, yellowish when
young), slowly glabrescent, going pale to deep brown later ; epidermis not
or scarcely flaking or peeling ; lenticels pale, dot-like. Stipules spinescent,
straight, on mature shoots (2–)3·5–6·5 cm. long (? more) ; " ant-galls "
and other prickles absent. Leaves : rhachis (1·5–)4–7·5(–12) cm. long,
densely pubescent ; pinnae (4–)8–14(–23) pairs ; leaflets 8–30 pairs,
2·5–5 mm. long, 0·5–1·25 mm. wide, ciliate, paler beneath than above.
Flowers white, in heads on axillary, tomentose or densely pubescent, eglandular peduncles 1·5–3 cm. long ; involucel basal or in lower half of peduncle.
Calyx 3–3·75 mm. long. Corolla 4–5·5 mm. long, lobes densely white-
tomentose outside. Pods straight or nearly so, about 10–22 cm. long and
0·9–1·6 cm. wide, attenuate at both ends, densely grey-velvety, cylindrical
when ripe. Seeds large, blackish, ellipsoid, about 12 mm. long and 10 mm.
wide ; areole about 10 × 5 mm.

KENYA. Northern Frontier Province: 88 km. N. of Wajir, Jan. 1949, *Dale* K752! & 23 km. NE. of Wajir, 27 May 1952, *Gillett* 13364! & Wajir, Jan. 1955, *Hemming* 461! & 21 km. NE. of Garissa, 5 Jan. 1958, *Hemming* 1298!
DISTR. **K**1; not known elsewhere
HAB. Dry scrub; 150–270 m.

Insufficiently known species
64. A. sp. C

Dwarf, decumbent shrub about 0·4 m. high. Young branchlets pubescent with spreading hairs up to about 0·75 mm. long, slowly glabrescent, grey to grey-brown, becoming much branched but remaining rather slender. Stipules spinescent, slender, mostly grey, straight or the shorter ones slightly curved, 0·2–2·0 cm. long ; " ant-galls " and other prickles absent. Leaves small ; rhachis pubescent, only up to about 1 cm. long ; pinnae 2–4 pairs, short ; leaflets 4–11 pairs, 1·5–3 mm. long, 0·7–1 mm. wide, glabrous or slightly ciliate, lateral nerves invisible beneath. Flowers and pods unknown.

TANGANYIKA. Lushoto District: between Buiko and Lake Manka, 8 June 1926, *Peter* 41063! & Mkomazi, Oct. 1946, *Yussif bin Mohamedi* 3!
DISTR. **T**3; not known elsewhere
HAB. Dry scrub with trees, and semi-desert scrub; *Peter* 41063 occurred at about 400 m.

65. A. sp. D

Branchlets persistently pubescent, dark grey-brown. Stipular spines short, to about 5 mm., straight, conical, pointing diagonally upwards ; " ant-galls " and other prickles absent. Leaves : rhachis about 1–3·5 cm. long, pubescent ; pinnae 2–4 pairs ; leaflets 8–14 pairs, 5–8 mm. long, 1·5–2·5 mm. wide with spreading pubescence on both surfaces ; lateral nerves (except sometimes a basal one) not visible. Flowers in heads on axillary pubescent and very glandular peduncles ; involucel at or above the middle. Corolla glabrous outside. Pods unknown.

TANGANYIKA. Moshi District: Sanya, Aug. 1928, *Haarer* 1546!
DISTR. **T**2; not known elsewhere
HAB. Unknown; about 1220 m.

NOTE. The glandular peduncles suggest a possible relationship with 33, *A. hockii*, of which *Haarer* 1546 might conceivably be a very aberrant form. The larger leaflets are, however, distinctive.

66. A. sp. E

Small sparsely branched tree 2·1 m. high, from a short stump. Young branchlets minutely and ± densely puberulous, eglandular. Stipular spines straight or nearly so, some short, up to about 1 cm. long, others up to 5 cm. long and going ashen-grey ; some of the long pairs of spines fused at base into deeply bilobed mostly grey " ant-galls," the lobes attenuate-conical. Leaves : rhachis usually 2–4 cm., puberulous ; pinnae 6–11 pairs ; leaflets 15–20 pairs, 2–3·5 mm. long, 0·75–1 mm. wide, glabrous except for slight and inconspicuous puberulence on margins, subacute or acute at apex ; lateral nerves invisible. Flowers and pods unknown.

TANGANYIKA. Iringa District: on the Kilosa road near the Ruaha R., 26 Oct. 1936, *B. D. Burtt* 6039!
DISTR. **T**7; not known elsewhere
HAB. " *Acacia-Commiphora* desert bush "; 1070 m.

NOTE. The deeply bilobed " ant-galls " of *Burtt* 6039 recall those of *A. seyal* var. *fistula*. There is, however, probably no close affinity between them, as *Burtt* 6039 has puberulous, eglandular twigs and no sign of powdery bark.

67. A. sp. F

Shrub or tree to 7·5 m. high. Young branchlets ± densely pubescent ; epidermis here and there falling away to expose a rusty-red layer. Stipules spinescent, straight, up to 7 cm. long, some of them fused and inflated below into bilobed " ant-galls " up to about 2·5 cm. across. Leaves with 3–9 pairs of pinnae ; leaflets 10–17 pairs, similar to those of *A. gerrardii*. Flowers similar to those of *A. gerrardii;* involucel at or near base of peduncle. Pods unknown.

TANGANYIKA. Kondoa District: Sambala, 12 Mar. 1929, *B. D. Burtt* 1963! & North Sambala Hills, 28 Mar. 1929, *B. D. Burtt* 2001 !

DISTR. T5; not known elsewhere

HAB. Along valley in woodland; about 1550 m.

NOTE. This except for the " ant-galls ", is very similar to *A. gerrardii* var. *gerrardii* (see p. 119), and may indeed be a form of that species. However the pods are still unknown, and it seems preferable to keep this distinct for the present, especially as, although *A. gerrardii* is common and widespread, I have seen nothing to match this " galled " acacia elsewhere.

Although the gradually inflated bases to the spines make the " galls " ± bilobed, the central longitudinal furrow separating the lobes is very slight or absent.

17. ALBIZIA

Durazz., Magazz. Tosc. 3 (4) (vol. 12) : 10, 13, illust. (1772) *

Trees, sometimes shrubs, very rarely climbing (not so in Africa) ; prickles or spines absent in the African species (except for a very small prickle beneath the node in *A. harveyi* and that in *A. anthelmintica* some branchlets may be sharp and spinescent at ends) ; sharp hooks apparently representing petiole-bases present in a very few extra-African species. Leaves bipinnate ; pinnae each with one to many pairs of leaflets. Inflorescences of round heads, or (not in native African species) spikes or spiciform racemes, pedunculate, axillary and solitary or much more often fascicled, often aggregated near ends of branchlets, which may be lateral and much shortened, sometimes paniculately arranged. Flowers ☿ or occasionally ♂ and ☿; 1–2 central flowers in each head frequently larger, different in form from the others and apparently ♂. Calyx gamosepalous, with normally 5 teeth or lobes (rarely 4, 6 or 7). Corolla gamopetalous, infundibuliform or campanulate, with normally 5 lobes (rarely 4 or 6, or in *A. coriaria* and *A. tanganyicensis* the lobes may be irregularly connate among themselves). Stamens numerous (19–50), fertile, their filaments united in their lower part into a slender tube sometimes projecting from, sometimes shorter than the corolla. Pods oblong, straight, flat, usually dehiscent, not septate inside, the valves papery to rigidly coriaceous but not thickened or fleshy. Seeds usually ± flattened.

A genus of about 100–150 species throughout the tropics, a few in the subtropics.

The generic name is often misspelt *Albizzia;* for the reasons for rejecting this spelling see Little in Amer. Midl. Nat. 33: 510 (1945).

KEY TO EXOTIC SPECIES

Several exotic species have been introduced into our area, which may be keyed as follows. I have seen East African material of all except *A. caribaea*, which is taken from U.O.P.Z. : 113.

Inflorescences spicate :

Flowers pedicellate ; spikes axillary .   . *A. lophantha* (Willd.) Benth. (*A. distachya* (Vent.) Macbr.) (Native of Australia)

Flowers sessile ; spikes paniculate .   . *A. falcata* (L.) Backer (Native of the Malay Islands)

*I have not seen this very rare work. Reference from Little in Amer. Midl. Nat. 33: 510 (1945).

Inflorescences capitate :
  Midrib of leaflets running along or near
their upper margin to base ; leaflets
usually ± pubescent or puberulous beneath, acute at apex, but not apiculate
or acuminate ; midrib marginal ; stipules large and conspicuous, but
quickly falling ; flowers mostly
creamy .   .   .   .   .   *A. chinensis* (Osbeck) Merr.
(Native of tropical Asia)

  Midrib of leaflets not marginal :
    Leaflets very small and numerous, about
1–1·5 mm. wide, glabrous or sub-
glabrous (margins puberulous)   . *A. caribaea* (Urb.) Britton & Rose (Native of
West Indies and central
America)

    Leaflets 3·5 mm. or more wide :
      Pairs of leaflets 2–4 ; leaflets elliptic . *A. saponaria* (Lour.) Bl. ex
Miq. (Native of East
Indies)

      Pairs of leaflets 5 or more :
        Calyx glabrous or nearly so, 2·5–3 mm.
long ; leaflets mostly 12 mm. or
more wide   .   . *A. procera* (Roxb.) Benth.
(Native of tropical Asia)

        Calyx densely pubescent, about 1 mm.
long ; leaflets mostly 0·35–
12 mm. wide .   .   . *A. odoratissima* (L.f.) Benth.
(Native of tropical Asia)

Key to native and naturalized species

  In the following key the descriptions of floral parts must not be taken to
apply to the 1–2 larger modified flowers commonly present in the centre of
the heads ; in these the staminal tube is not or scarcely exserted even when
it is longly exserted in the others.

Staminal tube not or scarcely projecting beyond the corolla :
  Leaflets small or very small, 0·5–3(–4) mm.
wide, usually in numerous pairs (pairs
usually 12–48, but sometimes as few as
5) ; pinnae 2–48 pairs :
  The leaflets mostly very small, 0·5–1·5
mm. wide and 2–6 mm. long :
    Apex of leaflets acute, asymmetrical,
the point turned towards pinna-
apex ; stamen-filaments about
1·5–2 cm. long ; bracteoles persistent till flowers open ; pods
glabrous or nearly so except for a
little pubescence near base and
along margins ; lateral nerves of
leaflets ± raised and visible beneath   .   .   .   . 9. *A. harveyi*

Apex of leaflets obtuse or sometimes subacute, symmetrical ; stamen-filaments about 0·5–1·2 cm. long ; bracteoles already fallen when flowers open ; pods minutely puberulous over surface ; lateral nerves of leaflets usually not distinct beneath . . . 11. *A. amara*

The leaflets mostly 1·25–4 mm. wide and (4–)6·5–12 mm. long, rounded to subacute at apex ; bracteoles normally already fallen when the flowers open ; pods normally puberulous over surface ; lateral nerves of leaflets usually ± raised and visible beneath :

Calyx 3–5 mm. long ; pods closely transversely venose, the veins ± parallel and about 2–4 mm. apart ; seeds 4·5–6·5 mm. wide, nearly twice as long as broad ; flowers sessile or almost so ; leaflets asymmetrical at apex . . . 10. *A. forbesii*

Calyx 1–2·5 mm. long ; seeds 7 mm. wide or more :

Pedicels of flowers 1·5–6 mm. long :

Leaflets 1·25–4 mm. wide ; pedicels 1·5–1·75 mm. long ; indumentum on outside of calyx and corolla grey . . . 12. *A. isenbergiana*

Leaflets mostly 4 mm. wide or more ; pedicels 2–6 mm. long ; indumentum on outside of calyx and corolla grey or, much more commonly, rusty . 14. *A. schimperiana*

Pedicels of flowers 0–1·5 mm. long :

Leaflets 3–8 mm. wide ; indumentum on outside of calyx and corolla ± rusty (when dry) ; pods normally with very prominent transverse veins which are raised in centre, and sometimes almost wing-like . . . . 13. *A. zimmermannii*

Leaflets up to 2·5 mm. wide ; indumentum on outside of calyx and corolla grey to golden (when dry) ; pods not prominently veined as above . 11. *A. amara*

Leaflets medium to large (3·5–)4–45 mm. wide, in 1–20(–23) pairs ; pinnae 1–8 (–11) pairs :

Rhachides of leaves and pinnae of all or most leaves projecting at ends as a short rigid persistent deflexed or downwards-bent hook or claw ; often a single stipel similarly bent near

base of pinnae ; flowers usually
precocious on almost or quite leafless
shoots, with calyces and corollas
glabrous or sparsely puberulous outside ; pinnae 1–4 pairs ; leaflets
usually about 3 pairs (range 1–5
pairs); [if indumentum on flowers
rusty, compare 5, *A. versicolor* and
3, *A. tanganyicensis*, below]　.　.　8. *A. anthelmintica*

Rhachides of leaves and pinnae not projecting at ends, or else projections
straight, not hooked or deflexed
(except rarely and casually in 5, *A.
versicolor*) and usually caducous ;
calyces and corollas usually ± densely
puberulous to tomentose outside, or
if glabrous then flowers not precocious
(except sometimes in 3, *A. tanganyicensis*) :

Stamen-filaments 1·5–5 cm. long ; calyx
normally 3–7 mm. long (in 6, *A.
lebbeck* sometimes only 2·5 mm.) ;
corolla frequently more than
8 mm. long ; pods glabrous and
often glossy on surface, or with
a few hairs along margins and at
base only (in 4, *A. malacophylla*
rarely puberulous all over) :

Filaments of stamens red above and
white below ; leaflets mostly in
6–11 pairs, subglabrous or thinly
puberulous, oblong to elliptic-
or ovate-oblong　.　.　.　1. *A. coriaria*

Filaments of stamens white to green
or greenish-yellow, not red :

Leaflets beneath grey to whitish
and very glaucous, glabrous
on both sides, ovate to rhombic-ovate or elliptic-oblong ;
flowers sessile or up to 2 mm.
pedicellate　.　.　.　.　7. *A. antunesiana*

Leaflets not very glaucous beneath, or if so then leaflets ±
pubescent to subtomentose :

Indumentum on outside of calyx
and corolla conspicuously
rusty (when dry), at least
on the lobes :

Leaflets in 3–5 (occasionally 6)
pairs, densely pubescent
to tomentose beneath,
mostly broadly obovate
to suborbicular ; young
branchlets densely rusty-
tomentose ; pods chest-
nut-brown to crimson ;
bark rough　.　.　.　5. *A. versicolor*

Leaflets (at least of the distal pinnae) always in 7 or more pairs :

The leaflets oblong, not more than 10 mm. wide ; flowers not precocious ; bracteoles often persistent at flowering time ; pods with rather thin valves abruptly narrowed near base ; bark rough, not peeling .   .   .   2. *A. ferruginea*

The leaflets ovate-elliptic to ovate-oblong, 6–29 (–32) mm. wide ; flowers precocious, usually produced before the young leaves appear ; bracteoles already fallen when the flowers open ; pods with rather thick and stiff valves gradually narrowed near base ; bark smooth, peeling off in papery pieces .   3. *A. tanganyicensis*

Indumentum on outside of calyx grey to whitish, not rusty :

Pedicels 1·5–4·5(–7·5) mm. long ; leaflets usually not glaucous beneath, subglabrous or rarely pubescent ; stamen-filaments green or greenish-yellow in upper part ; pods straw-coloured .   .   .   6. *A. lebbeck*

Pedicels 0–0·75. mm. long ; leaflets glaucescent beneath and sparsely pubescent to subtomentose ; stamen-filaments white .   .   4. *A. malacophylla*

Stamen-filaments 0·5–1·3 cm. long ; calyx 1–2·5 (–3) mm. long ; corolla 3–7·5 mm. long ; pods puberulous over their surface and not glossy :

Pedicels 1·5–7 mm. long ; pods not especially prominently veined :

Indumentum on outside of calyx and corolla grey (when dry) :

Leaflets in 3–6 (rarely 8) pairs, 9–40 mm. wide   .   .   15. *A. glaberrima*

Leaflets in 6–17 pairs, 4–6 mm. wide .   .   .   .   14. *A. schimperiana* var. *tephrocalyx*

Indumentum on outside of calyx and corolla brown (when dry) ;

leaflets on the two distal pairs of pinnae 8–23 pairs, 3·5–8·5 mm. wide .   .   .   .   14. *A. schimperiana* var.
schimperiana

Pedicels 0–1 mm. long ; pods normally with very prominent transverse veins, raised in centre and sometimes almost wing-like ; leaflets rounded at apex, 4–8 mm. wide .   .   .   .   13 *A. zimmermannii*

Staminal tube projecting beyond the corolla for a length of about 0·7–2·5 cm. (usually more than 1 cm.), usually red, pink or greenish, at least partly :

Calyx and corolla ± puberulous to pubescent outside, the former (2–) 2·5–5 mm. long :

3–6 pairs of leaflets per pinna (of the 2 distal pairs), the terminal pairs larger than the others, the pinnae thus broadening upwards :

Leaflets not auriculate at base on the proximal side :

The leaflets normally in 3–5 pairs, rarely 6 ; pinnae normally in 1–3 pairs, rarely 4 :

Bracts at base of peduncles, also stipules, broadly ovate, suborbicular or reniform, 8–20 mm. wide ; leaflets ± thinly pubescent all over beneath ; terminal leaflets acute or subacute at apex ; young branchlets densely pubescent .   .   19. *A. grandibracteata*

Bracts at base of peduncles, also stipules, very small and inconspicuous, narrowly lanceolate, about 1–2 mm. wide ; leaflets glabrous beneath except on midribs and margins ; terminal leaflets obtuse at apex ; young branchlets minutely puberulous .   .   .   .   18. *A. zygia*

The leaflets rarely as few as 6 pairs, and then normally accompanied by pinnae with more pairs on other leaves ; pinnae in 3–6 pairs .   .   .   .   16 × 18. *A. gummifera* × *zygia*

Leaflets auriculate at base on the proximal side ; bracts and stipules ± broadly ovate ; leaflets beneath with sparse hair between midrib and margins, often acute or subacute at apex ; peduncles ± densely pubescent .   .   19 × 16. *A. grandibracteata* × *gummifera*

7 or more pairs of leaflets per pinna (of the
2 distal pairs) :
The leaflets not auriculate at base
on the proximal side, though
the proximal margin may be ±
rounded to the insertion of the
petiole :
Terminal pairs of leaflets rather
larger than the rest, the pinnae
thus ± broadening upwards ;
pinnae in 3–6 pairs ; leaflets in
(6–)7–9 pairs :
Leaflets very thinly pubescent on
lower surface between midrib
and margins ; bracts and
stipules usually broadly ovate,
4–8 mm. wide ; uppermost
leaflets acute . . 19 × 16. *A. grandibracteata* ×
gummifera

Leaflets glabrous beneath except
on midrib and margins; bracts
and stipules narrowly lanceo-
late . . . . 16 × 18. *A. gummifera* × *zygia*
Terminal pairs of leaflets rather
smaller than the rest, the pinnae
thus ± narrowing upwards ;
pinnae in 4–9 pairs (rarely as
few as 3 pairs, and then only on
occasional reduced leaves) ;
leaflets in 9–17 pairs :
Young branchlets and rhachides of
leaves and pinnae finely and
shortly pubescent ; leaflets
pubescent beneath only on
midrib and margins, glabrous
between or rarely, especially
when young, with some occasional hairs on primary lateral
nerves ; stipules lanceolate,
up to 6–7 × 2–2·5 mm. ; pods
glabrescent . . . . 16. *A. gummifera* var.
ealaënsis

Young branchlets and rhachides
of leaves and pinnae densely
fulvous-pubescent ; leaflets ±
pubescent all over lower surface ; stipules ovate, about
5–12 × 3–8 mm. ; pods ±
densely and persistently pubescent . . . . . 17. *A. adianthifolia*
The leaflets markedly auriculate at base
on the proximal side :
Terminal pairs of leaflets rather
larger than the rest, the pinnae
thus ± broadening upwards ;
bracts and stipules ± broadly
ovate ; leaflets in not more

than 10 pairs, with sparse hair
beneath between midrib and
margins, often acute or subacute at apex ; pinnae in 2–5
pairs   .   .   .   . 19 × 16. *A. grandibracteata* ×
                                                    gummifera

Terminal pairs of leaflets rather
smaller than the rest, the pinnae
thus ± narrowing upwards ;
bracts and stipules normally
lanceolate ; leaflets in 9–17
pairs, pubescent beneath only on
midrib and margins, glabrous between or rarely, especially when
young, some occasional hairs on
primary lateral nerves ; pinnae
in 5–7 pairs (rarely as few as 3
pairs and then only on occasional
reduced leaves)   .   .   . 16. *A. gummifera*
Calyx (except for margins) and corolla
glabrous outside, the former 1–2 mm.
long :
Leaflets in 5–12 pairs, 2·5–13(–17) mm.
wide   .   .   .   .   .   . 20. *A. petersiana*
Leaflets in 1–3 pairs, 15–35 mm. wide or
" somewhat narrower "   .   .   . 21. *A. euryphylla*

1. **A. coriaria** [*Welw. ex*] *Oliv.*, F.T.A. 2 : 360 (1871) ; L.T.A. : 861
(1930) ; T.S.K. : 72 (1936) ; Bogdan in Nature in E. Afr., No. 2 : 17 (1947) ;
T.T.C.L. : 342 (1949) ; Gilb. & Bout. in F.C.B. 3 : 187 (1952) ; I.T.U.,
ed. 2 : 217 (1952) ; Consp. Fl. Angol. 2 : 291 (1956) ; F.W.T.A., ed. 2, 1 :
502 (1958). Types : Angola, Cuanza Norte, Golungo Alto, *Welwitsch* 1762
(BM, K, isosyn. !) & 1764 (BM, isosyn. !) & 1764b (BM, isosyn. !) & Cazengo,
Welwitsch 1763 (BM, isosyn. !)

Tree 6–36 m. high ; crown spreading, flat ; bark rough, flaking off.
Young branchlets puberulous or shortly pubescent, later glabrescent.
Leaves : rhachis ± thinly crisped-puberulous or shortly pubescent ; pinnae
(2–)3–6(–8, *fide* I.T.U.) pairs ; leaflets (4–)6–11(–12) pairs, oblong to
elliptic- or ovate-oblong, 13–33 mm. long, 5–14(–17) mm. wide, rounded at
apex, subglabrous except for a few hairs on midrib beneath, or sometimes ±
thinly puberulous beneath especially towards base. Flowers subsessile or
on pedicels 0·5–2 mm. long ; bracteoles already fallen by flowering time,
minute, mostly 1·5–2 mm. long. Calyx 3·5–6·5 mm. long, not slit unilaterally, puberulous outside, with a few shortly stipitate glands (× 20 lens
necessary) principally on outside of lobes. Corolla 8–13·5 mm. long, white
or whitish, puberulous outside. Staminal tube not or scarcely exserted
beyond corolla ; filaments 1·7–4 cm. long, red above, white below. Pod
oblong, (10–)14–21 cm. long, (2·3–)3·2–3·7(–4·8) cm. wide, glabrous or
nearly so, ± glossy, obscurely venose, brown or purplish-brown, usually ±
tapered and acute at base and sometimes apex. Seeds about 9–12 mm. long
and 8–9 mm. wide, flattened.

UGANDA. Karamoja District: Kakumongole, 7 Jan. 1937, *A. S. Thomas* 2203 !; Teso
   District: Serere, Feb. 1932, *Chandler* 637 !; Mengo District: Mawokota, Feb. 1905,
   Brown 168 !
KENYA. N. Kavirondo District: Bukura, 8 Dec. 1943, *M. D. Graham* 27 !; Central Kavi-
   rondo District: Ukwala, 29 Aug. 1944, *Davidson* 228 *in Bally* 4412 !; Kisumu-Londiani
   District: Kibigori, *Green* 22 !

TANGANYIKA. Lake Victoria, Maboko Is., 27 Dec. 1939. *Hornby* S1058!
DISTR. U1–4; K5; T1; the Ivory Coast eastwards to the Sudan and southwards to Angola;
   apparently absent from south-eastern and south-central Africa
HAB. Riverine forest and wooded grasslands; 850–1680 m.

NOTE. *A. coriaria* is closely related to *A. ferruginea* (below) and their separation has
   sometimes given difficulty. *A. coriaria* is distinguished by:—
      (1) Indumentum much less dense; in particular the leaflets are never densely pubes-
   cent or puberulous all over beneath, and very often only on midrib.
      (2) Calyx and corolla finely and rather sparingly and openly puberulous outside, not
   densely so. The minute stipitate glands on the calyx-lobes of *A. coriaria* (see description above) are, if present at all in *A. ferruginea*, effectively cpncealed by the indumentum.
      (3) Bracteoles smaller (about 1·5–2 mm. long, not about 3–7 mm.) and often falling
   earlier.
      (4) Stamen-filaments red or rosy above, not white throughout.
      (5) Pods usually more tapered especially below, their valves more coriaceous, the seeds
   normally showing as darker circles on the outside of the valves, which are concolorous
   in *A. ferruginea*.

2. **A. ferruginea** (*Guill. & Perr.*) *Benth.* in Hook., Lond. Journ. Bot. 3 :
88 (1844) ; L.T.A. : 861 (1930) ; Gilb. & Bout. in F.C.B. 3 : 185, fig. 9 A–B
(1952) ; I.T.U., ed. 2 : 219 (1952) ; Consp. Fl. Angol. 2 : 290 (1956) ;
F.W.T.A., ed. 2, 1 : 502 (1958). Type : Gambia, Albreda, *Leprieur* (G,
holo. !)

   Tree 6–45 m. high ; crown rounded ; bark rough. Young branchlets ±
densely rusty-pubescent or sometimes subtomentose. Leaves : rhachis
clothed like the young branchlets or less densely ; pinnae 3–6(–9 *fide* I.T.U.)
pairs ; leaflets (of the 2 distal pairs of pinnae) (9–)10–14(–17) pairs, though
sometimes as few as 5 pairs on the basal pinnae, slightly oblique, oblong (the
topmost pair obovate), 11–23 mm. long, 4·5–10 mm. wide, rounded at apex,
± densely pubescent or puberulous beneath, ± sparsely so above. Flowers
white or greenish-white, subsessile or up to 3 mm. pedicellate ; bracteoles
often persistent till flowering time, to about 7 mm. long, mostly oblanceolate,
densely rusty-pubescent. Calyx (3–)4–6 mm. long, densely and shortly
rusty-pubescent outside, not slit unilaterally. Corolla shortly rusty-pubescent outside, 9–12 mm. long.   Staminal tube not or scarcely exserted
beyond corolla ; filaments 3–5 cm. long. Pod oblong, usually 15–24 cm.
long, 3–5 cm. wide, glabrous (or with a very few hairs on stipe and margins
only), often ± glossy, venose, brown, obtuse or abruptly narrowed near base
and apex. Seeds about 7–10 mm. long and 4·5–8 mm. wide, flattened.

UGANDA. West Nile District: Zoka Forest, June 1933, *Eggeling* 1240!; Bunyoro District:
   Budongo Forest, Mar. 1932, *Harris* H72 *in F.H.* 638! & Budongo Forest, Sonso R.,
   Nov. 1932, *Harris* 156 *in F.H.* 1120!; Mengo District: Mabira Forest, Oct. 1904,
   Dawe 16/175!
DISTR. U1, 2, 4; Senegal, Ubangi-Shari and Uganda to Angola
HAB. Lowland rain-forest; 790–1220 m.

SYN. *Inga ferruginea* Guill. & Perr. in Fl. Seneg. Tent. 1: 236 (1832)

NOTE. Guillemin & Perrottet describe the leaflets as up to 20-paired, but I cannot
   confirm this, and their description of the stamen-filaments as deep red appears to be
   erroneous.

3. **A. tanganyicensis** *Bak. f.* in J.B. 67 : 199 (1929) ; L.T.A. : 862 (1930) ;
T.T.C.L. : 342 (1949) ; Consp. Fl. Angol. 2 : 293 (1956). Type : Tanganyika, Kondoa District, Simbo Hills, *B. D. Burtt* 716 (BM, holo., K, iso. !)

   Tree (3–)9–20 m. high, deciduous and usually flowering when quite
leafless ; trunk smooth except at base where burned, with old bark peeling
off in brown papery pieces, the young bark creamy-white to ochre-yellow
or yellow-green ; crown flat or rounded. Young branchlets glabrous to
pubescent. Leaves : rhachis clothed like young branchlets, not hooked

or clawed at ends ; pinnae 3–6 pairs ; leaflets (4–)7–13(–17) pairs, ovate-elliptic or ovate-oblong, 11–55 mm. long, 6–29(–32) mm. wide, rounded to subacute at apex, glabrous to ± crisped-pubescent on both sides. Flowers white, usually produced before the young leaves, sessile or to 1 mm. pedicellate ; bracteoles spathulate, about 2 mm. long, already fallen when the flowers open. Calyx 4–6 mm. long, sometimes slit unilaterally, densely brown-tomentellous on lobes ; tube glabrous to ± pubescent, occasionally with sessile glands. Corolla 7–11 mm. long, ± brown-tomentellous on lobes ; tube glabrous to ± pubescent, occasionally with sessile glands. Staminal tube not or scarcely exserted beyond corolla, filaments 1·5–4 cm. long. Pod oblong, 15 -30 cm. long, 2·5–5 cm. wide, glabrous, brown, ± glossy, not or only obscurely venose. Seeds about 10–17 mm. long and 8–12 mm. wide, flattened.

TANGANYIKA. Shinyanga, *Koritschoner* 3025!; Singida plateau and upper rift escarpment, Sept. 1935, *B. D. Burtt* 5282!; Kilosa District: Elpon's Pass, Oct. 1951, *Eggeling* 6314!

DISTR. **T**1, 4–6, 8; Portuguese East Africa, Northern and Southern Rhodesia, Bechuanaland, Angola and the Transvaal

HAB. *Brachystegia-Julbernardia* deciduous woodland, especially on rocky hills and outcrops; 760–1680 m.

SYN. *A. rhodesica* Burtt Davy, Man. Fl. Pl. Transv. 2: xviii, 348 (1932); Coates Palgrave, Trees Centr. Afr.: 269–271 (1956). Types: Rhodesia, Victoria Falls, *Allen* 174, *Rogers* 5319 (K, syn.!) & Southern Rhodesia, Matopos, *Galpin* 7082 (PRE, syn.)

VARIATION. The leaflets may be glabrous or nearly so, with calyx and corolla glabrous except on the lobes (e.g. Mpwapwa, *Hornby* 536, EA!, K!), or the leaflets may be pubescent on both sides, with the corolla-tube rather densely pubescent, and with some pubescence on the calyx-tube (e.g. Tabora District, Sikonge, *Lindeman* 787, EA!, K!). The type of *A. tanganyicensis* is about midway between these extremes. Since flowers and foliage are rarely collected together, these correlations of indumentum are based on limited evidence and require further checking.
    A specimen from Masasi, *Gillman* 1189 (EA!) has smaller leaflets than usual, up to about 2 × 1 cm.; a similar form occurs at the Victoria Falls in Rhodesia.
    Hornby 536 (see above) is unusual in showing numerous sessile glands on the calyx and corolla.
    At present it seems better to treat all these variations as falling within the ordinary range of variability of the species, and not to name them.

NOTE. The remarkable papery peeling bark of *A. tanganyicensis* seems most distinctive, and the seeds are usually larger than in other any East African *Albizia*.
    In the majority of gatherings examined, the flowers in the capitula (other than the central modified ones) have a minute ovary and style only about 1 mm. long in all, and apparently non-functional, the capitula being thus apparently ♂. *Eggeling* 6314 shows normal hermaphrodite flowers with well-developed ovaries and elongate styles. Observers in the field are asked to find out whether it is that certain trees are ♂ and others hermaphrodite, or whether the ♂ capitula are produced by hermaphrodite trees at a certain season, or whether the sex of the capitula varies with their position on the tree.

4. **A. malacophylla** (*A. Rich.*) *Walp.*, Ann. 2 : 457 (1851–2) ; L.T.A. : 860 (1930). Type : Ethiopia, between Shire and Sana, *Schimper* 521 (P, lecto., K, isolecto.!)

Tree up to 6 m. high, sometimes to 12 m. ; bark rough, pale brown to grey. Young branchlets densely grey- to pale brown-pubescent. Leaves : rhachis clothed as the young branchlets ; pinnae 2–8 pairs ; leaflets 4–15 pairs, ± obliquely oblong-elliptic to obovate-elliptic or sometimes nearly oblong, 10–35(–40) mm. long, 4–22(–30) mm. wide, rounded and mucronate or slightly emarginate at apex, ± pubescent above, glaucescent and densely pubescent to subtomentose, sometimes only sparsely appressed-pubescent, beneath. Flowers white, sweetly scented, sessile or to 0·75 mm. pedicellate ; bracteoles already fallen at flowering time. Calyx 3–5 mm. long, ± densely and shortly grey-pubescent outside, not slit unilaterally. Corolla clothed like the calyx, 5–7 mm. long. Staminal tube not or scarcely exserted beyond corolla ; filaments about 2–2·5 cm. long. Pod oblong, about 10–21 cm. long, 2–4 cm. wide, subglabrous, slightly pubescent towards base and margins, sometimes puberulous or pubescent all over, ± glossy, transversely venose, brown. Ripe seeds not seen.

var. **ugandensis** *Bak. f.*, L.T.A. : 860 (1930) ; I.T.U., ed. 2 : 221 (1952) ; F.W.T.A., ed. 2, 1 : 502 (1958). Types : Uganda, Bunyoro District, *Bagshawe* 1473, 1503 (BM, syn. !)

Pinnae 2–6 pairs. Leaflets 3–9(–11) pairs, mostly 15–40 mm. long and 10–22(–30) mm. wide, often nearly straight or slightly rounded on proximal side at base and markedly " shouldered " on the distal side.

Uganda. West Nile District: Terego, Mt. Ite, Feb. 1934, *Eggeling* 1514 *in F.H.* 1444! & Koboko, Feb. 1934, *Eggeling* 1530 *in F.H.* 1455!; Teso District: Serere, Feb. 1932, *Brasnett in F.H.* 431 ! & *Chandler* 524 !
Distr. U1, 3, 4; French Sudan, Ivory Coast, Nigeria, Ubangi-Shari and the Sudan
Hab. Wooded grassland; 1100–1310 m.

Syn. (of var.). *A. boromoënsis* Aubrév. & Pellegr. in Not. Syst. 14: 56 (1950); Aubrév., Fl. Forest. Soudano-Guin. 300, t. 57, fig. 17 (1950). Types: Ivory Coast, *Aubréville* 1830, 2142*, 2614, 2772, 2864 & French Sudan, *Waterlot* 1075, *Vuillet* 519, *Dubois* 159 & Ubangi-Shari, Archambault, *S. de Ganay* 100 (all P, syn. !; isosyn. of 1830 at K !)

Distr. (of species). As for the var. but also in Ethiopia and Eritrea

Syn. (of species). *Inga malacophylla* A. Rich., Tent. Fl. Abyss. 1: 235 (1847)
     A. elliptica Fourn. in Ann. Sci. Nat., ser. 4, 14: 374 (1860); L.T.A.: 860 (1930). Type: Ethiopia, Laegga, *Schimper* 1087 (of 1854 consignment) (P, holo. !)

Note. Typical *A. malacophylla* has smaller, more numerous and more nearly oblong leaflets, but its range of variation is at present not certain. It, and other forms with more numerous leaflets from Ethiopia and Eritrea, are not known from our area or from West Africa, and the var. *ugandensis* is therefore maintained. Within our area var. *ugandensis* appears to vary little. *Eggeling* 1530 has the leaflets closer together on the pinnae and more elliptic than in other specimens; the indumentum on the lower surface of the leaflets is dense and almost tomentose, contrasting thus with *Eggeling* 1514, where it is much shorter and sparser. The indumentum of the pod requires further investigation. In typical *A. malacophylla* and related forms the pods are often puberulous all over, but in West African material of *A. boromoënsis* they are nearly glabrous.

5. **A. versicolor** [*Welw. ex*] *Oliv.* in F.T.A. 2 : 359 (1871) ; L.T.A. : 863 (1930) ; T.S.K. : 72 (1936) ; Bogdan in Nature in E. Afr., No. 2 : 16 (1947) ; T.T.C.L. : 343 (1949) ; Gilb. & Bout. in F.C.B. 3 : 182, fig. 7 (1952) ; I.T.U., ed. 2 : 222 (1952) ; Consp. Fl. Angol. 2 : 293 (1956). Type : Angola, Cuanza Norte, Golungo Alto, *Welwitsch* 1760 (LISU, lecto., BM, K, isolecto. !)

Tree 5–15 m. high (to 20 m., *fide* F.C.B.), deciduous ; crown spreading, ± flat ; bark rough, greyish-brown. Young branchlets densely rusty-tomentose. Leaves : rhachis clothed like the young branchlets ; pinnae 1–4 (–5) pairs ; leaflets 3–6 pairs, broadly and obliquely obovate to suborbicular, sometimes broadly oblong, 14–63 mm. long, 12–49 mm. wide, rounded and mucronate to emarginate at apex, rarely subacute, becoming coriaceous, pubescent above, densely tomentose or pubescent beneath. Flowers white to greenish-yellow ; pedicels 0–2(–2·5) mm. long ; bracteoles at flowering-time present or already fallen. Calyx 4·5–7 mm. long, densely rusty-pubescent or -tomentose outside, not slit unilaterally. Corolla clothed like calyx, 8–12 mm. long. Staminal tube not or scarcely exserted beyond corolla ; filaments 2·5–4·5 cm. long (to 5·5 cm. *fide* F.C.B.). Pod oblong, 10–27(–30 *fide* F.C.B.) cm. long, 3·2–6·5 cm. wide, glabrous (or with a very few hairs on stipe and margins only), ± glossy, obscurely venose,

*The sheet of 2142 at Paris is indicated as the type.

chestnut-brown or crimson. Seeds about 9–13 mm. long and 8–11 mm. wide, flattened.

UGANDA. Ankole District: Kagera R. between Nsongezi and Ruborogoto, Oct. 1932, *Eggeling* 685 *in F.H.* 1059 ! & Nsongezi, May 1950, *Eggeling* 5877 !
KENYA. Kwale District: Mwachi, *St. Barbe Baker* 43 ! & Kwale, *C. W. Elliot* Q87 *in F.H.* 1677 & *in C.M.* 14009 !; Kilifi/Kwale Districts: Mazeras, *R. M. Graham* N331 *in F.H.* 1724 !
TANGANYIKA. Lushoto District: Longuza–Sengale, 7 Feb. 1934, *Greenway* 3709 !; Tanga District: Tengeni, 4 Dec. 1929, *Greenway* 1930 !; Morogoro, 30 Nov. 1950, *Wigg* 931 !; Rungwe District: Masukulu, 18 Nov. 1912, *Stolz* 1690 !
DISTR. **U**2; **K**7; **T**1, 3–8; southwards to Angola, the Transvaal and Natal
HAB. Deciduous woodland and bushland, wooded grassland ; locally frequent and found throughout the *Brachystegia* areas (*fide* B. D. Burtt); from probably near sea-level to 1680 m.

NOTE. A very distinct and easily recognized species, marked by the combination of tomentum or coarse pubescence, usually ± rust-coloured, over the vegetative parts, and the comparatively few and broad leaflets. The ± glossy, glabrous or subglabrous pods are also characteristic. There is not much variation: the leaflets vary somewhat in size, and the indumentum beneath may be comparatively short and coarsely pubescent, or longer and tomentose.
    There is a resemblance between *A. versicolor* and *Samanea saman* (Jacq.) Merr., which is sometimes planted in our area. The latter may be separated by its indumentum being yellowish (when dry), not rusty; by the gland on the petiole being smaller and at or near the insertion of the lowest pair of pinnae, not well below them; by the leaflets being more glabrescent above, and by the different pods. In habit the two are very different, *A. versicolor* being deciduous with a lightly branched crown, while *Samanea* is evergreen with a dense heavy crown.

6. **A. lebbeck** (*L.*) *Benth.* in Hook., Lond. Journ. Bot. 3 : 87 (1844) ; L.T.A. : 862 (1930) ; Bogdan in Nature in E. Afr., No. 2 : 16 (1947) ; Lawrence in Gentes Herbarum 8 : 44–5 (1949) ; T.T.C.L. : 342 (1949) ; U.O.P.Z. : 111, 112 (fig.) (1949) ; Gilb. & Bout. in F.C.B. 3 : 187 (1952) ; Consp. Fl. Angol. 2 : 292 (1956) ; F.W.T.A., ed. 2, 1 : 502 (1958). Type : Egypt, *Herb. Linnaeus* 1228.16 (LINN, syn. !)

Tree 2·5–15 m. high ; bark grey, rough. Young branchlets puberulous, sometimes pubescent. Leaves : rhachis subglabrous, puberulous or sometimes pubescent ; pinnae (1–)2–4(–5) pairs ; leaflets 3–11 pairs, 15–48 (–65) mm. long, (6–)8–24(–33) mm. wide, oblong or elliptic-oblong (terminal leaflets ± obovate), somewhat asymmetric with midrib nearer upper margin, rounded at apex, glabrous or rarely thinly pubescent above, beneath subglabrous or puberulous, rarely pubescent. Flowers pedicellate ; pedicels 1·5–4·5(–7·5) mm. long, puberulous ; bracteoles minute, about 2–3 mm. long, falling in early bud. Calyx (2·5–)3·5–5 mm. long, not slit unilaterally, ± puberulous outside. Corolla 5·5–9 mm. long, glabrous outside except for puberulence on outside of lobes. Staminal tube not or scarcely exserted beyond corolla ; filaments 1·5–3 cm. long, pale green or greenish-yellow in upper part, white below. Pod oblong, (12–)15–33 cm. long, (2·4–)2·9–5·5 (–6) cm. wide, glabrous, or almost so except near base, coriaceous, glossy, ± venose, pale straw-coloured, not (or rarely to 5 mm.) stipitate at base. Seeds 7–11·5 mm. long and 7–9 mm. wide, flattened, marked by bumps on the outside of the valves, the alternate ones more projecting on each valve.

UGANDA. West Nile District: Arua golf links, Dec. 1931, *Brasnett* 317 !
KENYA. Kisumu-Londiani District: Kisumu, Oct. 1944, *Harger in C.M.* 11915 !
TANGANYIKA. Lushoto District: Magila Road, 17 Nov. 1929, *Greenway* 1885 !; Rufiji District: Mafia Is., Chole Is., 19 Sept. 1937, *Greenway* 5264 !; Rungwe District: Mwakalele, 21 Oct. 1932, *R. M. Davies* 645 !
ZANZIBAR. Zanzibar Is.: Zanzibar, Mnazi Moja, 3 Oct. 1926, *Vaughan* 11 !
DISTR. **U**1; **K**5; **T**1–3, ?4, 6, 7; **Z**; **P**; pantropical now, but probably nowhere native in Africa and originating from tropical Asia. I have not found it possible to distinguish in our area between those localities where it is more or less naturalized and those where it is only planted and not established; all the provinces are therefore given from which I have seen specimens

HAB. Planted for shade in streets, gardens etc. and becoming naturalized, but always associated with human habitations (T.T.C.L.: 349 (1949)); from near sea-level to about 1280 m.

SYN. *Mimosa lebbeck* L., Sp. Pl.: 516 (1753)

NOTE. *Vaughan* 11 (EA !) is an extreme, with indumentum more developed than usual, the lower surface of the leaflets and the rhachis of the pinnae in particular being pubescent. This plant corresponds to var. *pubescens* Benth. in Hook., Lond. Journ. Bot. 3: 87 (1844); intermediates, however, occur between this and the commoner sub-glabrous plants.

The pods of *A. lebbeck*, with their included seeds, are said when agitated by the breeze to make an incessant rattle that has been variously compared with women's chatter and the sound of fish being fried.

The epithet of *A. lebbeck* has often been misspelt " *lebbek* ".

7. **A. antunesiana** *Harms* in E.J. 30 : 317 (1901) ; L.T.A. : 861 (1930) ; T.T.C.L. : 342 (1949) ; Gilb. & Bout. in F.C.B. 3 : 189, fig. 10 C, D (1952) ; Consp. Fl. Angol. 2 : 291 (1956) ; Coates Palgrave, Trees Centr. Afr. : 261–4 (1956). Types : Tanganyika, Mbeya District, Unyika, Iyunga village, *Goetze* 1372 (B, holo.†) & Angola, Huila, *Antunes* 330 (B, holo.†)

Tree (1·5–)6–18 m. high ; branches spreading ; bark rough, reticulate. Young branchlets glabrous or nearly so, or very shortly pubescent. Leaves : rhachis glabrous or subglabrous ; pinnae (1–)2–3(–4) pairs ; leaflets 4–8 (–9) pairs, oblique, ovate to rhombic-ovate or elliptic-oblong (15–)23–50 (–63) mm. long, (7–)11–25(–41) mm. wide, rounded or slightly emarginate at apex, glabrous, papery to subcoriaceous, venose, very glaucous beneath. Flowers greenish-yellow, ochre when over, with whitish filaments ; flowers sessile or up to 2 mm. pedicellate ; bracteoles already fallen by flowering time, minute, up to 1·7 mm. long. Calyx (3–)3·5–5·5 mm. long, rather densely puberulous or minutely pubescent outside, not slit uni-laterally. Corolla (5–)5·5–11 mm. long, densely minutely appressed-pubescent outside. Staminal tube not or scarcely exserted beyond corolla ; filaments about 1·5–3 cm. long. Pod oblong, 12–23 cm. long, 2·7–4·4 cm. wide, glabrous except for some hairs near base and near margins, slightly venose, ± transversely plicate, thin, usually pale brown. Seeds about 7–9 mm. in diameter, flattened.

TANGANYIKA. Bukoba District: Nshamba, Oct. 1935, *Gillman* 555 !; Ufipa District: Mbisi escarpment, 1 Nov. 1933, *Michelmore* 708 !; Dodoma District: Kazikazi, 20 Mar. 1933, *B. D. Burtt* 4540 !

DISTR. T1, ? 2, 4, 5, 7; Belgian Congo, Portuguese East Africa, Nyasaland, the Rhodesias, Angola and South West Africa

HAB. Deciduous woodland and wooded grasslands (*Brachystegia*, *Brachystegia-Julber-nardia* and *Stereospermum-Markhamia-Dombeya*); 1200–1530 m.

8. **A. anthelmintica** *Brongn.* in Bull. Soc. Bot. France 7 : 902 (1860) ; L.T.A. : 859 (1930) ; T.S.K. : 71 (1936) ; Bogdan in Nature in E. Afr., No. 2 : 16 (1947) ; T.T.C.L. : 341 (1949) ; I.T.U., ed. 2 : 217 (1952). Type : Ethiopia, Add'erbati, *Quartin Dillon & Petit* (P, holo., K, iso. !)

Bush or tree 3–9(–12) m. high, deciduous ; bark smooth, grey to brown. Young branchlets glabrous or sometimes shortly pubescent ; twigs often with short divaricate lateral branches. Leaves : rhachides of leaves and pinnae glabrous to shortly pubescent, in all or most leaves projecting at ends in a short rigid persistent deflexed or downwards-bent hook or claw ; often a single stipel similarly bent near base of pinnae ; pinnae 1–2(–4) pairs ; leaflets 1–4(–5) pairs, obliquely obovate to suborbicular, (7–)10–36 (–42) mm. long, (4–)6–31 mm. wide, mucronate at apex, venose, glabrous to ± sparsely shortly pubescent. Flowers usually on leafless twigs, on pedi-cels 0·5–5·5 mm. long. Calyx pale greenish, (very rarely 2–)3–5 mm. long, glabrous to sparsely finely pubescent outside, irregularly denticulate at apex and usually slit unilaterally to about 1–2·5 mm. Corolla pale greenish,

6–12 mm. long, glabrous, or puberulous on and near lobe-margins. Staminal tube not or scarcely exserted beyond corolla ; filaments about 1·5–2 cm. long, white. Pod oblong, 7–18 cm. long, 1·5–2·9 cm. wide, quite glabrous or occasionally puberulous all over, straw-coloured. Seeds 9–13 mm. in diameter, flattened, round.

UGANDA. Acholi District: Agora, *Eggeling* 840 *in F.H.* 1212 !; Karamoja District: Lokapeliethe, 29 Oct. 1939, *A. S. Thomas* 3116 !
KENYA. Northern Frontier Province: Moyale, 19 July 1952, *Gillett* 13612 !; Machakos District: Kibwezi, 6 Oct. 1908, *Scheffler* 205 !; Kitui District: Mutha plains, Aug. 1938, *Joana in C.M.* 7436 !; Masai District: Magadi road, 8 Aug. 1943, *Bally* 2677 !
TANGANYIKA. Shinyanga District: Seseku, 10 June 1931, *B. D. Burtt* 2527 !; Moshi District: Himo-Ruvu R. bridge, 17 Aug. 1935, *R. M. Davies* 996 !; Mpwapwa, 25 Sept. 1930, *Hornby* 303 !; Morogoro District: without locality, 30 Nov. 1932, *Wallace* 510 !
DISTR. U1; K1, 2, 4, 6, 7; T1–6; the Sudan and Eritrea southwards to Bechuanaland, the Transvaal and Zululand
HAB. Deciduous and evergreen bushland and dry scrub with trees, according to Burtt (MS.) especially along seasonal rivers and in termite-mound clump-thickets; 80–1520 m.

SYN. *Besenna anthelmintica* [A. Rich., Tent. Fl. Abyss. 1: 253 (1847), *nom. provis.*]; [A. Rich. ex] Walp., Ann. 2: 461 (1851–2)
    Albizzia conjugato-pinnata Vatke in Oesterr. Bot. Zeitschr. 30: 278 (1880). Type: Kenya, Kilifi District, Ribe, *Hildebrandt* 1937 (? B, holo.†, K, iso. !)
    A. anthelmintica Brongn. var. *pubescens* Burtt-Davy, Man. Fl. Pl. Transv. 2: xvii, 348 (1932). Types: Transvaal, Messina, *Rogers* 19347 (syn., location not known ?, K, isosyn. !) & Waterpoort, *Rogers* 21504 (syn., location not known ?)

VARIATION. In our area the plant is usually glabrous. Burtt-Davy has separated (see above) a var. *pubescens* with sparse pubescence on the peduncles, calyces and petals, and pubescent fruits. Similar forms occur in East Africa: *Koritschoner* 1699 (K !) from Tanganyika, Shinyanga District, Usule, shows pubescence on the inflorescences, and *Hornby* 303 ! and *B. D. Burtt* 1761 ! both have puberulous fruits. Pubescent forms became commoner to the south of our area, where the hairiness may extend over the surface of the leaflets. I find, however, so many insensible gradations connecting the glabrous and the relatively strongly pubescent forms that var. *pubescens* does not seem worth recognizing as a distinct entity. It seems to be a rather weak clinal trend from north to south.

9. **A. harveyi** *Fourn.* in Bull. Soc. Bot. France 12 : 399 (? 1866) ; L.T.A. : 865 (1930) ; Bogdan in Nature in E. Afr., No. 2 : 16 (1947) ; T.T.C.L. : 341 (1949) ; Gilb. & Bout. in F.C.B. 3 : 173 (1952). Type : Bechuanaland, near Lake Ngami, *M'Cabe* (K, holo. !)

Tree 1·5–15 m. high, deciduous ; crown flat or compressed-rounded ; bark fissured, reticulate. Young branchlets with grey to pale brown (when dry) spreading pubescence, not silvery. Leaves : gland on upper side of petiole (often absent) prominent and sometimes shortly stalked, (0·25–)0·5–1 mm. high ; pinnae 6–20(–22) pairs ; leaflets (7–)12–27(–30) pairs, 2–6(–7) mm. long, (0·6–)1–1·25(–2) mm. wide, ± falcate, apex asymmetrical, acute, the point turned towards apex of pinna, ± appressed-pubescent on both sides, or glabrous or nearly so above even when young ; midrib nearer upper margin ; lateral nerves ± raised and visible beneath ; lower surface of leaflet paler. Flowers white, sessile or up to 0·5 mm. pedicellate ; bracteoles persistent during flowering time. Calyx 1·5–2·5 mm. long, densely pubescent outside. Corolla 3·5–6 mm. long, densely pubescent outside. Staminal tube not or scarcely exserted beyond corolla ; filaments about 1·5–2 cm. long. Pod oblong, 8–18(–25 *fide* F.C.B.) cm. long, (1·5–)2·5–3·5 cm. wide, glabrous or nearly so except for a little pubescence near base and along margins, brown to purple. Seeds 8–12 mm. long, 6–9 mm. wide, flattened. Fig. 20, p. 150.

KENYA. Machakos District: Kiboko, 16 Feb. 1949, *Bogdan* 2248 !; Masai District: Nguruman Hills, Lenyora, 26 Sept. 1944, *Bally* 3823 !
TANGANYIKA. Mwanza, *Rounce* 235 !; Tabora District: Sagara near Nguruka, 16 Nov. 1949, *Vesey-FitzGerald* 419 !; Masasi District: Lissekesse, 12 Dec. 1942, *Gillman* 1197 !

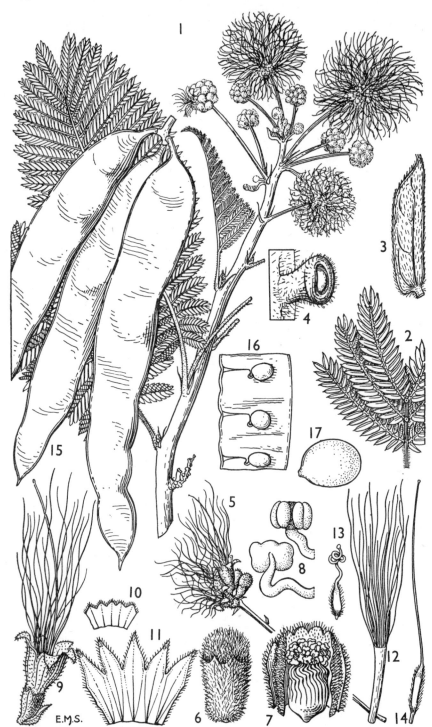

FIG. 20. *ALBIZIA HARVEYI*—**1**, part of flowering branch, × 1; **2**, part of leaf, showing glands on rhachis, × 2; **3**, leaflet, × 8; **4**, gland on rhachis, × 24; **5**, young flower-head, × 2; **6**, flower-bud, × 8; **7**, flower-bud opened, showing arrangement of stamens, × 8; **8**, anthers from bud, front and back views, × 40; **9**, open flower, × 4; **10**, calyx, opened out, × 4; **11**, corolla, opened out, × 4; **12**, stamen-filaments and tube, × 4; **13**, ovary from bud, × 8; **14**, ovary from mature flower, × 4; **15**, pods, × ⅓; **16**, part of valve of pod, seen from inside, × ⅔; **17**, seed, × 2. 1–3, from *B. D. Burtt* 3809; 4–14, from *B. D. Burtt* 5037; 15, from *B. D. Burtt* 1661; 16, 17, from *Legat* 65.

DISTR. **K**4, 6, 7; **T**1–8; extending southwards to Bechuanaland and the Transvaal
HAB. Deciduous bushland and woodland, dry scrub with trees; 80–1520(–2130) m.

SYN. *A. pallida* Harv. in Harv. & Sond., Fl. Cap. 2: 284 (1862), *non* Fourn. (1860),
  nom. illegit. Type: as *A. harveyi* Fourn.
  A. hypoleuca Oliv., F.T.A. 2: 356 (1871), *nom. illegit.* Type: as *A. harveyi*
  Fourn.
  A. pospichilii Harms in N.B.G.B. 1: 183 (1896). Type: Kenya, Machakos,
  Pospichil (B, holo. †). Syn. *fide* L.T.A.

10. **A. forbesii** *Benth.* in Hook., Lond. Journ. Bot. 3 : 92 (1844). Type :
Portuguese East Africa, Delagoa Bay, *Forbes* (K, holo.!)

Tree 2–21 m. high ; bark grey to blackish, thick, rough. Young branchlets densely grey-pubescent. Leaves : rhachis ± densely grey-pubescent ;
pinnae (2–)3–7(–8) pairs ; leaflets 5–16(–20) pairs, 4–8·5 mm. long, 1·5–
4 mm. wide, obliquely oblong to oblong-elliptic (the terminal pair obovate),
with the midrib nearer upper margin, rounded and mucronate to subacute
at apex, which is turned towards apex of pinna, glabrous or sometimes
pubescent above, beneath glabrous except for pubescence on midrib and
recurved margins, sometimes pubescent all over. Flowers creamy-white,
sessile or almost so ; bracteoles linear or oblanceolate, 1·5–2 mm. long,
falling before the flowers open. Calyx 3–5 mm. long, densely and shortly
pubescent outside, not or slightly slit unilaterally. Corolla 5–8 mm. long,
densely and shortly ± appressed-pubescent outside. Staminal tube not or
scarcely exserted beyond corolla ; filaments 1–1·5 cm. long. Pod oblong,
9–19 cm. long (including 1–2 cm. long stipe), 3·2–5 cm. wide, ± puberulous
over surface (use × 10 lens), sometimes nearly glabrous except on margins
and stipe, dark brown, closely and prominently transversely venose, the
veins ± parallel and about 2–4 mm. apart. Seeds slightly flattened, oblong-
ellipsoid or ellipsoid, 11–12 mm. long and 4·5–6·5 mm. wide.

TANGANYIKA. Probably Lindi District, but without locality, *Busse* 2338 !, 2985 !;
Lindi District: Sudi, *Gillman* 1468 !
DISTR. **T**8; Portuguese East Africa, Transvaal and Zululand
HAB. " Coastal scrub on white sandy soil " (*Gillman* 1468)

NOTE. More observations about this species in Tanganyika are desired. The pods are
most distinctive, and the seeds unusually narrow in proportion to their length.

11. **A. amara** (*Roxb.*) *Boiv.* in Encycl. XIX me Siècle, ed. 1, 2 : 34 (? 1834)*.
Type : India, *Roxburgh* (K, painting of holotype material, No. 486 !)

Tree, rarely shrubby, 1·5–12(? 15) m. high, deciduous ; crown rounded
or flat ; bark fissured, rough. Young branchlets with rather short dense
spreading grey to golden pubescence. Leaves : gland on upper side of
petiole low, sessile, up to about 0·25 mm. high ; pinnae 4–46 pairs ; leaflets
12–48 pairs, oblong to linear, 2–5(–7·5) mm. long, 0·5–1·5(–2·5) mm. wide,
symmetrical and obtuse or sometimes subacute at apex, ± appressed-pubescent on one or both sides or on margins only, glabrescent or not later ; midrib
nearly central (except at base) ; lateral nerves not distinct beneath, rarely
slightly raised. Flowers white or flushed pink, subsessile or up to 1·5 mm.
pedicellate ; bracteoles very caducous, fallen by flowering time. Calyx
1–2 mm. long, puberulous or pubescent outside. Corolla 3·5–7 mm. long,
puberulous or pubescent outside. Staminal tube not or scarcely exserted
beyond corolla ; filaments about 0·5–1·2 cm. long. Pod oblong, 10–28 cm.
long, 2–5 cm. wide, puberulous over surface, brown. Seeds 8–13 mm. long,
7–8 mm. wide, flattened.

subsp. **amara**; Brenan in K.B. 1955: 189 (1955)

Pinnae and leaflets comparatively few, former 4–12(–16) pairs, latter 10–29 pairs;
leaflets 3–7·5 mm. long, 0·75–1·5(–2) mm. wide, often glaucous beneath when mature
(not in Africa), lateral nerves especially the basal rarely slightly raised beneath.

*I have not personally seen this work.

KENYA. Northern Frontier Province: Sololo, 4 Sept. 1952, *Gillett* 13776!
TANGANYIKA. Lindi District: *Gillman* 1531! & Ruponda, 24 July 1947, *Semsei* 2081!
DISTR. K1; T8; India, Ceylon, see also note below
HAB. *Acacia-Commiphora* scrub; 820 m. See note below

SYN. *Mimosa amara* Roxb., Corom. Pl. 2: 13, t. 122 (1798)
       Albizzia gracilifolia Harms in N.B.G.B. 8: 146 (1922); L.T.A.: 866 (1930);
       T.T.C.L.: 341 (1949). Types: Tanganyika, Kilwa District, Kibata, *Brulz* 3132
       (B, syn. †) & without locality, *Koerner* 2278 (B, syn. †). See note below.

NOTE. The above specimens closely resemble Indian material of subsp. *amara*; judging
from dried specimens, the glaucousness of the lower side of the leaflets is not constant,
even in India. The following specimens are *amara* on the balance of characters, but the
leaflets are persistently appressed-pubescent on both sides and mostly smaller than
usual in *amara*. They may prove a separable African variant. *Hornby* 607 and 608
are described as smooth-barked.

KENYA. Northern Frontier Province: Takabba, Jan. 1949, *Dale* K703! & Dandu,
   18 Mar. 1952, *Gillett* 12576!; Kitui District: Mutha plains, Aug. 1938, *Joana in
   C.M.* 7388!
TANGANYIKA. Dodoma District: Saranda to Kilimatinde, 25 Dec. 1925, *Peter* 33519!;
   Mpwapwa, 27 Dec. 1931, *Hornby* 427! & 2 Dec. 1934, *Hornby* 607! & 24 Nov. 1934,
   Hornby 608!
DISTR. K1, T5
HAB. In *Acacia-Commiphora* scrub; 610–1220 m.

   Hornby 427, cited above, has pinnae in 14–22 pairs, leaflets in 17–27 pairs; in
number of pinnae it is near subsp. *sericocephala*, while in other characters it agrees with
the other Mpwapwa specimens cited above.
   A. gracilifolia, known to me only from the description, is probably *A. amara* subsp.
amara; the only possible discrepancies are that the calyx and corolla are said to be
puberulous or subglabrous; they are not subglabrous in *A. amara*.
   See also note under *A. zimmermannii* (p. 154).

   subsp. **sericocephala** *(Benth.) Brenan* in K.B. 1955: 190 (1955); Coates Palgrave, Trees
Centr. Afr.: 258–60 (1956). Types: Sudan, Sennar, *Kotschy* 244 & Kordofan, Milbes,
Kotschy 294 & Ethiopia, Gapdia, *Schimper* 818 & Dscheladscheranne [Jelajeranne],
Schimper 883 (all K, syn.!)

   Pinnae and leaflets comparatively many, the former (7–)14–46 pairs, the latter (12–)
21–48 pairs; leaflets 2–4(–4·5) mm. long, 0·5–1(–1·25) mm. wide, usually green beneath;
lateral nerves not distinct.

UGANDA. Acholi District: Agoro, *Eggeling* 795 *in F.H.* 1176!; Karamoja District:
   Kakumongole, 5 Jan. 1937, *A. S. Thomas* 2178!; Teso District: Serere, Feb. 1932,
   Chandler 525!
KENYA. Nairobi, 8 Apr. 1945, *Bally* 4401!; Machakos, 2 Jan. 1932, *van Someren* 1641!
TANGANYIKA. Shinyanga District: foot of Mantini Hills, Sept. 1935, *B. D. Burtt* 5298!;
   Ufipa District: ridge E. of Lake Kwela, 7 Nov. 1950, *Bullock* 3474!; Masasi, *Gillman*
   1023!
DISTR. U1–3; K1, 3–5; T1–5, 7, 8; Sudan and Eritrea southwards to Bechuanaland and
   the Transvaal (Zoutpansberg)
HAB. Wooded grassland, thickets and scrub, often on or near rock; 450–1890 m.

SYN. *A. sericocephala* Benth. in Hook., Lond. Journ. Bot. 3: 91 (1844); Milne-Redhead
       in K.B. 1934: 301 (1934); T.T.C.L.: 341 (1949); I.T.U., ed. 2: 222 (1952)
       [*A. amara* sensu Oliv., F.T.A. 2: 356 (1871); L.T.A.: 865 (1930); Gilb. & Bout. in
       F.C.B. 3: 172 (1952), *non* (Roxb.) Boiv. *sensu stricto*]
       A. struthiophylla Milne-Redhead in K.B. 1933: 144 (1933). Type: Northern
       Rhodesia, Mazabuka, *Milne-Redhead* 1207 (K, holo.!)

NOTE. There seems a frequent tendency, especially towards the south, for the indumentum of the inflorescences and young parts to be golden rather than grey, at least
when dried.

12. **A. isenbergiana** *(A. Rich.) Fourn.* in Ann. Sci. Nat., Bot. sér. 4, 14 :
373 (1860) ; Pichi-Serm., Miss. Stud. Lago Tana, Ric. Bot. 1 : 50 (1951) ;
I.T.U., ed. 2 : 221 (1952). Type : Ethiopia, Tigré, Adowa, *Schimper* 275
(P, holo., K, iso.!)

   Tree up to 12 m. high ; crown flat or umbrella-shaped. Young branchlets
densely and rather shortly pubescent. Leaves : gland on upper side of

petiole low and sessile, or somewhat raised, 0·25–0·75 mm. high ; pinnae (3–)6–12 pairs ; leaflets (12–)16–33 pairs, slightly falcate to oblong, (5·5–) 6·5–12(–14·5) mm. long, 1·25–4 mm. wide, with apex usually slightly asymmetric, obtuse to subacute, ± appressed-pubescent on both surfaces or glabrescent above ; lateral nerves ± raised and visible beneath ; lower surface of leaflet paler. Flowers white, on pedicels 1·5–1·75 mm. long ; bracteoles soon caducous, fallen by the time the flowers open. Calyx 1·5–2 mm. long, densely short-pubescent outside. Corolla 4–5 mm. long, densely short-pubescent or puberulous outside. Staminal tube not or scarcely exserted beyond corolla ; filaments about 1 cm. long or less. Pod oblong, 10–18 cm. long, 2–3 cm. wide, puberulous over surface. Seeds about 12 mm. long and 7 mm. wide, flattened.

UGANDA. Karamoja District: Mt. Morongole, June 1946, *Eggeling* 5641 !
TANGANYIKA. Rungwe District: Bulambia, 13 Nov. 1912, *Stolz* 1668 !
DISTR. **U**1; **T**7; Ethiopia, Northern Rhodesia
HAB. Unknown; 1680–1830 m. (*fide* I.T.U.)

SYN. *Inga isenbergiana* A. Rich., Tent. Fl. Abyss. 1: 236 (1847)
   [*A. julibrissin* sensu L.T.A.: 866 (1930), pro parte, *non* Durazz.]

NOTE. More material of this species is desirable, particularly from Tanganyika, from which area the solitary gathering is inadequate.
   The West African *A. chevalieri* Harms, which might occur in the drier parts of Uganda, is a very close relative indeed, but has the leaflets decidedly acute and mucronate, and occurs in dry lowland savannah.
   A. isenbergiana much resembles certain forms of *A. schimperiana* with small leaflets and grey indumentum on outside of calyx and corolla, but the leaflets of the former are smaller and the pedicels shorter.

13. **A. zimmermannii** *Harms* in E.J. 53 : 455 (1915) ; L.T.A. : 864 (1930) ; T.T.C.L. : 343 (1949). Types : Tanganyika, Lushoto District, Amani, *Zimmermann* 2807 (B, syn.†, K, isosyn.!) & Morogoro, *Holtz* 1271 (B, syn.†)

Tree 6–15 m. high ; crown flat, spreading ; bark smooth, finely fissured, grey-brown. Young branchlets sparsely to densely rusty-puberulous or -pubescent, sometimes nearly glabrous. Leaves : rhachis clothed like young branchlets, not hooked or clawed at end ; pinnae 3–6 (rarely to 10) pairs ; leaflets 8–17 pairs, oblong-elliptic, slightly oblique, 7–15 mm. long, (3–) 4–8 mm. wide, rounded at apex, beneath paler, ± glaucous and appressed-puberulous (occasionally glabrous or nearly so). Heads often aggregated on short leafless branchlets ; pedicels 0–1 mm. long ; bracteoles very minute, but sometimes present when the flowers open. Flowers white and pink. Calyx 1–2·5 mm. long, densely rusty-puberulous outside, not slit uni-laterally. Corolla 3·5–5 mm. long, densely rusty-puberulous outside. Staminal tube not or scarcely exserted beyond corolla ; filaments 0·7–1·3 cm. long. Pod oblong, 15–32 cm. long, 3·8–7 cm. wide, ± densely rusty-puberulous, crimson near maturity, turning brown with age (Greenway), normally with very prominent transverse veins much raised, particularly in centre, sometimes almost wing-like, and anastomosing. Seeds about 10 mm. long and 7 mm. wide, flattened.

KENYA. Teita District: Mbololo, Mar. 1938, *Joana in C.M.* 8905 !
KENYA/TANGANYIKA. Teita/Moshi Districts: Lake Chala, 21 Jan. 1936, *Greenway* 4447 !
TANGANYIKA. 8 km. W. of Moshi, by R. Njoro, 3 Nov. 1955, *Milne-Redhead & Taylor* 7217 !; Lushoto District: Amani, Msindi, 12 July 1932, *Greenway* 2991 !; Mpwapwa, 24 Nov. 1934, *Hornby* 611 !; Morogoro District: Kikundi, 25 Oct. 1934, *E. M. Bruce* 25 !
DISTR. **K**7; **T**2, 3, 5, 6; specimens apparently referable to this species also from Nyasa-land, Portuguese East Africa and Southern Rhodesia
HAB. Riverine forest, secondary bush near lowland rain-forest, and evergreen or semi-evergreen bushland; 580–1130 m.

NOTE. The large very conspicuously cross-veined pods are quite different from those of the other East African *Albizia* spp.

A. zimmermannii and *A. amara* subsp. *amara* appear to grow together in riverine forest at Mpwapwa. The much larger leaflets and of course the pods will easily separate the former from the latter, but when, as happens, both species may produce flowers without foliage, separation, at any rate in the herbarium, may be very hard. The indumentum of *A. zimmermannii* is usually browner when dry than in *A. amara;* but the possibility of the two species hybridizing should be investigated.

Portuguese East African and Nyasaland specimens have pods less prominently cross-veined than usual; further material may show this to be inconstant, especially as *Holst* 2393 (K!), Tanganyika, Lushoto District, Bwiti, has pods similarly less venose than usual. *A. nyasica* Dunkley in K.B. 1937: 469 (1937) appears to be the same.

14. **A. schimperiana** *Oliv.*, F.T.A. 2 : 359 (1871) ; L.T.A. : 866 (1930) ; Gilb. & Bout. in F.C.B. 3 : 183 (1952). Type : Ethiopia, Tigré or Begemdir, *Schimper* 1396 (K, holo. !, BM, iso. !)

Tree 5–23(–30) m. high ; crown flat or not ; bark smooth, grey, or sometimes brownish and rough. Young branchlets densely, sometimes sparsely, and shortly brown-pubescent (grey to golden in var. *tephrocalyx*), later glabrescent. Leaves : rhachis shortly and densely to sparsely pubescent ; pinnae 2–7 pairs ; leaflets (of the 2 distal pairs of pinnae) 6–21(–23) pairs (sometimes as few as 5 pairs on lower pinnae), variable in shape and size, obliquely oblong, or rhombic to falcate-oblong, acute to rounded and mucronate at the apex, which is turned towards the pinna-apex, with diagonal midrib, 7–21(–30) mm. long, 3·5–8 5(–16) mm. wide, ± appressed-pubescent beneath and often whitish when dry, glabrescent above. Flowers white or pale yellow, pedicellate ; pedicels 2–6 mm. long, densely and shortly brown- (grey in var. *tephrocalyx*) pubescent or sometimes puberulous, as are the calyces and corollas. Calyx 1·5–2·5 mm. long, not slit unilaterally. Corolla 4–7·5 mm. long. Staminal tube not or scarcely exserted beyond corolla ; filaments about 0·7–1·2 cm. long. Pod oblong, 18–34 cm. long, (2–)2·8–5·9 cm. wide, puberulous (sometimes sparsely so), not glossy, venose, brown. Seeds 9–11 mm. long and 6·5–8 mm. wide, flattened.

KEY TO INTRASPECIFIC VARIANTS

Indumentum on outside of calyx and corolla grey   .   var. **tephrocalyx**
Indumentum on outside of calyx and corolla brown :
  Each pinna somewhat narrowing towards apex ;
    leaflets mostly 3·5–8·5 mm. wide   .   .   .   var. **schimperiana**
  Each pinna somewhat broadening towards apex ;
    leaflets 5–16 mm. wide .   .   .   .   .   var. **amaniënsis**

var. **schimperiana**; Brenan in K.B. 1955: 190 (1955)

Young shoots brown-pubescent. Each pinna somewhat narrowing towards its apex; leaflets (2 distal pairs of pinnae) 8–21(–23) pairs with the terminal pair somewhat smaller than the rest, 7–21 mm. long, 3·5–8·5 mm. wide, rarely more, 2–3 times as long as wide. Indumentum on pedicels, calyx and corolla brown.

UGANDA. Acholi District: Agoro, Mar. 1935, *Eggeling* 1723 !; Karamoja District: Lozut, 5 Nov. 1939, *A. S. Thomas* 3200 !
KENYA. Kiambu District: Kabete, 5 Nov. 1950, *Bogdan* 2845 !; Nairobi: Thika Road House, 4 Dec. 1950, *Verdcourt* 394 !; Masai District: Ngong Hills, pipe-line on eastern foot-slopes, 13 Mar. 1933, *van Someren* 2575 *in C.M.* 5166 !
TANGANYIKA. Moshi District: Lyamungu, 26 Nov. 1943, *Wallace* 1144 !; Lushoto, 10 Nov. 1930, *Milne* 4 !; N. Mpwapwa, 11 Dec. 1935, *Hornby* 733 !; Rungwe District: Mwakalele Mission, 24 Oct. 1932, *R. M. Davies* 300 !
DISTR. U1; K4, 6; T2, 3, 5, 7; the Sudan, Ethiopia and Somalia southwards to Portuguese East Africa and Southern Rhodesia
HAB. Upland dry evergreen forest, upland rain-forest and evergreen bushland, riverine forest; 1130–2130 m.

SYN. ? *A. maranguënsis* [Taub. ex] Engl. in Abh. Preuss. Akad. Wiss. 1891: 241 (1892); L.T.A.: 867 (1930); Bogdan in Nature in E. Afr., No. 2: 17 (1947); T.T.C.L.:

341 (1949). Types: Tanganyika, Moshi District, between Marangu and Mashame, *Meyer* 359, 367 (B, syn. †).

[*A. sassa* sensu Chiov., Racc. Bot. Miss. Consol. Kenya: 40 (1935), pro parte, quoad *Balbo* 210, 371, *non* (Willd.) Chiov.]

VARIATION. *A. schimperiana* var. *schimperiana* is remarkably variable in its foliage, so that the extremes do not appear conspecific until the full range of variation has been examined. I have not, however, found it possible to split into any sharply defined entities this range, which may be analysed as follows:—

(1) The ratio of length to breadth of leaflet varies from about 2 to 3 or slightly over. *A. S. Thomas* 3200! (length/breadth about 3), *Napier* 3615! from Nairobi (length/breadth 2·3–3), *Bogdan* 1403! from Nairobi (length/breadth 2·5–2·75), *van Someren* 2575 *in C.M.* 5166! (length/breadth 2–2·5), *Verdcourt* 394! (length/breadth about 2) are examples of this.

(2) The apex of the leaflet varies from acute to rounded. Narrow leaflets tend to be acute, while the broader ones tend to be obtuse, but this is quite inconstant. The amount of falcateness also varies.

(3) The number of pairs of leaflets varies from 8–21(–23).

(4) The lower surface of the leaflet when dry varies from pale green to whitish.

These trends do not appear correlated with each other or with geography. Typical *A. schimperiana* has rather numerous, narrow, acute to obtuse leaflets.

It should be emphasized that the whole range of variation does not occur on one tree, and probably not even in one population.

A. schimperiana appears prone to a malformation affecting a part or the whole of single pinnae, whose leaflets are ± confluent into a lobed lamina resembling a fern pinnule.

A. maranguensis may in part be synonymous with var. *amaniensis* rather than with var. *schimperiana*, since the leaflets were described as 15–25 mm. long and 7–15 mm. wide, although in up to 14 pairs.

var. **tephrocalyx** *Brenan* in K.B. 1955: 191 (1955). Type: Uganda, Payida, *Eggeling* 1484 *in F.H.* 1424 (K, holo!., ENT, iso.!)

Young shoots golden- to grey-pubescent. Each pinna somewhat narrowing towards its apex; leaflets in 6–17 pairs, with the terminal pair somewhat smaller than the rest, 10–21 mm. long, 4–6 mm. wide, mostly about 3 times as long as wide. Indumentum on pedicels, calyx and corolla grey, very short.

UGANDA. West Nile District: Payida, Feb. 1934, *Eggeling* 1484 *in F.H.* 1424! & Mar. 1935, *Eggeling* 1925!; Acholi District: Lututuru, Feb. 1938, *Eggeling* 3495!
DISTR. U1; the Sudan
HAB. Riverine forest; 1520 m.

SYN. *A. maranguensis* Taub. forma; I.T.U., ed. 2: 221 (1952), excl. *Eggeling* 1723

NOTE. The leaflets of var. *tephrocalyx* are in 6–17 pairs, mostly falcate-oblong, usually about thrice as long as broad, but some only twice, and usually acute or subacute at apex.

This is close to *A. isenbergiana;* see note under that species (p. 153).

var. **amaniensis** (*Bak. f.*) *Brenan* in K.B. 1955: 191 (1955). Type: Tanganyika, Lushoto District: Amani, *Zimmermann* G3026 (BM, holo.!, EA, K, iso.!)

Indumentum as in var. *schimperiana*. Each pinna ± broadening towards apex; leaflets in 5–11 pairs, with the terminal pair somewhat larger than rest, mostly 10–30 mm. long, 5–16 mm. wide; leaflets about twice as long as wide.

TANGANYIKA. Moshi District: Rau & Kilimanjaro, 11 Dec. 1924, *Lewis* 224! & Kilimanjaro, Moshi–Marangu road, 29 Dec. 1934, *R. M. Davies* 985!; Lushoto District: Amani, 18 Jan. 1910, *Zimmermann* G3026!
DISTR. T2, 3; not known elsewhere
HAB. Lowland rain-forest and ground-water forest

SYN. *A. amaniensis* Bak.f. in J.B. 70: 255 (1932); T.T.C.L.: 341 (1949)

NOTE. The status of this group is doubtful. The material is rather heterogeneous, but apparently distinct from the other varieties by the pinnae somewhat broadening upwards, with the terminal leaflets somewhat larger than the others.

In some ways var. *amaniensis* is similar to *A. glaberrima* var. *mpwapwensis*, but differs in the indumentum of the inflorescence being brown not grey, and in the apex of the leaflets not being minutely emarginate. The possibility of var. *amaniensis* being the product of crossing between *A. glaberrima* var. *glabrescens* and *A. schimperiana* var. *schimperiana* should, I feel, be borne in mind.

A. maranguensis [Taub. ex] Engl. may in part be synonymous with var. *amaniensis* rather than with var. *schimperiana* (see note above).

15. **A. glaberrima** (*Schumach. & Thonn.*) *Benth.* in Hook., Lond. Journ. Bot. 3 : 88 (1844), *non* F.W.T.A. 1 : 362 (1928), syn. *Mimosa glaberrima* Schumach. & Thonn. except., vide Keay in K.B. 1953 : 489–490 (1954) ; F.W.T.A., ed. 2, 1 : 502 (1958). Type : Ghana, between Asiama and Jadofa, *Thonning* (C, holo. !)

Tree 9–24 m. high ; crown ± flattened ; bark smooth, grey. Young branchlets finely and usually ± sparingly puberulous or shortly pubescent. Leaves : rhachis clothed like the young branchlets ; pinnae 1–3(–4) pairs ; leaflets 3–6(–8) pairs, obliquely rhombic-ovate to rarely -obovate or slightly falcate-oblong or asymmetrically elliptic, with the midrib often ± diagonal, 18–70(–90) mm. long, 9–33(–40) mm. wide, narrowed to an obtuse (or, but rarely in E. Africa, subacute or acute) apex, glabrous or nearly so above, the midrib puberulous to almost glabrous, beneath glabrous or nearly so except for sparse puberulence on midrib, rarely finely appressed-puberulent over surface, often pale beneath when dry. Flowers white or whitish, pedicellate ; pedicels 1·5–7 mm. long, densely covered with grey puberulence or very short fine pubescence, as are the outsides of the calyces and corollas. Calyx 1·5–2·5(–3) mm. long, occasionally slit unilaterally. Corolla 3–5·5 mm. long. Staminal tube not or scarcely exserted beyond corolla ; filaments about 0·6–1·0 cm. long. Pod oblong, 12–26 cm. long, 3–4·2 cm. wide, puberulous over surface, not or only slightly glossy, somewhat venose, usually brown. Seeds about 9–11 mm. long and 6–7 mm. wide, ± flattened.

var. **glaberrima**; Brenan in K.B. 1955: 192 (1955); Consp. Fl. Angol. 2: 292 (1956)

Leaflets with minute appressed puberulence over lower surface; apex obtuse to subacute or almost acute; midrib above puberulent to apical part of leaflet.

UGANDA. West Nile District: East Madi, Zoka Forest, June 1933, *Eggeling* 1249 *in F.H.* 1345 !; Acholi District: Chua, *Eggeling* 3512 !; Toro District: S. Kibale Forest, 14 Dec. 1938, *Loveridge* 249 !
DISTR. U1, 2; from Ghana eastwards to the Sudan, extending southwards in West Africa to the French Cameroons and, apparently, Angola
HAB. Forest; 790–1370 m.

SYN. *Mimosa glaberrima* Schumach. & Thonn., Beskr. Guin. Pl. : 321 (1827)
      A. warneckei Harms in E.J. 30: 75 (1901); F.W.T.A. 1: 363 (1928). Type: Togo, near Lome, Feb. 1900, *Warnecke* 57 (B, holo. †, K, iso. !)
      A. eggelingii Bak. f. in J.B. 76: 237 (1938); I.T.U., ed. 2: 219 (1952). Type: Uganda, Acholi District, Chua, *Eggeling* 3512 (BM, holo. !)
      Pithecellobium glaberrimum (Schumach. & Thonn.) Aubrév. in Not. Syst. 14: 57 (1950), pro parte quoad syn., excl. nota.
      [*Albizia zygia* sensu I.T.U., ed. 2 : 222 (1952), pro parte, quoad *Eggeling* 1249, non (DC.) Macbr.]

NOTE. The two cited specimens other than 3512, having neither flowers nor fruits, are somewhat doubtfully placed here.

var. **mpwapwensis** *Brenan* in K.B. 1955: 192 (1955). Type: Tanganyika, Mpwapwa, B. D. Burtt 5016 (K, holo. !, EA, iso. !)

Leaflets with minute appressed puberulence over lower surface; apex obtuse; midrib above glabrous or nearly so, with puberulence only near base.

TANGANYIKA. Mpwapwa, 13 Dec. 1931, *Hornby* 402 ! & Mpwapwa, Kikombo streams, 30 Nov. 1933, *B. D. Burtt* 5016 !
DISTR. T5; not known elsewhere
HAB. Riverine forest; 1070–1130 m.

NOTE. This variety is intermediate between var. *glabrescens* and var. *glaberrima*, sharing with the latter the minute puberulence on the lower surface of the leaflets, to see which a lens is essential, but differing in the leaflet apex, indumentum of midrib on upper side, and in the glaucescence of the lower surface of the leaflets.
      For the distinction between var. *mpwapwensis* and *A. schimperiana* var. *amaniënsis*, see note under the latter (p. 155).

var. **glabrescens** (*Oliv.*) *Brenan* in K.B. 1955: 192 (1955). Types: Tanganyika, Uzaramo District, Dar es Salaam, *Kirk* (K, syn. !) & Portuguese East Africa, Kongone R., *Kirk* (K, syn. !)

Leaflets with puberulence on midrib, otherwise glabrous or nearly so beneath; apex obtuse; midrib above glabrous or nearly so, with puberulence only near base, or sometimes puberulent to apical part of leaflet.

KENYA. Kitui/Machakos Districts: Athi Camp below Yatta Gap, 29 Jan. 1942, *Bally* 1764!; Kwale District: Gogoni, Gazi, Sept. 1936, *Dale in F.H.* 3540!
TANGANYIKA. Pare District: Kisiwani, 7 Nov. 1955, *Milne-Redhead & Taylor* 7248!; Lushoto District : Sigi–Kwamkuyu Rivers, 15 Jan. 1931, *Greenway* 2805!; Utete District : Mafia Island, Kipandeni, 26 Sept. 1937, *Greenway* 5322!; Newala District : Ngunja, 12 Dec. 1942, *Gillman* 1102!
ZANZIBAR. Zanzibar Is., Kombeni Caves, 16 Aug. 1930, *Vaughan* 1443! & near Chaani, 22 Jan. 1929, *Greenway* 1123!
DISTR. K4, 7; T2, 3, 6–8; Z; Belgian Congo, Nyasaland, Portuguese East Africa, Northern and Southern Rhodesia
HAB. Lowland rain-forest, riverine forest, coastal evergreen bushlands; also recorded from open *Brachystegia* woodland in Handeni District; from near sea-level to 760 m.

SYN. *A. glabrescens* Oliv., F.T.A. 2: 357 (1871); L.T.A.: 862 (1930); Bogdan in Nature in E. Afr. No. 2: 16 (1947); T.T.C.L.: 342 (1949); Gilb. & Bout. in F.C.B. 3: 184 (1952), saltem pro parte, sed excl. distrib. quoad Ghana, Ivory Coast, Angola

NOTE. The var. *glabrescens* differs from the other two varieties in being almost or quite glabrous on the lower surface of the leaflets (apart from the midrib).

16. **A. gummifera** (*J.F. Gmel.*) *C.A. Sm.* in K.B. 1930 : 218–9 (1930), pro parte, excl. syn. *Mimosa adianthifolia* [" *adiantifolia* "], *Zygia fastigiata* et *Albizzia fastigiata;* Dale in T.S.K. : 72 (1936), pro parte, excl. " Coast form (*A. sassa*) " ; F.P.N.A. 1 : 392 (1948), saltem pro parte ; T.T.C.L. : 339–40 (1949), pro parte, quoad " Species A " tantum ; Gilb. & Bout. in F.C.B. 3 : 181 (1952) ; I.T.U., ed. 2 ; 220 (1952) ; Brenan in K.B. 1952 : 511 (1953) ; Coates Palgrave, Trees Centr. Afr. 265–8 (1956). Type : description and two plates of *Sassa*, from Ethiopia, in Bruce, Travels 5 : 27 (1790) ; no herbarium-specimens extant

Medium or large tree up to 30 m. high ; crown flattened ; bark smooth, very rarely rough, grey. Young branchlets finely and shortly brownish-pubescent, soon glabrescent and usually deep or blackish-purple, ultimately grey-barked. Leaves : pinnae (Fig. 22/1, p. 163) 5–7(–8) pairs (rarely only 3 on occasional reduced leaves), each pinna more or less narrowing upwards ; leaflets of 2 distal pairs of pinnae 9–16(–17 *fide* I.T.U., ed. 2) pairs, obliquely rhombic-quadrate to rhombic-subfalcate, mostly about 10–20(–25) mm. long, 4–8(–13) mm. wide, auricled or sometimes not on proximal side, obtuse to acute at apex, subglabrous or somewhat pubescent on midrib and margins, rarely, especially when young, some occasional hairs on the primary lateral nerves ; raised venation beneath lax. Stipules and bracts at base of peduncles lanceolate, up to about 6–7 mm. long and 2–2·5 mm. wide. Peduncles puberulous or finely pubescent ; bracteoles mostly caducous, linear, inconspicuous, (1–)2–6 mm. long, normally shorter than the flower-buds except when extremely young. Flowers subsessile ; pedicels puberulous or sometimes glabrous, 0·25–0·75(–1) mm. long. Calyx 2·5–5 (very rarely indeed 1·5–2) mm. long, minutely, shortly, and rather appressedly brownish-pubescent to subglabrous outside. Corolla 6·5–12 mm. long, minutely pubescent outside, white. Staminal tube exserted about 1·5–2·8 cm. beyond corolla, white below, crimson above. Pod oblong, flat or slightly transversely plicate, (8–)10–21 cm. long, 2–3·4 cm. wide (to 4 cm. wide, *fide* I.T.U. ed. 2), glabrescent, glossy, eglandular, less prominently and closely venose than in *A. adianthifolia*, pale brown to reddish-brown or purplish. Seeds 9–12 mm. long, 10 mm. wide, flattened.

var. **gummifera** ; F.W.T.A., ed. 2, 1 : 503 (1958)

Leaflets conspicuously auriculate at base on proximal side. Figs. 21/1–4, p. 159, & 22/1, p. 163.

UGANDA. Toro District: Mpanga R., 1905, *Dawe* 500 !; E. side of Ruwenzori, May 1939, *Sangster* 540 !; Mbale District: Suam R., 12 Apr. 1956, *Kimera* K30 !
KENYA. Elgon, 23 Jan. 1931, *Lugard* 511 !; Nandi, *E. C. M. Green* 34 !; Masai District: Ngong, Dec. 1931, *van Someren* 1629 !
TANGANYIKA. Lushoto District: W. Usambara Mts., *Gillman* 1001 !; Mafia Island: Mchangani, 26 Sept. 1937, *Greenway* 5325 !; Rungwe District: Rungwe Mt., 13 Oct. 1913, *Stolz* 2252 !
ZANZIBAR. Zanzibar Is., Fumba, 11 Oct. 1930, *Vaughan* 1657 ! and Chukwani, 1 Oct. 1950, *R. O. Williams* 70 !
DISTR. U2–4; K3–7; T2–8; Z; mainly eastern Africa from the Sudan and Ethiopia southwards to Southern Rhodesia and Portuguese East Africa, westwards to the region of Lakes Kivu and Edward in the Belgian Congo, and then in the British Cameroons (Bamenda) and SE. Nigeria; also in Madagascar
HAB. Lowland and upland rain-forest, riverine forest, and in open habitats near forest; near sea-level to 2440 m. In Southern Rhodesia said to be " a pioneer in the natural expansion of the forest "

SYN. *Sassa gummifera* J.F. Gmel., Syst. 2 (2) : 1038 (1791)
     Inga sassa Willd., Sp. Pl. 4 (2): 1027 (1806). Type as for *A. gummifera*
     Albizzia sassa (Willd.) Chiov., Monogr. Rapp. Colon. Roma (Etiopia: Osserv. Bot. Agrar. Indust.) 24: 103 (1912); L.T.A.: 868 (1930), pro parte, quoad syn. *Sassa* Bruce, *Mimosa sassa* et *Albizzia mearnsi* [" *mearnsii* "] tantum; Chiov., Racc. Bot. Miss. Consol. Kenya: 40 (1935), pro parte, quoad *Mearns* 1197, *Balbo* 209
     A. mearnsi De Wild., Pl. Bequaert. 3: 51–2 (1925); L.T.A.: 869 (1930), sub *A. sassa*. Type: Kenya, Fort Hall, *Mearns* 1197 (BR, holo. !, K, photo. !, FI, iso. !)
     A. laevicorticata Zimm., *nomen nudum*; T.T.C.L.: 343 (1949)

var. **ealaënsis** (*De Wild.*) Brenan in K.B. 1952: 518 (1953); Consp. Fl. Angol. 2: 294 (1956) ; F.W.T.A., ed. 2, 1 : 503 (1958). Type : Belgian Congo, Eala, *Laurent* 665 (BR, holo. !, K, photo. !)
Leaflets not auriculate at base on proximal side, at most with the proximal margin minutely rounded at its junction with the pulvinus. Fig. 21/5, p. 159.

UGANDA. Mengo District: Mabira Forest, Mar. 1947, *Kigundu* 54 !
TANGANYIKA. Bukoba District: Kaagya, 1935, *Gillman* 364 ! & Mansira Is., Aug. 1953, *Procter* 70 !
DISTR. U4; T1; Nigeria, French Cameroons, Belgian Congo and Angola
HAB. Lowland rain-forest; altitude range uncertain, about 1140 m.
SYN. *Albizzia ealaënsis* De Wild. in Ann. Mus. Congo, sér. 5, 2: 126–7 (1907); L.T.A.: 869 (1930), sub *A. sassa* [ut " A. ealensis "]; Gilb. & Bout. in F.C.B. 3: 177 (1952)

NOTE. The var. *ealaënsis* seems always to have short bracteoles not exceeding 2·5 mm. long; however, var. *gummifera* may also have them as short.
    In spite of its treatment in F.C.B. as a species, I still do not consider the characters of var. *ealaënsis* sufficient for that. The other slight differential characters in the F.C.B. description do not work satisfactorily. It is there said that the filaments of var. *ealaënsis* are greenish, but although precise field-notes on this have not been seen by me, Gillman's specimen cited above shows evident signs of their having been red, as in var. *gummifera*. According to F.C.B. var. *ealaënsis* can attain 35 m. in height.
    Vaughan 47, from Zanzibar Is., Muyuni, 4 Nov. 1926 (BM !) is a form of *A. gummifera* with very reduced auricles to the leaflets, thus approaching var. *ealaënsis*.
    Purseglove 3118, from Uganda, Kigezi District, Kachwekano Farm, Sept. 1949, is very near *A. gummifera*, but with stronger indumentum, including some hair on the lower surface of the leaflets; it may well be a hybrid with *A. adianthifolia*.

16 × 18. **A. gummifera** (*J. F. Gmel.*) C. A. Sm. × **zygia** (*DC.*) *Macbr.*; Brenan in K.B. 1952 : 531 (1953) ; Consp. Fl. Angol. 2 : 296 (1956)

Differs from *A. gummifera* in the often fewer pinnae (3–6 pairs), each pinna broadening towards apex, and in the normally fewer leaflets in 6–9 pairs, without auricles at base on the proximal side, the terminal leaflets being more or less larger than the others ; from *A. zygia* in the normally more numerous pinnae and leaflets. Fig. 22/7, 8, p. 163.

UGANDA. Masaka District: Sese Islands, Bugala Is., Nov. 1931, *Eggeling* 81 *in F.H.* 254 !
DISTR. U4; Ubangi, Gaboon, Belgian Congo and Angola
HAB. Presumably where those of the parents overlap; 1190 m.

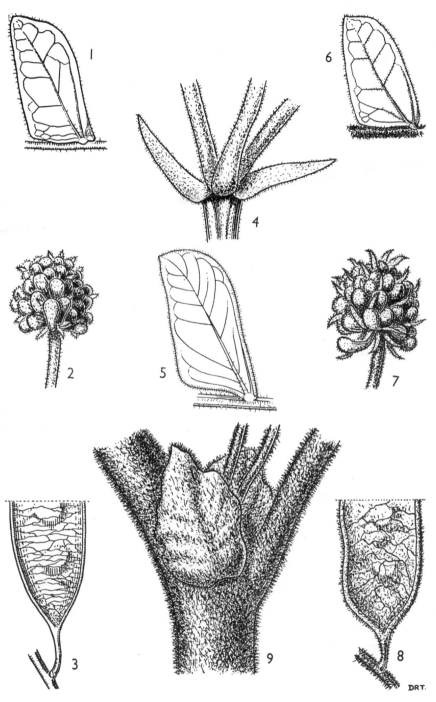

FIG. 21. *ALBIZIA GUMMIFERA* var. *GUMMIFERA*—**1**, leaflet, underside, × 2; **2**, young flower-head, × 4; **3**, basal part of pod, × 1; **4**, stipules, × 4. *A. GUMMIFERA* var. *EALAENSIS*—**5**, leaflet, underside, × 4. *A. ADIANTHIFOLIA*—**6**, leaflet, underside, × 2; **7**, young flower-head, × 4; **8**, basal part of pod, × 1; **9**, stipules, × 4. 1, 2, 4, from *Lugard* 511; 3, from *Jackson* 336; 5, from *Louis* 3506; 6, 8, 9 from *B. D. Burtt* 2894; 7, from *Greenway* 3376.

NOTE. As suggested in K.B. 1952: 532 (1953), it is quite likely that *A. gummifera* var. *ealaënsis* rather than *A. gummifera* var. *gummifera* is one parent of this hybrid.

17. **A. adianthifolia** (*Schumach.*) *W. F. Wight* in U.S. Dept. Agric. Bur. Pl. Industry, Bull. 137 : 12 (1909) ; I.T.U., ed. 2 : 217 (1952) ; Gilb. & Bout. in F.C.B. 3 : 178 (1952) ; Brenan in K.B. 1952 : 520 (1953) ; Consp. Fl. Angol. 2 : 295 (1956) ; F.W.T.A., ed. 2, 1 : 502, fig. 160 (1958). Type : ? Ghana, Bligusso, *Thonning* (C, holo. !, K, photo !)

Tree 4–30 m. high ; crown flattened ; bark grey to yellowish-brown and rough (rarely smooth in our area). Young branchlets densely, rather coarsely and persistently rusty- to fulvous-pubescent. Leaves : pinnae 5–8 pairs (rarely only 3 on occasional reduced leaves), each pinna more or less narrowing upwards ; leaflets of 2 distal pairs of pinnae 9–17 pairs, obliquely rhombic-quadrate or -oblong, mostly about 7–17(–20) mm. long and 4–9(–11) mm. wide ; proximal margin at base usually more or less rounded into the pulvinus but not auriculate ; apex of leaflet usually obtuse and mucronate, sometimes subacute, surface of leaflet thinly pubescent above, rather plentifully pubescent all over beneath, raised venation beneath close. Stipules and bracts at base of peduncles ovate, about 5–12 mm. long and 3–6(–8) mm. wide. Peduncles clothed as the young branchlets ; bracteoles variably persistent, linear-spathulate to oblanceolate, 5–8 mm. long, exceeding the flower-buds. Flowers subsessile ; pedicels pubescent, 0·5–1 mm. long. Calyx 2·5–4 (rarely only 2) mm. long, pubescent outside. Corolla 6–11 mm. long, pubescent outside, white or greenish-white. Staminal tube exserted about 1·3–2 cm. beyond corolla, red to wholly greenish or pink. Pod oblong, flat or slightly transversely plicate, 9–19 cm. long, 1·9–3·2 (–? 4) cm. wide, more or less densely and persistently pubescent, not glossy, prominently venose, usually pale brown. Seeds 7–9·5 mm. long, 6·5–8·5 mm. wide, flattened. Figs. 21/6–9, p. 159, & 22/2, p. 163.

UGANDA. Ankole District: Igara, Kasambia, 27 May 1929, *Snowden* 1364 ! & between Nsongezi and Ruborogoto, Oct. 1932, *Eggeling* 679 *in F.H.* 1053 !; Masaka District: Koki and Ankole, 1905, *Dawe* 395 !
KENYA. Mombasa, Jan. 1876, *Hildebrandt* 1936 !; Kwale District: Shimba, *Gardner* 1443 ! & Kwale, *R. M. Graham* 312 !
TANGANYIKA. Mwanza District: Geita, 6 June 1937, *Burtt* 6583 !; Lushoto District: Amani, 28 Mar. 1933, *Greenway* 3376 !; Newala District: Mahuta, 12 Dec. 1942, *Gillman* 1050 !
ZANZIBAR. Pemba, Makongwe Is., 16 Dec. 1930, *Greenway* 2733 !
DISTR. U2, 4; K7; T1–4, 6–8; Z; P; widespread in tropical and extending to South Africa, from Gambia and Kenya southwards to Angola and Pondoland
HAB. Lowland rain-forest, deciduous woodland and wooded grassland, also in upland grassland; 30–1680 m.

SYN. *Mimosa adianthifolia* Schumach. in Beskr. Guin. Pl.: 322 (1827)
   Zygia fastigiata E. Mey., Comm. Pl. Afr.: 165 (1836). Types: South Africa, between Rivers Umkomaas and Umzinkulu, *Drège* (B, holo. †), & Port Natal [Durban], *Drège* (B, holo. †, K, iso. !)
   Albizzia fastigiata (E. Mey.) Oliv., F.T.A. 2: 361 (1871)
   [*A. sassa* sensu L.T.A.: 868–9 (1930), pro parte, excl. syn. *Sassa* et *Mimosa sassa*, non (Willd.) Chiov.]
   [*A. gummifera* sensu C. A. Sm. in K.B. 1930: 218–9 (1930), pro parte; T.T.C.L.: 339–40 (1949), pro parte, quoad " Species B " tantum ; U.O.P.Z.: 113 (1949); non (J. F. Gmel.) C. A. Sm.]

Further details of the synonymy will be found in K.B. 1952: 520–1 (1953). *A. adianthifolia* has a wide range of habitat, and ecotypes may be recognizable. It is not, however, a particularly variable plant. The colour of the staminal tube varies (see the description), and also the surface of the bark. For notes on the occurrence of rough and smooth barked trees in Tanganyika and elsewhere see K.B. 1952: 527–8 (1953).

Milne-Redhead & Taylor 8970B ! from Tanganyika, Songea District, Matengo Hills, about 1·5 km. E. of Ndengo, 4 Mar. 1956, has auriculate leaflets (although not so markedly so as in *A. gummifera*), and short sparse indumentum over both upper and lower surfaces of the leaflets. It was growing with typical *A. gummifera*, and is probably a hybrid between that species and *A. adianthifolia*. The specimen is unfortunately sterile, and it would therefore be preferable to await more complete material before claiming this hybrid with certainty for our area. See also last paragraph of note under *A. gummifera* (p. 158).

18. **A. zygia** (*DC.*) *Macbr.* in Contrib. Gray Herb., n.s. 59 : 3 (1919) ; L.T.A. : 868 (1930) ; T.S.K. : 72 (1936) ; F.P.N.A. 1 : 394 (1948) ; T.T.C.L. : 340 (1949) ; Gilb. & Bout. in F.C.B. 3 : 176 (1952) ; I.T.U., ed. 2 : 222 (1952) ; Consp. Fl. Angol. 2 : 294 (1956) ; F.W.T.A., ed. 2, 1 : 502 (1958). Type : " Antilles " (? erroneous label), *unknown collector* (G–DC, holo. (fragment !), K, photo. !)

Tree 4·5–30 m. high ; crown spreading ; bark rough or smooth. Young branchlets densely to very sparsely clothed with minute crisped puberulence, usually soon disappearing but sometimes persistent. Leaves : pinnae (Fig. 22/6, p. 163) (1–)2–3(–4, even—*fide* I.T.U. –5) pairs, each pinna broadening towards apex ; leaflets of 2 distal pairs of pinnae 2–5 (very rarely 6) pairs, obliquely rhombic to obovate, with the distal pair largest, 29–72 mm. long, 16–43 mm. wide ; lower pairs smaller, down to about 12 × 8 mm. ; apex obtuse, rarely subacute ; base not auricled ; surface beneath glabrous except for puberulence on midrib and margins. Stipules and bracts at base of peduncles very caducous, narrowly triangular-lanceolate and acute, 2–7 mm. long, 1–2 mm. wide. Peduncles clothed as branchlets ; bracteoles very caducous, subulate-linear to oblanceolate, 1·75–3 mm. long. Flowers subsessile ; pedicels puberulous, 0·25–0·5(–0·75) mm. long. Calyx 2–4 mm. long, puberulous outside. Corolla 6–9(–10) mm. long, densely and minutely pubescent or puberulous outside, white or pink. Staminal tube red, exserted about 1–1·8 cm. beyond corolla. Pod oblong, flat or somewhat transversely plicate, mostly 10–18 cm. long and 2–4 cm. wide, usually pubescent or puberulous on stipe, otherwise glabrous or nearly so. Seeds about 7·5–10 mm. long and 6·5–8·5 mm. wide, flattened.

UGANDA. Acholi District: Gulu, Mar. 1934, *Tothill* 2491 !; Bunyoro District: Budongo Forest, Busingiro Hill, Mar. 1932, *Harris* 66 *in F.H.* 636 !; Teso District: Serere, Feb. 1933, *Chandler* 1090 !
KENYA. Central Kavirondo District: Gem Location, Yala R., 21 Sept. 1944, *Davidson* 216 *in Bally* 4428 !
TANGANYIKA. Mwanza District: Lake Victoria, Luwondo & Maisome Isls., July 1932, *Herring in F.H.* 1897 ! & Geita Gold Mine, 14 Apr. 1937, *B. D. Burtt* 6522 !; Mpanda District: Mpangwa R., Uvinza Forest Reserve, Nov. 1954, *Procter* 302 !
DISTR. U1–4; K5; T1, 4; widespread in tropical Africa from Senegal and the Sudan to Gaboon, Cabinda and the Belgian Congo
HAB. Lowland rain-forest, riverine forest, but also occurs in wooded grassland; 915–1370 m.

SYN. *Inga zygia* DC., Mém. Fam. Légum. 440, t. 65 (1825)
    Zygia brownei Walp., Rep. 1: 928 (1842). Type: as *Inga zygia* DC.
    Albizzia brownei (Walp.) Oliv., F.T.A. 2: 362 (1871)
    A. welwitschioïdes [Schweinf.] ex Bak. f., L.T.A.: 867 (1930). Types: French Cameroons, Dengdeng to Kongola, *Mildbraed* 8915 (BM, syn. !, K, isosyn. !) & the Sudan, Dar Fertit, *Schweinfurth* Ser. 2, 103 (BM, syn. !, K, isosyn. !)

19. **A. grandibracteata** *Taub.* in P.O.A. C : 193 (1895) ; L.T.A. : 867 (1930) ; T.S.K. : 72 (1936) ; Bogdan in Nature in E. Afr., No. 2 : 16 (1947) ; F.P.N.A. 1 : 393 (1948) ; T.T.C.L. : 339 (1949) ; Gilb. & Bout. in F.C.B. 3 : 175 (1952) ; I.T.U., ed. 2 : 220, t. 40 (1952). Type : Uganda, " Menjo " [Mengo], *Stuhlmann* 1288 (B, holo.†)

Tree, 6–30 m. high, deciduous ; crown rounded or flat ; bark smooth, or pock-marked at base, lenticels frequently coalescing in vertical columns ; heartwood pinkish with red streaks. Young branchlets at first with short dense spreading pubescence brownish when dry, slowly glabrescent. Leaves :

pinnae (Fig. 22/4, opposite) (1–)2–3 pairs, each pinna broadening towards apex ; leaflets of 2 distal pairs of pinnae 3–6 pairs, obliquely rhombic to obovate, with the distal pair largest, 29–72(–100) mm. long, 16–32 mm. (–42) mm. wide ; lower pairs smaller, down to about 12 × 8 mm. ; apex acute or sometimes subacute; base not or scarcely auricled on proximal side ; surface beneath ± pubescent between midrib and margins. Stipules and bracts at base of peduncles ± caducous, broad, suborbicular to reniform, 7–18 mm. long, 8–20 mm. wide. Peduncles clothed as branchlets ; bracteoles ± persistent, oblanceolate or the lower obovate, 3·5–7 mm. long, 0·25–4 mm. wide, pubescent. Flowers subsessile ; pedicels puberulous, 0·25–0·75 mm. long. Calyx 3–5 mm. long, or minutely pubescent outside. Corolla about 7–10 mm. long, densely and minutely pubescent outside, pink to white. Staminal tube pink or red, exserted about 1·2–2 cm. beyond corolla. Pod oblong, transversely plicate, 7–15 cm. long, 1·5–3 cm. wide, very finely and minutely pubescent (use × 20 lens) especially near base, eglandular, venose. Seeds about 8–11 mm. long and 6·5–8 mm. wide, flattened.

UGANDA. Toro District: Ruwenzori, lower slopes, *Dawe* 434 !; Mengo District: Kiagwe, Namanve Forest, Apr. 1932, *Eggeling* 406 *in* F.H. 688 !; Mubende, 19 June 1945, *A. S. Thomas* 4127 !
KENYA. Trans-Nzoia District: Caves of Elgon, 18 Apr. 1943, *Bally* 2498 !; Nandi District: Kaimosi, 3 June 1933, *C. G. Rogers* 723 !; Nakuru District: Rongai, Feb. 1932, *Cooper in* F.H. 2751 !
TANGANYIKA. North Mara District: Tarime, 16 Dec. 1937, *Bancroft in* F.H. 1243 !
DISTR. U1–4; K3, 5; T1; Belgian Congo and the Sudan
HAB. Lowland and upland rain-forest, riverine forest; the most ubiquitous first-stage colonizer in the Mabira Forest, Uganda; recorded from grassland in Uganda, Ankole District, Igora, 1650 m., and Kenya, Nakuru District, Rongai, 1830 m., where it is possibly relict; 1160–2130 m.

19 × 16. **A. grandibracteata** *Taub.* × **gummifera** (*J. F. Gmel.*) *C. A. Sm.*; I.T.U., ed. 2 : 221 (1952) ; Brenan in K.B. 1952 : 530 (1953).

Differs from *A. grandibracteata* in the often narrower stipules and bracts, the pinnae (Fig. 22/5, opposite) often more numerous (2–3 pairs, as against 1–3), the more numerous leaflets of the two distal pairs of pinnae (6–10, as against 3–6 pairs), usually ± clearly auriculate at base on the posticous side ; from *A. gummifera* in the broad stipules and bracts, the pinnae in 2–5 pairs (as against (3–)5–7), the fewer leaflets sparsely pubescent beneath on surface between midrib and margins, the terminal ones rather larger than the rest, the pinnae thus ± widening upwards.

UGANDA. Acholi District: SE. Imatong Mts., Lomwaga Mt., 4 Apr. 1945, *Greenway & Hummel* 7271 !; Bunyoro District: Budongo Forest edge, June 1932, *Harris* 127 *in* F.H. 842 !; Masaka District: Kabula near Ankole District boundary, Oct. 1932, *Eggeling* 690 *in* F.H. 1064 !
KENYA. Uasin Gishu District: Lamok R., 3 May 1951, *G. R. Williams* 149 !; N. Kavirondo District: Kakamega, comm. *Cons. of Forests* H 235/31 !
DISTR. U1, 2, 4; K3, 5; Belgian Congo and the Sudan
HAB. Presumably where those of the parents overlap

NOTE. *Eggeling* 1906 (West Nile District: Zeio, Mar. 1935, BM !) is unusual in being much on the *A. grandibracteata* side, having most of the leaflets non-auricled as in that species.
    Mr. H. C. Dawkins (E. Afr. Agric. Journ. 17: 99 (1951)) remarks that in Mengo District " there are so many intermediates between the above two species [*A. grandibracteata* and *A. gummifera*] that without the inflorescences it is often impossible to decide to which a specimen tends, much less belongs ".

20. **A. petersiana** (*Bolle*) *Oliv.*, F.T.A. 2 : 362 (1871) ; Gilg in P.O.A. B : 299 (1895) ; L.T.A. : 867 (1930) ; T.T.C.L. : 340 (1949). Type : Portuguese East Africa, Boror and Sena, 16–18° S. lat., *Peters* (B, holo.†)

Tree, sometimes shrubby, 3–21 m. high, deciduous ; crown rounded or flat ; bark smooth unless fire-scarred ; heartwood red. Young branchlets

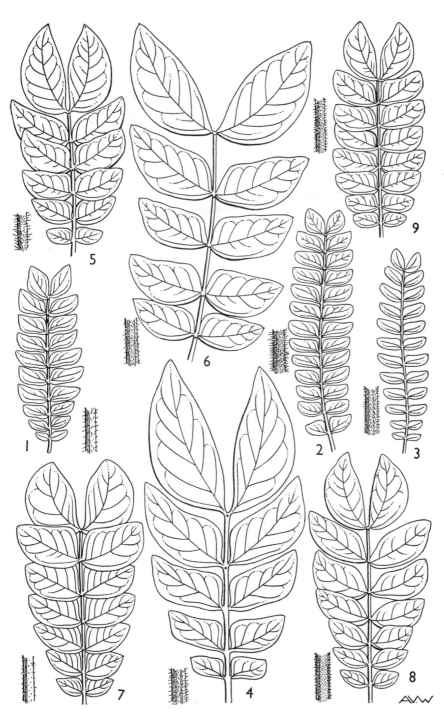

FIG. 22. *ALBIZIA*—Pinnae and parts of their rhachides, × ⅔. **1,** *A. gummifera* var. *gummifera;* **2,** *A. adianthi-folia;* **3,** *A. intermedia* (not so far found in our area); **4,** *A. grandibracteata;* **5,** *A. grandibracteata* × *gummi-fera;* **6,** *A. zygia;* **7, 8,** *A. gummifera* × *zygia;* **9,** *A. adianthifolia* × *zygia* (not so far found in our area). 1, from *Green* 34;   2, from *Yates* 13;   3, from *Gossweiler* 8719;   4, from *Rogers* 723;   5, from *Harris* 842; 6, from *Brown* 81;   7, from *Welwitsch* 1770;   8, from *Eggeling* 81 *in F.H.* 254;   9, from *Vigne* 77.

shortly crisped-pubescent to almost glabrous. Leaves : pinnae 2–5 pairs, each pinna not broadening towards apex ; leaflets of 2 distal pairs of pinnae 5–12 pairs, oblong- to obovate-rhombic, 5–23 mm. long, 2·5–13(–17) mm. wide ; base auricled or not on proximal side ; apex obtuse to subacute, sometimes acute ; surface beneath subglabrous to rather densely appressed-pubescent, above glabrous to minutely appressed-pubescent. Stipules very quickly falling and usually absent, oblanceolate or triangular-acute, 1·75–3·5 mm. long, 0·6–1 mm. wide. Peduncles subglabrous to shortly crisped-pubescent ; bracteoles falling while flowers are still in bud, oblanceolate, 1 mm. long, pubescent. Flowers on glabrous pedicels 1–3 mm. long. Calyx 1–1·75 mm. long, glabrous except on margins. Corolla about 7–10 mm. long, glabrous outside, white to pink. Staminal tube red, exserted about 1·1–3 cm. beyond corolla. Pod oblong, transversely plicate, 4–15 cm. long, 1·7–3 cm. wide, glabrous, eglandular, venose, deep red-purple when ripe. Seeds 9–10 mm. long or in diameter.

UGANDA. Ankole District: Mulema, 14 Apr. 1903, *Bagshawe* 219 !
KENYA. Masai District: Mara Masai Reserve, Telek R., 13 Sept. 1947, *Bally* 5335 ! & Egerok, 17 Sept. 1947, *Bally* 5388 !; Teita District: Kasigao (Kisigau), Sept.–Oct. 1938, *Joana in C.M.* 8796 !
TANGANYIKA. Mwanza District: Ihumba, Mbaruka, 18 Oct. 1951, *Tanner* 337 !; Lushoto District: Kongei, 9 Feb. 1933, *Greenway* 3346 !; Tanga District: Ngomeni, 30 July 1953, *Drummond & Hemsley* 3547 !; Mpwapwa, 27 Oct. 1931, *Hornby* 324 !
DISTR. U2; K6, 7; T1–8; Portuguese East Africa and Nyasaland
HAB. Ground-water forest, riverine forest, and on termite-mounds and in ravines in deciduous woodland; a scattered but dominant tree on mountain-slopes and valleys cleared of forests in the Usambaras; 380–1700 m.

SYN. *Zygia petersiana* Bolle in Peters, Mossamb. Bot.: 1, t. 1 (1861)
　　Albizzia brachycalyx Oliv., F.T.A. 2: 361 (1871); L.T.A.: 868 (1930); Bogdan in Nature in E. Afr., No. 2: 17 (1947); T.T.C.L.: 339 (1949); I.T.U., ed. 2: 217 (1952). Type: Tanganyika, Biharamulo District, Usui, *Speke & Grant* 205 (K, holo. !)

VARIATION. *A. petersiana* and *A. brachycalyx* have been separated merely on foliage. The original plate of the former shows the two distal pairs of pinnae with 5–6 pairs of leaflets per pinna; the type of the latter shows 8–10 pairs. The following are other ranges, each based on a count of the leaflet pairs of 10 pinnae:—5–6 (*Burtt* 4471), 6–8 (*Lynes* P.g.49), 7–9 (*Rounce* 232), 8–9 (*Vesey-FitzGerald* 394), 7–10 (*Bally* 5388), 8–10 (*Bancroft* 42), (8–)11–12 (*Burtt* 5670). I cannot confirm the statement in L.T.A.: 868 (1930) that the leaflets of *A. brachycalyx* may be in as many as 15 pairs.
　　Both the size and spacing of the leaflets also vary a good deal, greater size and wider spacing going with fewer pairs. The indumentum also varies: at one extreme the lower surface of the leaflets is subglabrous with a very few hairs on midrib and margins only (*Burtt* 523, *Gillman* 1467); or the hairs may be still so confined though rather more frequent (*R. M. Davies* 984); or there may be sparse hairs over the actual surface between midrib and margins (*Rounce* 222); or the pubescence may be fairly dense all over the lower surface (*Burtt* 717, *Hornby* 324).
　　The variation is doubtless genetic, but does not show clear correlation with geography.

21. **A. euryphylla** *Harms* in E.J. 33 : 151 (1902) ; L.T.A. : 867 (1930) ; T.T.C.L. : 339 (1949). Type : Tanganyika, Dodoma District, Mapanga, *Busse* 254 (B, holo.†)

Shrub about 3 m. high, glabrous. Leaves : pinnae 1–3 pairs ; leaflets 1–3 pairs, obliquely suborbicular to broadly obovate or rhombic, 15–25 mm. long and as wide or somewhat narrower [in *Busse* 258 20–35 mm. long, about 18–35 mm. wide], rounded-obtuse or slightly emarginate at apex, glabrous, somewhat glaucous beneath. Flowers white, on pedicels 2–3 mm. long. Calyx 1–2 mm. long, glabrous. Corolla glabrous outside. Staminal tube long-exserted. Pod [described from *Busse* 258] more or less oblong and somewhat transversely plicate, 10–14 cm. long, 2·5–2·9 cm. wide, glabrous, eglandular, venose, brownish-purple. Seeds 8·5–11·5 mm. long, 8–10 mm. wide, flattened.

Tanganyika. Dodoma District: Mapanga, July [? Aug.] 1900, *Busse* 254 & 258 !
Distr. **T5**; not known elsewhere
Hab. " Bush-steppe "

Note. *Busse* 258 is said by Harms (see above reference) probably to belong to *A. euryphylla*. More material of this species is wanted. The above description, except where otherwise indicated, is adapted from that of Harms.

18. PITHECELLOBIUM

Mart. in Flora 20 (2), Beibl. : 114 (1837) (erroneously as *Pithecollobium*) ; Kosterm., Monogr. Asiatic etc. Sp. Mimos. Formerly Incl. in Pithecolobium Mart. : 8 (Bull. 20, Org. Sci. Res. Indonesia (1954))

Trees or shrubs, often armed with spinescent stipules. Leaves bipinnate ; petiole normally with a gland at its apex at the junction of the pinnae ; pinnae usually with 1–3 pairs of leaflets, rarely more. Inflorescences of heads or spikes often racemosely or paniculately aggregated. Flowers usually ⚥, usually sessile. Calyx gamosepalous with (4–)5(–6) teeth. Corolla (4–)5(–6)-lobed, puberulous to glabrous outside. Stamens many, fertile, their filaments connate below into a tube ; anthers eglandular at apex. Ovary usually puberulous. Pods compressed or convex, spirally twisted, circinate or curved, splitting into 2 twisting coriaceous or subcoriaceous valves. Seeds unwinged, without endosperm, often arillate.

In a restricted sense, as interpreted here, the genus is probably confined, as a native, to South America. The number of species cannot be estimated without fuller revision, but they are probably fairly numerous.

The generic name is often spelled as *Pithecolobium* or *Pithecollobium*. For a discussion on this see Merrill in Journ. Wash. Acad. Sci. 6: 43 (1916) and Sprague in K.B. 1929: 243 (1929).

P. unguis-cati (L.) Benth. from tropical America, which is near *P. dulce* but has flower-heads on longer peduncles, and *P. pruinosum* Benth. from Australia, which has pedicellate flowers and up to 9 mostly alternate leaflets per pinna, are both said to have been cultivated in Tanganyika (T.T.C.L.: 347 (1949)).

P. dulce (*Roxb.*) *Benth.* in Hook., Lond. Journ. Bot. 3 : 199 (1844) ; L.T.A. : 871 (1930) ; T.T.C.L. : 347 (1949) ; U.O.P.Z. : 415, 416, fig. (1949) ; Dale, Introd. Trees Uganda : 60 (1953) ; Kosterm., Monogr. Asiatic etc. Sp. Mimos. Formerly Incl. in Pithecolobium Mart. : 8 (Bull. 20, Org. Sci. Res. Indonesia (1954)). Type : India, Coromandel, cultivated, *Roxburgh in Wallich* 5282 D. (K, ? holo. ! & painting of type material, No. 488 !).

Shrub or tree 4–15 m. high, armed with spinescent stipules up to 12 mm. long. Bark smooth. Young branchlets puberulous to pubescent. Leaves : petiole 0·3–2·8(–5) cm. long, glandular at apex at the junction of the single pair of pinnae ; leaflets 1 pair, ± asymmetrically elliptic to obovate-elliptic, mostly 0·7–5 cm. long, 0·3–2·3 cm. wide, glabrous or inconspicuously hairy, rounded or emarginate to subacute at apex. Flowers creamy or yellow, in small heads 0·8–1 cm. wide on short peduncles arranged racemosely or paniculately in ± leafless inflorescences. Calyx 1–1·5 mm. long, puberulous. Corolla 3–4·5 mm. long, puberulous. Free part of stamen-filaments about 6·5–7 mm. long. Pods spirally twisted. Seeds black, glossy, 9–10 mm. long, 7–8 mm. wide, covered with a white to reddish fleshy edible aril.

Kenya. Kwale District: railway loop 4 km. S. of Mazeras, 10 Sept. 1953, *Drummond & Hemsley* 4243 !
Distr. **K7**; native of tropical America, but widely cultivated in other parts of the tropics and naturalized in many places
Hab. Naturalized on fringe of woodland with *Cussonia zimmermannii*, *Acacia* spp. and *Tamarindus indica;* about 150 m. Perhaps derived from an agricultural research station that formerly existed at Mazeras

SYN. *Mimosa dulcis* Roxb., Pl. Corom. 1: 67, t. 99 (1795)

NOTE. *P. dulce* is cultivated in Uganda, Kenya, Tanganyika and Zanzibar, and other instances of its naturalization besides the one mentioned above may thus well be found. It is very popular as a hedge-plant at the lower altitudes.

19. CATHORMION

Hassk., Retzia, 1 : 231 (1855) ; Kosterm., Monogr. Asiatic etc. Sp. Mimos. Formerly Incl. in Pithecolobium Mart. : 11 (Bull. 20, Org. Sci. Res. Indonesia (1954))

Trees or shrubs, unarmed. Leaves bipinnate, pinnae each with several to many pairs of leaflets ; gland on upper side of petiole present or absent ; glands also often present at insertion of pinnae and on upper part of pinna-rhachis. Inflorescences of round heads which in the African species are pedunculate and mostly solitary or paired (sometimes in threes) in axils. Flowers ♀, or said to be rarely ♂ and ♀ ; 1 to several central flowers in each head often modified and different in form from the others, and sometimes at least ♀. Calyx gamosepalous, shortly (4–)5-dentate. Corolla gamosepalous, infundibuliform, (4–)5-lobed. Stamens numerous (about 16–22), fertile, their filaments united in their lower part into a slender tube not or scarcely projecting from the corolla (or very shortly so in the modified central flowers). Pods oblong, ± falcate or spirally curved, compressed, with their margins straight or lobed, ± constricted between the seeds, at maturity breaking up into coriaceous or hard 1-seeded joints. Seeds ± compressed.

Confined to the tropics. One species in Asia and Australia, about four in Africa, and at least seven in tropical America.

Cathormion altissimum (*Hook. f.*) *Hutch.* & *Dandy* in F.W.T.A. 1 : 364 (July 1928) & in K.B. 1928 : 401 (Dec. 1928) ; I.T.U., ed. 2 : 223 (1952) ; Consp. Fl. Angol. 2 : 298 (1956) ; F.W.T.A., ed. 2, 1 : 504 (1958). Types : Ghana, Cape Coast, *T. Vogel* (K, syn.!) & Nigeria, Ibo country, *T. Vogel* (K, syn.!)

Shrub or tree 5–35 m. high, unarmed or (I.T.U., ed. 2 : 223) often spinous on juvenile and sucker shoots ; crown spreading. Young branchlets puberulous. Leaves : rhachides ± densely and shortly crisped-pubescent or puberulous ; pinnae 5–7 pairs ; leaflets 11–22(–25 *fide* I.T.U.) pairs, somewhat obliquely oblong, and widest near the base which is slightly auricled on both sides, 7–15 mm. long, 2·5–6 mm. wide, narrowed to a usually obtuse apex, glabrous on both surfaces except for ciliolate margins at base, with the lateral nerves rather close and fine; the basal pair of leaflets on each pinna is characteristically represented by a single leaflet only, on the lower side, that on the upper side being replaced by a minute stipel. Inflorescences on peduncles 1·2–4·5 cm. long. Flowers white, sessile or subsessile. Calyx 3–3·5 mm. long, glabrous or subglabrous except on teeth. Corolla 5–7 mm. long, glabrous outside. Pods about 10–28 cm. long, 1·3–2 cm. wide, ± regularly lobed along one or both sutures which are puberulous, the pod otherwise subglabrous or sparingly puberulous. Seeds ± flattened, brown, 6–9 mm. long, 6·5–7 mm. wide. Fig. 23.

UGANDA. Bunyoro District: Siba R. area, Nov. 1932, *Harris* 166 *in F.H.* 1126 !
DISTR. U2; from Sierra Leone and the Sudan southwards to Angola, Northern Rhodesia and Uganda
HAB. Fresh-water swamp-forest; altitude range unknown

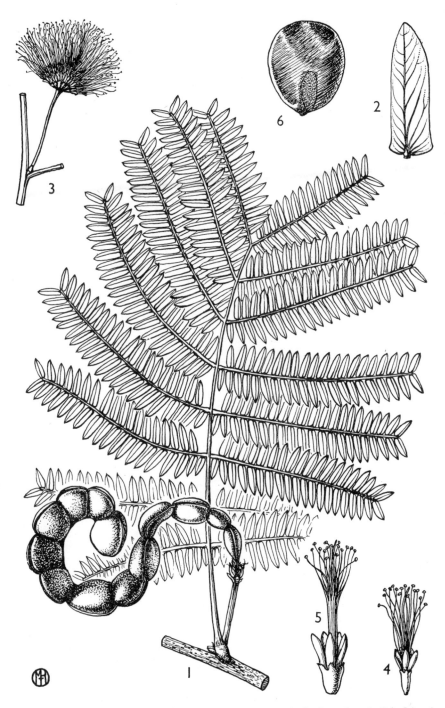

FIG. 23. *CATHORMION ALTISSIMUM*—**1**, part of fruiting branch, showing pod, × ⅔; **2**, leaflet, × 3; **3**, flower-head, × 1; **4**, flower from side of head, × 2; **5**, central flower of head, × 2; **6**, seed, × 3. 1, 2, from *Harris* 1126; 3–5, from *Linder* 933; 6, from *Unwin* 25.

Syn.   *Albizzia altissima* Hook. f. in Niger Fl.: 332 (1849)
    Pithecolobium altissimum (Hook. f.) Oliv., F.T.A. 2: 364 (1871); L.T.A.: 870 (1930)
    Arthrosamanea altissima (Hook. f.) Gilb. & Bout. in B.J.B.B. 22: 182 (1952) & in F.C.B. 3: 193 (1952)

Note.   ? *Pithecolobium stuhlmannii* Taub. (in P.O.A. C: 193 (1895); L.T.A.: 870 (1930); T.T.C.L.: 347 (1949)), based on *Stuhlmann* 2773 from Bataibo on the R. Duki (B, holo. †), was not collected in Tanganyika as stated in L.T.A., but in the Belgian Congo west of Lake Albert. It is certainly a *Cathormion*, and most probably synonymous with *C. altissimum*.

INDEX TO MIMOSOIDEAE

169